Heat and Mass Transfer

Series Editors: D. Mewes and F. Mayinger

For further volumes:
http://www.springer.com/series/4247

Preface

This book presents recent developments in my research on heat transfer of laminar forced convection and its film condensation. It is a monograph on advanced heat transfer, provided for university postgraduate students as textbook, students in self-study, researchers and professors as an academic reference book, and engineers and designers as a scientific handbook. A primary goal is to present a new similarity analysis method to replace the traditional Falkner-Skan type transformation for a deep investigation in this book. This method is so important that it becomes a theoretical basis of this book. A secondary goal is to report a system of research developments for heat and mass transfer of laminar forced convection and its two-phase film condensation, based on the present new similarity analysis method.

The book includes three parts: (1) theoretical foundation, including presentation of basic conservation equations for laminar convection, review of Falkner-Skan transformation for laminar forced convection boundary layer, and creation of the new similarity analysis method related to laminar forced convection and its two-phase film condensation; (2) laminar forced convection, including a system of complete similarity mathematical models based on the new similarity analysis method, rigorous numerical results of velocity and temperature fields, and advanced research results on heat transfer with ignoring variable physical properties or viscous thermal dissipation, considering viscous thermal dissipation, and considering coupled effect of variable physical properties; and (3) laminar forced film condensation, including a system of complete similarity mathematical models based on the new similarity analysis method, rigorous numerical results of velocity, temperature, and concentration fields, and advanced research achievements on condensate heat and mass transfer. In the research of this book, the following difficult issues have been resolved. They are (i) creating the new similarity analysis method, (ii) developing the system of complete similarity mathematical models, (iii) rigorously dealing with the variable physical properties, (iv) obtaining the reliable numerical results with satisfying whole interfacial physical matching conditions for the forced convection film condensation, (v) clarifying the effect of noncondensable gas on interfacial vapour saturation temperature as well as heat and mass transfer of the film condensation of vapour–gas mixture, (vi) clarifying the coupled effects of temperature-dependent physical properties on heat transfer of laminar forced convection, (vii) clarifying the coupled effect of concentration and temperature- dependent physical properties on heat and mass transfer of forced convection film condensation, (viii) creating a series of reliable prediction equations on heat and mass transfer for practical application, etc. Having resolved all above difficult issues demonstrates the scientific contributions of this book to extensive research on heat and mass transfer of laminar forced convection and its film condensation.

Obviously, all above difficult issues resolved in this book are big challenges encountered in my recent research. While the most significant challenge is how to clarify heat and mass transfer and create the related reliable prediction equations with consideration of various complicated physical factors including the coupled effect of variable physical properties, in order to realize theoretically reliable and convenient prediction of heat and mass transfer. For this purpose, all related variable physical properties are based on the experimental data, and rigorously dealt with in the complete similarity mathematical models in the system of related investigations of this book. On this basis, the theoretical research results of this book have their reliable application values.

In this book, all flows are taken on a horizontal flat plate for a research systemization. However, the present new similarity analysis method is suitable for general type of laminar forced convection and its film condensation. The research of this book has a spread space.

Here, I am very sincerely grateful to Professor Liangcai Zhong, Northeastern University, China, who provided his great help for development of the related mathematical models and numerical calculation. This book involves his significant contribution. I am heartily thankful to Professor B.X. Wang, my supervisor of my Ph.Dr. study from 1988 to 1991 in Tsinghua University, China, whose profound knowledge of the speciality, strict style of investigation, encouragement, and support enabled me to develop the challenging research. My respective friend, Mr. Sam Gelman gave up his holiday during the period of New Year going through parts of the manuscript. I would like to offer my sincere gratitude to him for all his help, support, hints, and valuable improvements.

I am so thankful for my wife Shihua Sun's sustained moral encouragements, which gave me a long-term power to persist in completing this book and carrying out the significant research work before that.

Ottawa, ON, Canada
June 2010

Deyi Shang

Contents

Nomenclature

A	area, m^2
C_{mg}	gas mass fraction in the vapour–gas mixture
C_{mv}	vapour mass fraction in the vapour–gas mixture
$C_{\mathrm{mv,s}}$	Interfacial vapour mass fraction
$C_{\mathrm{mv},\infty}$	bulk vapour mass fraction
$C_{x,f}$	local skin-friction coefficient with ignoring variable physical properties, $C_{x,f} = 0.664(\mathrm{Re}_{x,f})^{-1/2}$
$C_{x,w}$	local skin-friction coefficient with consideration of variable physical properties, $C_{x,w} = \sqrt{2}\frac{\nu_w}{\nu_\infty}(\mathrm{Re}_{x,\infty})^{-1/2}\left(\frac{\mathrm{d}W_x(\eta)}{\mathrm{d}\eta}\right)_{\eta=0}$
$\overline{C}_{x,f}$	average skin-friction coefficient with ignoring variable physical properties, $\overline{C}_{x,f} = 1.328\mathrm{Re}_{x,f}{}^{-1/2}$
$\overline{C}_{x,w}$	average skin-friction coefficient with consideration of variable physical properties, $\overline{C}_{x,w} = 2 \times \sqrt{2}\frac{\nu_w}{\nu_\infty}(\mathrm{Re}_{x,\infty})^{-1/2}\left(\frac{\mathrm{d}W_x(\eta)}{\mathrm{d}\eta}\right)_{\eta=0}$
C_{mh}	condensate mass–energy transformation coefficient
c_p	specific heat at constant pressure, J/(kg k)
c_{p_g}	gas specific heat, J/(kg K)
c_{p_v}	vapour specific heat, J/(kg K)
c_{p_m}	specific heat of vapour–gas mixture, J/(kg K)
D_v	vapour mass diffusion coefficient in the non-condensable gas, m^2/s
E	internal energy, J
e	internal energy per unit mass, J/kg
Ec	Eckert number, $\frac{w_{x,\infty}{}^2}{c_{p_f}(t_w - t_\infty)}$
$\mathrm{Ec_a}$	adiabatic Eckert number (corresponding to $q_x = 0$)
$E_{(Q_{x,f})_{\mathrm{Ec}=0}}$	deviation of heat transfer predicted by ignoring the viscous thermal dissipation to that by considering the viscous thermal dissipation, $\frac{(Q_{x,f})_{Ec=0} - (Q_{x,f})_{\mathrm{Ec}}}{(Q_{x,f})_{\mathrm{Ec}=0}}$
$\dot{E}$	internal energy per unit time in system, $\Delta\dot{E} = \dot{Q} + \dot{W}_{\mathrm{out}}$, W
ΔE	increment of internal energy in a system, J
$\Delta\dot{E}$	increment of internal energy per unit time in system, W

F	force, N
F_m	mass force acting on the control volume, N
F_s	surface force acting on the control volume, N
$\vec{F}$	surface force per unit mass fluid, N/kg
$\vec{\mathrm{F}}_m$	mass force per unit mass fluid, N/kg
$\vec{F}_{m,v}$	total mass force in the control volume, N
$f(\eta)$	intermediate dimensionless similarity variable on momentum field
$G_{\mathrm{increment}}$	momentum increment of the fluid flow per unit time in the control volume, N
g	gravity acceleration, m/s^2
g_x	local mass flow rate entering the fluid film at position x per unit area of the plate, kg/(m^2 s)
G_x	total mass flow rate entering the fluid film from position $x = 0$ to x with width of b of the plate, kg/s
H	enthalpy, $E + pV$, $c_p t$, J
h	specific enthalpy (enthalpy per unit mass), $e + pv$, J/kg
h_{fg}	latent heat of vapour condensation, J/kg
[J]	basic dimension for quantity of heat
[K]	basic dimension for absolute temperature
[kg]	basic dimension for mass
[m]	basic dimension for length
$\dot{m}_{\mathrm{increment}}$	mass increment per unit time in the control volume, kg/s
$\dot{m}_{\mathrm{in}}$	mass flowing per unit time into the control volume, kg/s
$\dot{m}_{\mathrm{out}}$	mass flowing per unit time out of the control volume, kg/s
n	number of independent physical variables
$\mathrm{Nu}_{x,w}$	local Nusselt number, with consideration of variable physical properties, $\frac{\alpha_{x,w} \cdot x}{\lambda_w}$
$\overline{\mathrm{Nu}}_{x,w}$	average Nusselt number with consideration of variable physical properties, $\frac{\overline{\alpha}_{x,w} \cdot x}{\lambda_w}$
$\mathrm{Nu}_{x,f}$	local Nussel number with ignoring variable physical properties, $\frac{\alpha_{x,f} \cdot x}{\lambda_f}$
$[\mathrm{Nu}_{x,f}]_{\mathrm{Ec}=0}$	local Nusselt number with ignoring variable physical properties and viscous thermal dissipation, $\frac{\alpha_{x,f} \cdot x}{\lambda_f}$
$[\overline{\mathrm{Nu}}_{x,f}]_{\mathrm{Ec}=0}$	average Nusselt number with ignoring variable physical properties and viscous thermal dissipation, $\frac{\overline{\alpha}_{x,f} \cdot x}{\lambda_f}$
$(\mathrm{Nu}_{x,f})_{\mathrm{Ec}}$	local Nusselt number with ignoring variable physical properties but consideration of viscous thermal dissipation, $\frac{(\alpha_{x,f})_{\mathrm{Ec}} \cdot x}{\lambda_f}$
$(\overline{\mathrm{Nu}}_{x,f})_{\mathrm{Ec}}$	average Nusselt number with ignoring variable physical properties but consideration of viscous thermal dissipation, $\frac{(\overline{\alpha}_{x,f})_{\mathrm{Ec}} \cdot x}{\lambda_f}$
n_λ	thermal conductivity parameter

n_{μ}	viscosity parameter
$n_{\mu\lambda}$	overall temperature parameter, $0.45n_{\mu} + 0.55n_{\lambda}$
$n_{\lambda,g}$	thermal conductivity parameter of gas
$n_{\lambda,v}$	thermal conductivity parameter of vapour
$n_{\mu,g}$	viscosity parameter of gas
$n_{\mu,v}$	viscosity parameter of vapour
p	pressure, N/m^2
Pr	Prandtl number at any temperature, $\frac{\mu \cdot c_p}{\lambda}$
Pr_f	average Prandtl number, $\frac{\mu_f \cdot c_{p_f}}{\lambda_f}$
Pr_g	gas Prandtl number
Pr_v	vapour Prandtl number
Pr_m	Prandtl number of vapour–gas mixture
$q_{x,f}$	local heat transfer rate at position x per unit area on the plate with ignoring variable physical properties, W/m^2
$Q_{x,f}$	total heat transfer rate from position $x = 0$ to x with width of b on the plate with ignoring variable physical properties, W
$q_{x,w}$	local heat transfer rate at position x per unit area on the plate with consideration of variable physical properties, W/m^2
$Q_{x,w}$	total heat transfer rate from position $x = 0$ to x with width of b on the plate with consideration of variable physical properties, W
$(q_{x,f})_{\mathrm{Ec}=0}$	local heat transfer rate at position x per unit area on the plate, with ignoring variable physical properties and viscous thermal dissipation, W/m^2
$(Q_{x,f})_{\mathrm{Ec}=0}$	total heat transfer rate from position $x = 0$ to x with width of b on the plate, with ignoring variable physical properties and viscous thermal dissipation, W
$(\overline{Q_{x,f}})_{\mathrm{Ec}=0}$	average heat transfer rate on plate area from $x = 0$ to x with width of b, with ignoring variable physical properties and viscous thermal dissipation, W
$(q_{x,f})_{\mathrm{Ec}}$	local heat transfer rate at position x per unit area on the plate, with ignoring variable physical properties but considering viscous thermal dissipation, W/m^2
$(Q_{x,f})_{\mathrm{Ec}}$	total heat transfer rate from position $x = 0$ to x with width of b on the plate, with ignoring variable physical properties but considering viscous thermal dissipation, W
$(\overline{Q_{x,f}})_{\mathrm{Ec}}$	average heat transfer rate on plate area from $x = 0$ to x with width of b, with ignoring variable physical properties but considering viscous thermal dissipation, W
r	number of basic dimension
Q	heat, J
$\dot{Q}$	heat entering the system per unit time, W
Q_{in}	heat transferred into the system from its surroundings, J
$\mathrm{Re}_{x,\infty}$	local Reynolds number with consideration of variable physical properties, $\frac{w_{x,\infty} x}{\nu_{\infty}}$

$\mathrm{Re}_{x,f}$	local Reynolds number with ignoring variable thermophysical properties, $\frac{w_{x,\infty}x}{\nu_f}$
$\mathrm{Re}_{xl,s}$	local Reynolds number of liquid film flow for film condensation of pure vapour, $\frac{w_{xv,\infty}x}{\nu_{l,s}}$, or that of vapour–gas mixture, $\frac{w_{xm,\infty}x}{\nu_{l,s}}$
$\mathrm{Re}_{xv,\infty}$	local Reynolds number of vapour film for forced film condensation of vapour, $\frac{w_{xv,\infty}x}{\nu_{v,\infty}}$
[s]	basic dimension for time
$Sc_{m,\infty}$	local Schmidt number
t	temperature, °C
t_{s}	vapour saturated temperature at the liquid–vapour interface for film condensation of pure vapour, or reference vapour saturation temperature (corresponding to $C_{\mathrm{mv,s}} = 1$) for film condensation of vapour–gas mixture, °C
$t_{\mathrm{s,int}}$	interfacial vapour saturation temperature for forced film condensation of vapour–gas mixture, °C
t_w	wall temperature, °C
t_∞	temperature in the bulk, °C
t_f	average temperature of boundary layer, $\frac{t_w+t_\infty}{2}$, °C
T	absolute temperature, K
T_{r}	reference temperature, K
T_w	wall temperature, K
T_∞	bulk temperature, K
$\frac{T_w}{T_\infty}$	boundary temperature ratio
V	volume, m^3
v	specific volume, m^3/kg
w_x, w_y, w_z	velocity components in x, y, z coordinates, respectively, m/s
w_{xl}, w_{yl}	velocity components of liquid film flow in the x and y directions, respectively, m/s
w_{xv}, w_{yv}	velocity components of vapour film flow in the x and y directions, respectively, m/s
w_{xm}, w_{ym}	velocity components of vapour–gas mixture in x and y coordinates, respectively, m/s
$w_{xl,s}$, $w_{yl,s}$	interfacial velocity components of condensate liquid film flow in x and y coordinates, respectively, m/s
$w_{xm,s}$, $w_{ym,s}$	interfacial velocity components of vapour–gas mixture film flow in x and y coordinates, respectively, m/s
$w_{x,\infty}$	bulk stream velocity. m/s
$w_{xv,\infty}$	vapour bulk velocity for film condensation of pure vapour, m/s
$w_{xm,\infty}$	vapour–gas mixture bulk velocity for film condensation of vapour–gas mixture, m/s
$W_x(\eta)$, $W_y(\eta)$	dimensionless velocity components in the x and y direction, respectively
$W_{xl}(\eta_l)$, $W_{yl}(\eta_l)$	dimensionless velocity components of liquid film flow in the x and y directions, respectively

$W_{xv}(\eta_v)$, $W_{yv}(\eta_v)$	dimensionless velocity components of vapour film flow in x and y directions, respectively
$W_{xl,s}(\eta_l)$, $W_{yl,s}(\eta_l)$	interfacial dimensionless velocity components of liquid film flow in x and y directions, respectively
$W_{xm}(\eta_m)$, $W_{yv}(\eta_m)$	interfacial dimensionless velocity components of vapour–gas mixture film flow in x and y coordinates, respectively
$W_{xv,s}$, $W_{yv,s}$	interfacial velocity components of vapour film flow in x and y coordinates, respectively, m/s
$\vec{W}$	velocity, $w_x i + w_y j + w_z k$, m/s
$\dot{W}$	work done per unit time, W
$\dot{W}_{\text{out}}$	work per unit time acting on the system, W
$\left(\frac{\partial W_{xl}}{\partial \eta_l}\right)_s$	dimensionless interfacial velocity gradient of condensate liquid film
x, y, z	coordinate variables

Greek Symbols

δ	boundary layer thickness, m
δ_l	liquid film thickness, m
$\eta_{l\delta}$	dimensionless liquid film thickness
δ_v	vapour film thickness, m
η_{v0}	dimensionless thickness of vapour film at the liquid–vapour interface
$\eta_{v,\infty}$	dimensionless thickness of vapour film
δ_c	concentration boundary layer thickness of vapour–gas mixture, m
δ_m	momentum boundary layer thickness of vapour–gas mixture, m
δ_t	temperature boundary layer thickness of vapour–gas mixture, m
η	dimensionless coordinate variable for boundary layer with consideration of the variable physical properties, $\frac{y}{x}\left(\frac{1}{2}\mathrm{Re}_{x,\infty}\right)^{1/2}$
η	dimensionless coordinate variable for boundary layer with ignoring variable physical properties, $\frac{y}{x}\left(\frac{1}{2}\mathrm{Re}_{x,f}\right)^{1/2}$
η_l	dimensionless coordinate variable of condensate liquid film flow, $\frac{y}{x}\left(\frac{1}{2}\mathrm{Re}_{xl,s}\right)^{1/2}$
η_v	dimensionless coordinate variable of vapour film flow, $\frac{y}{x}\left(\frac{1}{2}\mathrm{Re}_{xv,\infty}\right)^{1/2}$
$\eta_{l\delta}$	dimensionless thickness of liquid film
$\eta_{v\delta}$	dimensionless thickness of vapour film
η_m	dimensionless co-ordinate variable of vapour–gas mixture film flow, $\frac{y}{x}(\frac{1}{2}\mathrm{Re}_{xm,\infty})^{1/2}$
$\theta(\eta)$	dimensionless temperature, $\frac{t-t_\infty}{t_w-t_\infty}$

$\theta_l(\eta_l)$	dimensionless temperature of liquid film flow, $\frac{t_l - t_s}{t_w - t_s}$
$\theta_v(\eta_v)$	dimensionless temperature of vapour film flow, $\frac{t_v - t_\infty}{t_s - t_\infty}$
$\theta_m(\eta_m)$	dimensionless temperature of vapour–gas mixture film flow, $\frac{t - t_\infty}{t_s - t_\infty}$
α	angle, or local heat transfer coefficient, $W/(m^2k)$
$\alpha_{x,w}$	local heat transfer coefficient with consideration of variable physical properties, $W/(m^2K)$
$\overline{\alpha}_{x,w}$	average heat transfer coefficient with consideration of variable physical properties, $W/(m^2K)$
$[\alpha_{x,f}]_{\mathrm{Ec}=0}$	local heat transfer coefficient with ignoring variable physical properties and viscous thermal dissipation, W/($m^2 \times$°C)
$[\overline{\alpha}_{x,f}]_{\mathrm{Ec}=0}$	average heat transfer coefficient with ignoring variable physical properties and viscous thermal dissipation, W/(m^2K)
$(\alpha_{x,f})_{\mathrm{Ec}}$	local heat transfer coefficient with ignoring variable physical properties but consideration viscous thermal dissipation, W/(m^2K)
$(\overline{\alpha}_x)_{\mathrm{Ec}}$	average heat transfer coefficient with ignoring variable physical properties but consideration viscous thermal dissipation, W/(m^2K)
ρ	mass density, kg/m^3
ρ_f	average density, kg/m^3
ρ_g	gas density, kg/m^3
ρ_v	vapour density, kg/m^3
ρ_m	density of vapour–gas mixture, $\frac{\mathrm{kg}}{\mathrm{m}^3}$
ρ_{mg}	local density of gas in vapour–gas mixture, $\frac{\mathrm{kg}}{\mathrm{m}^3}$
ρ_{mv}	local density of vapour in vapour–gas mixture, $\frac{\mathrm{kg}}{\mathrm{m}^3}$
$\frac{1}{\rho}\frac{d\rho}{dx}$	density factor
$\frac{1}{\rho_g}\frac{d\rho_g}{d\eta_m}$	density factor of gas
$\frac{1}{\rho_v}\frac{d\rho_v}{d\eta_m}$	density factor of vapour
$\frac{1}{\rho_m}\frac{d\rho_m}{d\eta_m}$	density factor of vapour–gas mixture
λ	thermal conductivity, W/(m °C)
λ_f	average thermal conductivity, W/(m K)
λ_g	thermal conductivity of gas, W/(m K)
λ_v	thermal conductivity of vapour, W/(m K)
λ_m	thermal conductivity of vapour–gas mixture, W/(m K)
$\frac{1}{\lambda}\frac{d\lambda}{d\eta}$	thermal conductivity factor
$\frac{1}{\lambda_g}\frac{d\lambda_g}{d\eta_m}$	thermal conductivity factor of gas
$\frac{1}{\lambda_v}\frac{d\lambda_v}{d\eta_m}$	thermal conductivity factor of vapour
$\frac{1}{\lambda_m}\frac{d\lambda_m}{d\eta_m}$	thermal conductivity factor of vapour-gas mixture
μ	absolute viscosity, kg/(m s)
μ_f	average absolute viscosity, kg/(m s)
$\mu_{l,s}$	liquid absolute viscosity at the liquid–vapour interface

μ_g	gas absolute viscosity, kg/(m s)
μ_v	vapour absolute viscosity, kg/(m s)
μ_m	absolute viscosity of vapour–gas mixture, kg/(m s)
$\frac{1}{\mu}\frac{d\mu}{d\eta}$	viscosity factor
$\frac{1}{\mu_g}\frac{d\mu_g}{d\eta_m}$	viscosity factor of gas, (70)
$\frac{1}{\mu_v}\frac{d\mu_v}{d\eta_m}$	viscosity factor of vapour, (69)
$\frac{1}{\mu_m}\frac{d\mu_m}{d\eta_m}$	viscosity factor of vapour–gas mixture, (59)
ν	kinetic viscosity, m^2/s
ν_f	average kinetic viscosity, m^2/s
β	thermal volumetric expansion coefficient. K^{-1}
ψ	stream function, m^2/s
$-\left(\frac{d\theta}{d\eta}\right)_{\eta=0}$	wall dimensionless temperature gradient
$\left[-\left(\frac{d\theta}{d\eta}\right)_{\eta=0}\right]_{T_w/T_\infty\to 1}$	reference wall dimensionless temperature gradient with $T_w/T_\infty \to 1$ identical to the case with ignoring the variable thermophysical properties
$\left[-\left(\frac{d\theta}{d\eta}\right)_{\eta=0}\right]_{f,Ec=0}$	wall dimensionless temperature gradient with ignoring variable physical properties and viscous thermal dissipation
$\left[-\left(\frac{d\theta}{d\eta}\right)_{\eta=0}\right]_{f,Ec}$	wall dimensionless temperature gradient with ignoring variable physical properties but consideration of viscous thermal dissipation
$\left(\frac{d\theta_l}{d\eta_l}\right)_{\eta_l=0}$	liquid dimensionless temperature gradient on the plate (for short, wall temperature gradient)
Γ_{mv}	vapour relative mass fraction, $\frac{C_{mv}-C_{mv,\infty}}{C_{mv,s}-C_{mv,\infty}}$
Δt_w	wall subcooled temperature for film condensation of pure vapour, $t_s - t_w$, where t_s is vapour saturation temperature, °C
$\frac{\Delta t_w}{t_s}$	wall subcooled grade for film condensation of pure vapour, $\frac{t_s-t_w}{t_s}$ where t_s is vapour saturation temperature,
$(\Delta t_w)_s$	reference wall subcooled temperature for film condensation of vapour–gas mixture, $t_s - t_w$, where t_s is reference vapour saturation temperature, °C
$\frac{(\Delta t_w)_s}{t_s}$	reference wall subcooled grade for film condensation of vapour–gas mixture, $\frac{t_s-t_w}{t_s}$ where t_s is reference vapour saturation temperature
Δt_∞	vapour superheated temperature of film condensation of pure vapour, $t_\infty - t_s$, where t_s is vapour saturation temperature, °C

$\frac{\Delta t_\infty}{t_s}$	vapour superheated grade for film condensation of pure vapour, $\frac{t_\infty - t_s}{t_s}$ where t_s is vapour saturation temperature
$(\Delta t_\infty)_s$	reference vapour superheated temperature for film condensation of vapour–gas mixture, where t_s is reference vapour saturation temperature, °C
$\left(\frac{\Delta t_\infty}{t_s}\right)_s$	reference vapour superheated grade for film condensation of vapour–gas mixture, where t_s is reference vapour saturation temperature
$t_{s,\mathrm{int}} - t_w$	wall subcooled temperature for film condensation of vapour–gas mixture, °C
τ	time, s, or shear force, $\mathrm{N/m^2}$
τ_w	wall skin shear force, N/m^2, $\mu\left(\frac{\partial w_x}{\partial y}\right)_{y=0}$ (for Newtonian fluid flow)
$\vec{\tau}$	surface force acting on unit area , N/m^2
$\vec{\tau}_n$	surface force per unit mass fluid flow at per area of surface, $\mathrm{N/m^2}$
$\vec{\tau}_{n,A}$	surface force in the control volume, N
Φ	viscous dissipation function
Φ_s	mass flow rate parameter
ε	deformation rate
[ ]	symbol of tensor
{ }	symbol of quantity grade

Subscripts

Ec = 0	with ignoring viscous thermal dissipation
Ec	with consideration of viscous thermal dissipation
f	mean value
g	gas
l	liquid
m	mass force or vapour–gas mixture
s	saturation temperature for vapour, and reference vapour saturation temperature for vapour-gas fraction
v	vapour
w	at wall
∞	far from the wall surface

Chapter 1
Introduction

Abstract A new similarity analysis method is proposed in this book through a complete theoretical derivation. It leads to establishing systems of complete similarity governing mathematical models for deep investigations of laminar forced convection and its film flows for resolving the challenges in the research. The proposed novel similarity variables, the dimensionless velocity components, directly describe momentum field, which shows that the new similarity analysis method is different from the traditional Falkner–Skan transformation with the dimensionless function variable $f(\eta)$ for indirect description of the momentum field. With the new similarity analysis method, it is convenient to consider variable physical properties, especially to conveniently treat the interfacial physical matching conditions of two-phase film condensation and even to conveniently investigate the effect of noncondensable gas on the film condensation, compared to those with the traditional Falkner–Skan-type transformation. A series of results on rigorous analysis and calculation are reported for effects of the viscous thermal dissipation and variable physical properties on heat transfer of laminar forced convection, and a series of related prediction equations are provided. Furthermore, the complete similarity mathematical models for forced film condensation of pure vapour and vapour–gas mixture are developed, respectively, based on the new similarity analysis method, in which the coupled effect of the condensate liquid film flow and the induced vapour film flow is considered. The vapour film flow for the general forced film condensation involves vapour momentum and temperature boundary layers, while the laminar forced film condensation of vapour–gas mixture flow involves the additional concentration boundary layer. An even big challenge is resolved for rigorous calculation of the interfacial vapour saturation temperature, a decisive issue of heat and mass transfer for laminar film condensation of vapour–gas mixture. Then, it is realized to rigorously evaluate heat and mass transfer of the forced film condensation of vapour–gas mixture. Furthermore, a condensate mass–energy transformation equation is created under the new similarity analysis system, for a better clarification of the internal relations between heat and mass transfer of the forced film condensation.

D.Y. Shang, *Theory of Heat Transfer with Forced Convection Film Flows*,
Heat and Mass Transfer, DOI 10.1007/978-3-642-12581-2_1,

1.1 Scope

This book systematically presents my recent investigation developments on heat and mass transfer of laminar forced convection boundary layer and two-phase film condensation. The range of research in this book involves the following three parts.

In Part I, a new similarity analysis method is presented through a complete theoretical derivation. With the novel similarity variables, the dimensionless velocity components for direct description of momentum field, this new similarity analysis method is different from the traditional Falkner-Skan transformation with the dimensionless function variable $f(\eta)$ for indirect description of the momentum field. Such difference leads to more convenience, with the new similarity analysis method to consider variable physical properties, investigate heat and mass transfer of two-phase film condensation, and even clarify the effect of noncondensable gas on the film condensation than those with the traditional Falkner-Skan type transformation. This new similarity analysis method becomes a basis of whole investigation in this book for laminar forced convection and two-phase film flow condensation.

Part II is devoted to an extensive study of the laminar forced convection boundary layer. The effects of viscous thermal dissipation on heat transfer are investigated in depth. A series of results are reported on rigorous analysis and calculation of the effect of the viscous thermal dissipation on heat transfer. The related prediction equations on heat transfer are proposed with variations in a large range of the Eckert number and Prandtl number. On this basis, the rigorous equations are provided to describe the calculated deviation on heat transfer caused by ignoring the viscous thermal dissipation. Additionally, the studies are further conducted for comprehensive consideration of the coupled effects of variable physical properties on heat transfer. Since liquids and gases have quite different characteristics with temperature variation, the study developments of forced convection heat transfer of liquid and gas are presented separately. Subsequently, rigorous prediction equations on heat transfers are further created with consideration of the coupled effects of the temperature-dependent physical properties.

Part III of this book describes an extensive study of heat and mass transfer of laminar forced film condensation, corresponding to the coupled two-phase film flows. In this part, the laminar forced film condensations of pure vapour and vapour–gas mixture are respectively investigated in depth. The laminar forced film condensation of pure vapour consists of condensate liquid film flow and the induced vapour film flow. In fact, the vapour film flow involves vapour momentum and temperature boundary layers. While, the laminar forced film condensation of vapour–gas mixture flow consists of condensate film flow and vapour–gas mixture film flow. The vapour–gas mixture film flow involves vapour–gas mixture momentum, temperature, and concentration boundary layers. The temperature-dependent physical properties of liquid film flow and the concentration and temperature-dependent physical properties of vapour–gas mixture flow medium are considered for these phenomena. With these complex relationships, the study of laminar forced film condensation becomes very complicated and difficult for traditional analysis and calculation methods.

This complexity underlies the development of the new similarity analysis method. This methodology establishes a system of complete similarity governing mathematical models for clarification of these complexities. On this basis, the new systems of governing similarity mathematical models resolve the difficult point for satisfaction of whole interfacial physical matching conditions in numerical calculation. The key numerical results are used to derive rigorous prediction equations, which are further transformed to the prediction equations on condensate heat and mass transfer by combining them with the developed theoretical equations of heat and mass transfer.

The new similarity governing mathematical models also enable the interfacial vapour saturation temperature to be determined together with the system of numerical solutions of laminar forced film condensation of vapour–gas mixture. The system of rigorous numerical solutions on the interfacial vapour saturation temperature is formulated, and then the effect of the noncondensable gas on the interfacial vapour saturation temperature is clarified. In fact, the rigorous calculated result on the interfacial vapour saturation temperature is absolutely necessary for the reliable prediction of heat and mass transfer of laminar forced film condensation of vapour–gas mixture.

1.2 Application Background

Heat transfer with forced convection and its film condensation often involves large temperature differences. Its practical applications exist widely in various branches of industry, such as the metallurgical, chemical, mechanical, energy, electrical and food industries. The heat transfer on surfaces such as those of boilers, heating and smelting furnaces, heat exchangers, condensers, and other types of industrial and civic equipments is often caused by various forms of forced convection and its film condensation under large temperature differences. The heat transfer rate affects the related industrial heating process. While, in the electronic industry, the cooling process occurs with forced convection on the surface integrated circuits and tends to restrict the surface temperature to go below the allowable temperature. In addition, it is widely known that forced film condensation has significant applications in various condensers, which are widely used in various industries, especially nuclear power plants. The optimal design of the corresponding heating equipment and the optimal control of heat transfer processes of the heating devices depend on the correct prediction of heat transfer.

1.3 Previous Developments of the Research

1.3.1 Laminar Forced Convection Boundary Layer

Prandtl initiated the study of convection by means of boundary layer theory [1]. Its essential idea is to divide a flow into two major parts. The larger part concerns a stream of fluid far from any solid surface. The smaller part constitutes a thin layer next to a solid surface in which the effects of molecular transport properties (viscosity and thermal conductivity) are considered using some approximation. A short time later, Blasius [2] described the steady two-dimensional boundary layer of laminar forced convection on a semi-infinite plate parallel to a main constant flow. He transformed the partial differential equations related to the momentum boundary layer to a single ordinary differential equation by transforming the coordinate system. Then, Bohlhausen [3] applied the Blasius coordinate system, obtained the velocity field, and calculated the convective heat transfer with consideration of constant physical properties in the boundary layer on a horizontal flat plate.

Later on, the Blasius transformation approach was generalized by Falkner-Skan type transformation [4] for outer flows of the form $= cx^m$ results. Since then, laminar forced convection heat transfer by means of the boundary layer analysis has had widespread development, and only the studies undertaken in [5–29] are identified here for saving space. The research developments related to laminar forced convection and heat transfer have been collected and summarized in a series of books; see [30–37]. The above-mentioned studies are generally based on the boundary layer analysis. Meanwhile, the Falkner-Skan type analysis approach has had widespread application in the previous studies. Several studies have employed other analytical methods, for example, the asymptotic method and perturbation technique [13, 19, 25]. Additionally, in some studies, such as those in [6, 9, 10, 12, 19, 25, 26, and 29], the effect of the variable physical variables on laminar forced film convection and heat transfer were considered to a certain extent. To sum up, in these studies, the reference temperature method is taken as their general approach for investigation of the effect of variable physical properties.

1.3.2 Laminar Forced Film Condensation of Pure Vapour

Since the pioneer scientist Nusselt published his research [38] on film condensation, numerous studies have followed for laminar film condensation from free or forced pure vapour flows. There are numerous studies on the laminar forced convection film condensation (for short, laminar forced film condensation, the same below) so far, but only some of them, such as in [39–48], are listed here due to space limitations. Some detailed reviews on heat and mass transfer of laminar forced film condensation of vapour can be found in [49–53]. Generally, these studies are based on the approximate analysis on two-phase boundary layer equations to investigate the heat and mass transfer of the laminar forced film condensation. In these studies, there was a lack of consideration of the coupled physical relations between the condensate liquid film and the induced vapour film flows, such as interfacial temperature balance, velocity component balance, shear forced balance, mass transfer balance, and heat transfer balance. Meanwhile, there was a lack of studies on consideration of the coupled effects of variable physical properties on condensate heat and mass transfer.

1.3.3 Laminar Forced Film Condensation of Vapour–Gas Mixture

Although the laminar forced film condensation of pure vapour plays an important role in many industrial applications, the related condensation from vapour in the presence of noncondensable gas is equally important. This is due to its universal existence in general condensation. In the design of heat exchangers and condensers, as well as in the related optimal prediction of heat transfer processes in the chemical and power industries, including nuclear power plants, it is necessary to deal with the correct analysis and calculation of condensation from vapour–gas mixtures.

There were also numerous theoretical and experimental studies undertaken for laminar forced film condensation of vapour–gas mixture; however, only the studies with [54–68] are listed here for saving space. Some detailed reviews on heat and mass transfer of laminar condensation forced film condensation of vapour–gas mixture can be found in [69–73]. These studies demonstrated that the bulk concentration of noncondensable gas could have a decisive effect on the heat transfer rate, and their results indicated greater reductions in heat transfer. This is because of the fact that the presence of noncondensable gas lowers the partial pressure of the vapour and then reduces the condensate saturation temperature at liquid–vapour interface. Generally, in these previous studies, there was a lack of consideration of the coupled physical relations between the condensate liquid film and the induced vapour–gas mixture film flows, such as interfacial temperature, velocity component, shear force, mass flow, and energy rate matching conditions, as well as concentration matching condition corresponding to the interfacial vapour saturation temperature, and the balance matching conditions between the condensate liquid mass flow and vapour mass diffusion matching conditions. Meanwhile, there was a lack of the studies for clarification of the effect of noncondensable gas on the interfacial vapour saturation temperature. Additionally, there was a lack of studies on consideration of the coupled effects of concentration- and temperature-dependent physical properties on condensate heat and mass transfer of vapour–gas mixture.

1.4 Challenges Associated with Investigations of Laminar Forced Convection and Film Condensation

From the previous studies presented in Sect. 1.3, we can find the following challenges associated with investigations of laminar forced convection and film condensation.

1.4.1 Investigation of the Laminar Forced Convection Boundary Layer

Investigations of laminar forced convection are made difficultly by consideration of the coupled effect of variable physical properties. Due to the universality of forced convection with large temperature differences, consideration of the effect of variable temperature-dependent properties is very important in the corresponding studies. To sum up, in previous studies, the reference temperature method was often taken as the approach for investigation of the effect of variable physical properties, and there was an absence of reliable correlations of the variable physical properties used at any local temperature in numerical calculations. Additionally, the studies did not give consideration to the coupled effects of the whole temperature-dependent properties, such as viscosity, thermal conductivity, and density on heat transfer of laminar forced convection. In fact, based on the Falkner-Skan type analysis approach, it is not convenient to consider the coupled effect of these variable physical properties dependent on the local temperature for analysis and calculation of laminar forced convection.

1.4.2 Investigation of Laminar Forced Film Condensation of Pure Vapour

There are two challenges in studies of laminar forced film condensation of pure vapour. The first one is characterizing the variable physical properties of condensate liquid and vapour film fluids; the second one is to satisfy all interfacial physical matching conditions. The laminar forced film condensation of vapour consists of condensate liquid film flow and induced vapour film flow. Actually, at the liquid–vapour interface, the condensation follows five balance conditions: temperature balance, velocity component balance, shear forced balance, mass transfer balance, and heat transfer balance. Theoretically, the reliable calculated results of the related governing equations should satisfy the above five balance conditions. Therefore, satisfying these interfacial conditions is also the prerequisite for the reliable solutions of the governing equations of the two-phase forced film flows. However, satisfying all interfacial physical matching conditions is a difficult challenge, particularly in consideration of variable physical properties. So far, it is still a challenging work to develop a complete similarity model and a numerical approach for satisfying all the interfacial balance conditions for the laminar forced film condensation of vapour to further permit the reliable investigation of the velocity and temperature distributions, as well as condensate heat mass transfer.

Major challenges also exist in developing a complete similarity model and the numerical approach for assessing the coupled effect of temperature-dependent properties of condensate liquid and vapour film fluids of laminar forced film condensation of vapour. In fact, the traditional Falkner-Skan similarity model and some other models are inconveniently coupled with the models of the variable physical properties for the treatment of these variable properties.

1.4.3 Investigation of Laminar Forced Film Condensation of Vapour–Gas Mixture

There are three difficult challenges associated with investigations of laminar forced film condensation of vapour–gas mixture. The first one is to consider variable physical properties of condensate liquid and vapour film fluids; the second one is to satisfy all interfacial physical matching conditions; and the third one is to determine the interfacial vapour saturation temperature. For laminar forced film condensation of vapour–gas mixture, all seven interfacial physical matching conditions should be satisfied for a reliable solution. They include temperature, velocity, shear force, mass flow rate, and energy rate matching conditions, as well as concentration condition corresponding to the interfacial vapour saturation temperature and the balance matching conditions between the condensate liquid mass flow and the vapour mass flow with diffusion. With an additional concentration boundary layer existing in the vapour–gas mixture film flow, the investigation of laminar forced film condensation of vapour–gas mixture is an even more complicated issue. In this case, the concentration-dependent physical properties of the vapour–gas mixture should be further taken into account.

Since the wall subcooled temperature is the driving force of the condensate heat transfer, correct determination of the interfacial vapour saturation temperature is extremely important for reliable prediction of heat and mass transfer of laminar forced film condensation of the vapour–gas mixture. Since the interfacial vapour saturation temperature depends on the interfacial vapour mass fraction, it is difficult issue to correctly determine the interfacial vapour saturation temperature. Correct determination of the interfacial vapour saturation temperature is further complicated by the temperature-dependent physical properties of condensate liquid film fluid and concentration and temperature-dependent physical properties of vapour–gas mixture film fluids. A decline in the condensate saturation temperature in combination with increasing mass fraction of noncondensable gas in the vapour–gas mixture film flow should be dealt with seriously. This will enable clarification of the effects of vapour (or gas) concentration and temperature boundary conditions on velocity, temperature, and concentration fields of the two-phase flow phenomenon, as well as the condensate heat and mass transfer. For resolving these difficult points, it is necessary to develop the related advanced theoretical mathematical model and calculation approach. Developing an advanced mathematical model governing these difficult issues and obtaining the reliable solutions for investigations of film condensation of vapour–gas mixtures remains challenging work.

1.5 Limitations of Falkner-Skan Type Transformation

Theoretically, there are different numerical methods, which can be used for dealing with the calculation for the problem with laminar forced boundary layer and film condensation. However, an appropriate similarity analysis and transformation of the governing partial differential equations is always useful for simplification of the numerical calculation and further for heat and mass transfer analysis. So far, for similarity analysis and transformation of the governing partial differential equations of laminar forced convection boundary layer and two-phase film condensation, the popular approach is traditional Falkner-Skan type transformation. With this method, a stream function has to be induced at first, and then the dimensionless function variable $f(\eta)$ and its derivatives are produced for transformation of governing differential equations. However, with such dimensionless function variable $f(\eta)$, the traditional Falkner-Skan type transformation is not easily applied to treat forced convection boundary layer and film issues, such as those for consideration of variable physical properties, as well as for dealing with two-phase film flows condensation with three-point boundary value problems. Especially, with the governing similarity mathematical model of the laminar forced film condensation based on the Falkner-Skan type transformation, it is not convenient to treat and satisfy the whole set of interfacial physical matching conditions, in order to obtain reliable calculation results on condensate heat and mass and transfer. Up to now, these limitations of the Falkner-Skan type transformation have hampered extensive study on laminar forced convection with larger temperature differences and investigation of forced film condensation of vapour or vapour–gas mixture.

1.6 Recent Developments of Research in This Book

My recent research has focused on dealing with the difficult challenges associated with research on laminar forced convection boundary layer and two-phase film condensation. A number of significant developments have been achieved and are reported upon in this book. They are briefly summarized below.

1.6.1 New Similarity Analysis Method

In view of the challenges facing research on laminar forced convection boundary layer and forced film condensation, Chap. 4 presents a new similarity analysis method via a system of theoretical and mathematical derivations. With this method, the proposed expressions on the dimensionless similarity velocity components for laminar forced convection are proportional to the related velocity components. In this case, the solutions of the velocity components of the similarity model based on the new similarity analysis method can more directly express the real velocity field than the solution, the function variable $f(\eta)$ based on the Falkner-Skan type transformation. Although the present new similarity analysis method is derived by means of the governing partial differential equations with laminar forced convection boundary layers for ignoring the variable physical properties, it is suitable for the issues with consideration of variable physical properties. Furthermore, it is suitable for complete similarity analysis and transformation of the governing partial differential equations on laminar forced film condensation of vapour or vapour–gas mixture with the two-phase film flows. Meanwhile, by using the new similarity analysis method, a series of detailed derivations for creating the related complete similarity models are done in different chapters for the different issues. Also, with the new similarity analysis method, all variable physical properties become the related dimensionless physical factors existing in the derived dimensionless governing models. With the advanced approach on treatment of variable physical properties described in Chaps. 7 and 8, respectively, for laminar forced convection of gases and liquids, the system of the physical properties factor will be similarly transformed under the new similarity system. Thus, systems of complete similarity models are obtained respectively for laminar forced convection with and without consideration of variable physical properties, as well as for two-phase film condensation of vapour or vapour–gas mixture even with the consideration of concentration and temperature-dependent physical properties.

1.6.2 Treatment of Variable Physical Properties

In Chap. 7, the advanced temperature parameter method [74] is first applied for convenient and reliable treatment of variable thermophysical properties of gas laminar forced convection. In Chap. 8, the polynomial formulations suggested in [74] are first applied for that of liquid laminar forced convection. The physical property expressions based on the temperature parameter method of gases and polynomial equations of liquids respectively describe the local property values at any local temperature. The predicted values of the properties coincide with the related typical experimentally measurement values. In addition, in Chaps. 12, 13, and 14 for laminar forced film condensation of vapour–gas mixture, the concentration-dependent physical properties are further taken into account for the vapour–gas mixture film flow. Furthermore, the concentration-dependent physical properties are coupled with the temperature-dependent physical properties in any local position of the vapour–gas mixture film flow.

In the transformed similarity analysis models of laminar forced convection and two-phase film condensation of vapour or vapour–gas mixture, all thermal physical properties exist in the form of the physical property factor, which are beneficial in the treatment of the variable physical properties. Using the new similarity analysis method, these physical property factors become the functions of the dimensionless temperature and concentration. This is beneficial for simultaneous solution of the complete dimensionless similarity model of governing equations. Furthermore, the proposed similarity analysis method has functions effectively for the treatment of variable physical properties on laminar forced convection and film condensation of pure vapour or vapour–gas mixture. These attributes demonstrate that the new similarity method is a better alternative to the Falkner-Skan transformation for consideration of the variable physical properties.

1.6.3 Coupled Effect of Variable Physical Properties on Heat and Mass Transfer

In Chaps. 7, 8, 9, 10, 11, 12, 13, and 14, by using the developed complete similarity governing models, and the above advanced treatment of variable physical properties, the rigorous numerical solutions are obtained respectively for gas laminar forced convection, liquid laminar forced convection, laminar forced film condensation of pure vapour, and laminar forced film condensation of vapour–gas mixture. Due to the coupled effect of the variable physical properties, the numerical solutions not only depend on the Prandtl number but also depend on the temperature boundary conditions for laminar forced convection. For laminar forced film condensation of pour vapour, the coupled effect of variable physical properties is transformed to the effect of the wall subcooled grade and vapour superheated grade. While for laminar forced film condensation of vapour–gas mixture, the additional effect of bulk vapour mass fraction is also expressed, which exists in the coupled effect of variable physical properties on the numerical solutions.

By means of heat transfer analysis based on the present new similarity analysis method, a series of novel prediction equations on heat transfer and Nusselt number are contributed respectively for gas laminar forced convection, liquid laminar forced convection, laminar forced film condensation of pure vapour, and laminar forced film condensation of vapour–gas mixture. In these heat transfer prediction equations, only the wall dimensionless temperature gradient is the unknown variable for gas laminar forced convection, liquid laminar forced convection, and laminar forced film condensation of pure vapour. While for laminar forced film condensation of vapour–gas mixture, the interfacial vapours saturation temperature is also the unknown variable besides the wall dimensionless temperature gradient. The system of key numerical solutions on the wall dimensionless temperature gradient and interfacial vapour saturation temperature are rigorously formulated, and the related reliable formulated equations are created. Then, combined with the theoretical equations created based on the present new similarity analysis method, it is possible with these formulated equations of the key solutions to do the simple and reliable prediction of heat transfer on laminar forced convection and forced film condensation with consideration of the coupled effects of variable physical properties. Meanwhile, the coupled effects of variable physical properties on heat and mass transfer for the related issues are clarified.

1.6.4 Extensive Study of Effect of Viscous Thermal Dissipation on Laminar Forced Convection

In Chap. 6, the present new similarity analysis method is used for extensive study on heat transfer of laminar forced convection with consideration of viscous thermal dissipation. By means of the completely transformed governing similarity model created by the present new similarity method, a system of numerical solutions have been obtained in the wide ranges of the average Prandtl number and Eckert number. With these numerical solutions, the effect of Eckert number on heat transfer of laminar forced convection is demonstrated. The system of key numerical solutions on wall dimensionless temperature gradient is formulated with variation of the viscous thermal dissipation and average Prandtl number. The formulated equations are so reliable that they coincide with the related numerical results. Combined with the theoretical equations on heat transfer, the formulated equation on wall dimensionless temperature gradient becomes the simple and reliable prediction equation for heat transfer of laminar forced convection with consideration of viscous thermal dissipation.

Additionally, deviation equations of heat transfer predicted by ignoring the viscous thermal dissipation are obtained based on the provided formulated equations of the wall dimensionless temperature gradient. Then, it is easy to predict the deviation on heat transfer caused by ignoring the viscous thermal dissipation. Furthermore, a system of the adiabatic Eckert numbers is determined and formulated with variation of average Prandtl number, in order for further judgment of the heat flux direction. It is clear that the adiabatic Eckert number will increase at an accelerated pace with a decrease in the average Prandt number, especially in the region with lower Prandtl numbers.

1.6.5 Laminar Forced Film Condensation of Vapour

The laminar forced film condensation of vapour is extensively investigated in Chaps. 9, 10, and 11 with consideration of coupled effect of variable physical properties.

In Chap. 9, the new similarity analysis method is first successfully applied for the development of a complete similarity governing mathematical model on the laminar forced film condensation of superheated and saturated vapours for consideration of variable thermophysical properties. The laminar forced film condensation with two-phase film flows consists of the condensate liquid film and vapour film; and in both film flows, the variable thermophysical properties are considered in detail. For the model of the condensate liquid film, the determination of the similarity variables and consideration of the variable physical properties are based on the approach in Chap. 8 for liquid laminar forced convection. While for the model of the vapour film, the proposal of the similarity variables and consideration of the variable physical properties are based on the approach in Chap. 7 for gas laminar forced convection. The two models are combined with the five international matching balance equations, respectively, describing the velocity component balance, mass flow rate balance, the balance of shear force, the energy balance, and the temperature balance. On these bases, by using the new similarity analysis method, a system of complete similarity dimensionless governing model is created for rigorous investigation on laminar forced film condensation of saturated and superheated vapours.

In Chap. 10, the laminar forced film condensation of water vapour is taken as an example for treatment of variable physical properties. Meanwhile, the temperature parameter method for gases is used for treatment of vapour film variable physical properties, and the polynomial formulations are applied for treatment liquid film variable physical properties. Then, the physical property factors are further transformed under the present new similarity analysis model. An effective numerical procedure is developed for solution of the complete similarity governing mathematical model, satisfying the whole set of the interfacial matching conditions of the three-point boundary value problem. Thus, a system of the numerical solutions on the velocity and temperature fields of the two-phase film flows is rigorously determined. The system of numerical solutions expresses the effect of the wall subcooled grade on condensate liquid film thickness, vapour film thickness, wall temperature gradient, vapour film temperature gradient, and the interfacial velocity level. It is also found that compared with the effect of the wall subcooled grade on these physical phenomena, the effect of the vapour subcooled grade is very slight.

On the basis of Chaps. 9 and 10, heat and mass transfer analysis is illustrated in Chap. 11, and the theoretical equations on condensate heat and mass transfer are provided under the new similarity analysis system. From these theoretical equations it is found that the wall dimensionless temperature gradient and the induced condensate mass flow rate parameter, respectively, dominate condensate heat and mass transfer. Take the laminar forced film condensation of water vapour as an example, the systems of key numerical solutions on the dimensionless wall temperature gradient and mass flow rate parameter have been obtained and then formulated to rigorous equations. Combined with the theoretical equations on condensate heat and mass, the reliable formulated equations on the wall dimensionless temperature gradient and mass flow rate parameter are available for reliable prediction of condensate heat and mass transfer results.

It is also found that compared with the effect of the wall subcooled grade on the forced film condensation, the effect of the vapour subcooled grade is very slight in the relatively large range of bulk vapour superheated grade, and can be ignored in a large range (between 0 and 4.27) of the bulk vapour superheated grade for laminar forced film condensation of water vapour. The reason is that in a relatively large range of bulk vapour superheated grade, the heat capacity of superheated vapour is far smaller than the vapour condensate latent heat in this case.

At last, a condensate mass–energy transformation equation is created under the new similarity system, for a better clarification of the internal relations between heat and mass transfer balance of the forced film condensation.

1.6.6 Laminar Forced Film Condensation of Vapour–Gas Mixturer

In Chap. 12, the complete similarity mathematical model of two-phase laminar film condensation from vapour–gas mixture is developed based on the new similarity analysis method. The similarity models for the liquid and vapour–gas mixture film flows are coupled with seven interfacial matching conditions including the temperature condition, shear force balance, mass flow rate balance, temperature balance, heat transfer balance, concentration condition, as well as the balance between the condensate mass flow and the vapour mass diffusion. The focus of this chapter is to successfully apply the present new similarity analysis method in the laminar forced film condensation of vapour–gas mixture, for developing the complete similarity mathematical model. In the developed complete similarity model for liquid film flow, the temperature-dependent physical properties are considered, while for vapour–gas mixture film flow the concentration and temperature-dependent physical properties are taken into account. In the developed similarity governing mathematical model, these variable physical properties exist as the related dimensionless physical property factors. All these prove that the present new similarity method is a better alternative to the traditional Falkner-Skan type transformation for extensive investigation of laminar film condensation from vapour–gas mixture.

In Chap. 13, through a series of treatments on concentration- and temperature-dependent physical properties, the set of physical property factors in the developed complete similarity governing model becomes functions of the dimensionless temperature, dimensionless concentration, and their derivatives for convenient numerical calculation. By means of the available numerical procedure, the seven interfacial matching conditions are rigorously satisfied in the numerical calculation. Taking the laminar forced film condensation of water vapour–air mixture as an example, a formulation is provided for determination of the interfacial vapour saturation temperature at atmospheric pressure as the system pressure for iteration calculation. Then, the interfacial vapour saturation temperature is evaluated as the solution at the end of the numerical calculation. The correct determination of the interfacial vapour saturation temperature will lead to rigorous and reliable numerical results on the laminar forced film condensation of vapour–gas mixture. On the other hand, it is found the interfacial vapour saturation temperature is strongly reduced by the noncondensable gas existing in the vapour–gas mixture.

With the example of numerical calculation for laminar forced film condensation of water vapour–air mixture, a system of rigorous numerical results is successfully obtained, including velocity, temperature and concentration fields of the two-phase film flows. This facilitates identifying the effect of the bulk vapour mass fraction and the wall temperatures T_w on the velocity and temperature fields of the condensate liquid film and the induced vapour gas mixture film, as well as the concentration field of the vapour–gas mixture film.

Based on the rigorous numerical calculation, the interfacial vapour mass fraction and interfacial vapour saturation temperature, the decisive issues of heat and mass transfer for laminar film condensation of vapour–gas mixture are deeply investigated. A system of the numerical solutions on the interfacial vapour mass fraction and interfacial vapour saturation temperature is obtained for the laminar film condensation of water vapour–air mixture. On this basis, the formulated equation on reliable prediction of the interfacial vapour saturation temperature is rigorously created. Reliable determination of the interfacial vapour saturation temperature is necessary not only for reliable numerical solutions but also for correct prediction of condensate heat and mass transfer results.

Through the condensate heat and mass transfer analysis in Chap. 14, the theoretical equations of condensate heat and mass transfer rates are provided. These theoretical equations demonstrate that the unknown physical variables, the dimensionless wall temperature gradient, interfacial vapour saturation temperature, and induced mass flow rate parameter, respectively, dominate the condensate heat and mass transfer results. The defined mass flow rate parameter depends on condensate liquid film thickness as well as the interfacial condensate liquid velocity components.

The system of the rigorous key solutions on the wall dimensionless temperature gradient and mass flow rate parameter are formulated respectively to the simple and reliable equations for the laminar film condensation of water vapour–air mixture. Combined with the theoretical heat transfer equations, the formulated equations on the wall dimensionless temperature gradient and interfacial vapour saturation temperature are available for simple and reliable evaluation of condensate heat transfer on laminar film condensation of water vapour–air mixture. On this basis, with the condensate mass–energy equation provided in Chap. 11, the condensate mass transfer rate can be easily determined.

All above investigation developments demonstrate that the present new similarity analysis method is a better alternative to the traditional Falkner-Skan transformation for extensive study on laminar forced convection and film condensation.

Finally, the author believes the analysis and calculation methods reported in this book on horizontal flat plate can be extended to other different types of laminar forced convection and forced film condensation. The research results (including the novel similarity analysis method, the systems of the complete similarity governing mathematical models, the system of reliable numerical solutions, as well as the series of developed prediction equations on heat and mass transfer) are not only devoted to the theoretical investigations for laminar forced convection and forced film condensation but can be equally applied to practical industrial applications.

References

1. L. Prandtl, Uber Flussigkeits bewegung bei sehr kleiner Reibung. *Verhaldlg III Int. Math. Kong* (Heidelberg, Teubner, 1904), pp. 484–491, Also available in translation as: Motion of fluids with very little viscosity. NACA TM 452 (March 1928)
2. H. Blasius, Grenzschichten in Flussigkeiten mit kleiner Reibung. Z. Math. Phys. **56**, 1–37 (1908)
3. E. Pohlhausen, Der Wärmeaustausch zwischen festen Körpern und Flüssigkeiten mit kleiner Reibung und kleiner Wärmeleitung. Z. Angew. Math. Mech. **1**, 115–121 (1921)
4. V.M. Falkner, S.W. Skan, Some approximate solutions of the boundary layer equations. Phil. Mag. **12**, 865 (1931)
5. E.R.G. Eckert, O. Drewitz, The heat transfer to a plate in flow at high speed. NASA Tech. Mem. 1045 (1943), Transl. of Der Waermeuebergang an einer mit grosser Geschwindigkeit Laengs angestroemte platte, Forschung auf dem Gebiete des Ingenieurwesens **11**, 116–124 (1940)
6. E.R.van Driest, *The Laminar Boundary Layer with Variable Fluid Properties* (North Americaln Aviation rep. AL-1866, Los Angelos, 1954)
7. G.W. Morgan, W.H. Warmer, On heat transfer in laminar boundary layers at a high Prandtl numbers. J. Aeronaut. Sci. **23**, 937–948 (1956)
8. N.B. Cohen, Boundary-layer similar solutions and correlation equations for laminar heat transfer distribution in equilibrium air at velocities up to 41000 feet per second, NASA Tech. Rep. R-118 (1961)
9. R.E. Wilson, Real-gas laminar-boundary layer skin friction and heat transfer. J. Aerosp. Sci. **29**, 640–647 (1962)

10. R.A. Seban, *Laminar Boundary Layer of a Liquid with Variable Viscosity, in Heat Transfer Thermodynamics and Education*, ed. by H.A. Johnson. Boelter Annyversary Volume (Mcgtraw-Hiill, New York, NY, 1964), pp. 319–329
11. S.C.R. Dennis, N. Smith, Forced convection from a heated flat plate. J. Fluid Mech. **24**, 509–519, Part: Part 3 (1966)
12. G. Poots, G.F. Raggett, Theoretical results for variable property, laminar boundary layers in water. Int.J. Heat Mass Transfer **10**, 597–610 (1967)
13. R. Narasimh, N. Afzal, Laminar boundary layer on a flat plate at low Prandtl number. Int. J. Heat Mass Transfer, **14**(2), 279–292 (1971)
14. S. Churchill, H. Ozeo, Correlations for laminar forced convection in flow over an isothermal flat plate and in developing and fully developed flow in an isothermal tube. J. Heat Transfer Trans. ASME. **95**(4), 540–541 (1973)
15. V.M. Soundalgekar, H.S. Takhar, M. Singh, Velocity and temperature-field in MHD Falkner-Skan flow. J. Phys. Soc Japan, **50**(9), 3139–3143 (1981)
16. C.A. Forbrich, improved solutions to the Falkner-Skan boundary-layer equation. AIAA J. **20**, 1306–1307 (1982)
17. H. Chuang, improved solutions to the Falkner-Skan boundary-layer equation – comment. AIAA J. **23**, 2004–2005 (1985)
18. Lin HT, Lin LK, Similarity solutions for laminar forced-convection heat transfer from wedged to fluids of any Prandtl number. Int. J. Heat Mass Transfer **30**(6), 1111–1118 (1987)
19. H. Herwig, G. Wickern, The effect of variable properties on laminar boundary layer flow. Warme und Stoffubertragung **20**, 47–57 (1986)
20. L. Yu, L. Yili, New Series expansion method for the solution of the Falkner-Skan equations. AIAA J **27**, 1453–1455 (1989)
21. A. Kumar, B.B. Singh, Some aspects related to the asymptotic solutions of the laminar boundary-layer equations with heat-flux. Astrophys. Space Sci. **165**(1), 41–49 (1990)
22. J.S. Lim, A. Bejan, J.H. Kim, The optimal thickness of a wall with convection on one side. Int. J. Heat Mass Transfer. **35**(7), 1673–1679 (1992)
23. M. Vynnycky, S. Kimura, K. Kanev, et al., Forced convection heat transfer from a flat plate: The conjugate problem. Int. J. Heat Mass Transfer **41**(1), 45–59 (1998)
24. G. Polidori, M. Rebay, J. Padet, Revisited results about the theory of steady laminar forced convection for 2D external boundary layer flows. Int. J. Thermal Sci. **38**(5), 398–409 (1999)
25. M.A. Hossain, M.S. Munir, Mixed convection flow from a vertical flat plate with temperature dependent viscosity. Int. J. Thermal Sci. **39**(2), 173–183 (2000)
26. A. Postelnicu, T. Grosan, I. Pop, The effect of variable viscosity on forced convection flow past a horizontal flat plate in a porous medium with internal heat generation. Mech. Res. Commun. **28**(3), 331–337 (2001)
27. D.A. Nield, A.V. Kuznetsov, Boundary-layer analysis of forced convection with a plate and porous substrate. Acta Mechanica **166**(1–4), 141–148 (2003)
28. G.E. Cossali, Similarity solutions of energy and momentum boundary layer equations for a power-law shear driven flow over a semi-infinite flat plate. Eur. J. Mech B-Fluids, **25**(1), 18–32 (2006)
29. Weyburne DW, Approximate heat transfer coefficients based on variable thermophysical properties for laminar flow over a uniformly heated flat plate. Heat Mass Transfer **44**(7), 805–813 (2008)
30. M.W. Rubenson, H.A. Johnson, A summary of skin friction and heat transfer solution of the laminar boundary layer on a flat plate. Proc. 1948 Heat transfer fluid Mech. Inst.; also Trans. ASME **71**, 383–388 (1949)
31. E.R.G. Eckert, R.M. Drake, *Analysis of Heat and Mass Transfer* (McGraw-Hill, New York, NY 1972)
32. J.W. Rose, Boundary-Layer Flow on a Flat-Plate. Int. J. Heat Mass Transfer, **22**(6), 969–969 (1979)
33. B. Louis, et al., *Convective Heat Transfer,* 2nd edn. (Wiley, New York, NY, 1993)

34. S. Kakaç, et al., *Convective Heat Transfer*, 2nd edn. (CRC Press, Boca Raton, FL), p. 431 (1994)
35. I. Pop, D.B. Ingham, *Convective Heat Transfer – Mathematical and Computational Modelling of Viscous Fluids and Porous Media* (Elsevier, Amsterdam, 2001)
36. T. Cebeci, *Convective Heat Transfer,* 2nd edn (Springer, Heidelberg 2002)
37. H. Schlichting, *Boundary-Layer Theory* (Springer, Heidelberg 2004)
38. W. Nusselt, Die Oberflaechenkondensation des Wasserdampfes, Z VDI, **60**, 541–546, 569–575 (1916)
39. R.D. Cess, Laminar film condensation on a flat plate in the absence of a body force, Zeitschrift für Angewandte Mathematik und Physik (ZAMP), **11**, 426–433 (1960)
40. J.C.Y. Koh, Film condensation in a forced-convection boundary-layer flow. Int J Heat Mass Transfer **5,** 941–954 (1962)
41. I.G. Shekriladze, V.I. Gomelauri, Theoretical study of laminar film condensation of flowing vapor. Int. J. Heat Mass Transfer **9**, 581–591 (1966)
42. Jacobs HR, An integral treatment of combined body force and forced convection in laminar film condensation. Int. J. Heat Mass Transfer **9**, 637–648 (1966)
43. P. Beckett, G. Poots, Laminar film condensation in forced flows, The Quarterly J. Mech. Appl. Maths **25**(1), 125–152 (1972)
44. T. Fujii, H. Uchara, Laminar filmwise condensation on a vertical surface, Int.I. Heat Mass Transfer. **15**, 217–233 (1972)
45. H. Honda, T. Fujii, Condensation of a flowing vapor on a horizontal tube – numerical analysis on a conjugate heat transfer problem, Trans ASME. J Heat Transfer **106**, 841–848 (1984)
46. J.W. Rose, A new interpolation formula for forced-convection on a horizontal surface, Trans ASME. J Heat Transfer **111,** 818–819(1989)
47. S.B. Memory, V.H. Adams, P.J. Marto, Free and forced convection laminar film condensation on horizontal elliptical tubes, Int. J. Heat Mass Transfer, **40**(14), 3395–3406 (1997)
48. F. Méndez, C. Trevinõ, Analysis of a forced laminar film condensation including finite longitudinal heat conduction effects, Heat Mass Transfer **39**, 489–498 (2003)
49. D. Butterwoth, R.G. Sardesai, P. Griffith, A.E. Bergles, *Condensation, In: Heat Exchanger Design Handbook* (Hemisphere, Washington, DC, 1983) Chapter 2.6
50. P.J. Marto, Fundamemtals of condensation, in *Two-Phase Flow Heat Exchangers: Thermal-Hydraulic Fundamentals and Design*, eds. by S. Kakac, A.E. Bergles, E.Q. Fernandes, (Kluwer Academic, Dordrecht, 1988), pp. 221–291
51. T. Fujii, *Theory of Laminar Film Condensation* (Springer, Berlin, 1991)
52. J.G. Collier, J.R. Thome, in *Convective Boiling and Condensation*, 3rd edn (Oxford University Press, New York, NY 1996), pp. 430–487
53. J.W. Rose, Condensation heat transfer. *Heat Mass Transfer* **35**, 479–485 (1999)
54. W.J. Minkowycz, E.M. Sparrow, Condensation heat transfer in the presence of noncondensables, interfacial resistance, superheating, variable properties, and diffusion. Int. J. Heat Mass Transfer **9**, 1125–1144 (1966)
55. D.G. Kroger, W.M. Rohsenow, Condensation heat transfer in presence of non-condensable gas. Int. J. Heat Mass Transfer **11**(1), 15–26 (1968)
56. L. Slegers, R.A. Seban, Laminar film condensation of steam containing small concentration of air. Int. J. Heat Mass Transfer **13**, 1941–1947 (1970)
57. V.E. Denny, A.F. Mills, V.J. Jusionis, Laminar film condensation from a steam-air mixture undergoing forced flow down a vertical surface. J. Heat Transfer **93**, 297–304 (1971)
58. R.K. Oran, C.J. Chen, Effect of lighter noncond-Nsa31 -Gas on laminar film condensation over a vertical plate. Int. J. Heat Mass Transfer **18**, 993–996 (1975)
59. K. Lucas, Combined body forced and forced-convection in laminar-film condensation of mixed vapour-Integral and finite-difference treatment, Int. J. Heat Mass Transfer **19**(11), 1273–1280 (1976)
60. V. M. Borishanskiy, O.P. Volkov, Effect of uncondensable gases content on heat transfer in steam condensation in a vertical tube. Heat Transfer Soviet Res. **9,** 35–41 (1977)

61. J.W. Rose, Approximate equations for forced-convection in the presence of a non-condensing gas on a flat-plate and horizontal tube. Int. J. Heat Mass Transfer. **23**, 539–545 (1980)
62. S. Kotake, Effect of a small amount of noncondensable gas on film condensation of multicomponent mixtures. Int. J. Heat Mass Transfer **28**(2), 407–414 (1985)
63. P.J. Vernier and P. Solignac, A test of some condensation models in the presence of a noncondensable gas against the ecotra experiment. Nucl. Technol. **77**(1), 82–91 (1987)
64. S.M. Chiaasaan, B.K. Kamboj, S.I. Abdel-Khalik, Two-fluid modeling of condensation in the presence of noncondensables in two-phase channel flows. Nucl. Sci. Eng. **119**, 1–17 (1995)
65. V. Srzic, H.M. Soliman, S.J. Ormiston, Analysis of laminar mixedconvection condensation on isothermal plates using the full boundary layer equations: Mixtures of a vapor and a lighter gas. Int. J. Heat Mass Transfer **42**(4), 685–695 (1999)
66. E.C. Siow, S.J. Ormiston, H.M. Soliman, A two-phase model for laminar film condensation from steam–air mixtures in vertical parallel-plate channels. Heat Mass Transfer **40**(5), 365–375 (2004)
67. S.T.Revankar, D. Pollock, Laminar film condensation in a vertical tube in the presence of noncondensable gas. Appl. Math. Model. **29**(4), 341–359 (2005)
68. S. Oh, S.T. Revankar, Experimental and theoretical investigation of film condensation with noncondensable gas. Int. J. Heat Mass Transfer **49**(15–16), 2523–2534 (2006)
69. Tamasawa, *Advances in Condensation Heat Transfer, Advances in Heat Transfer* (Academic, New York, 1991), pp. 55–139
70. K. Stephan, *Heat Transfer in Condensation and Boiling* (Springer, New York, NY, 1992)
71. T. Fujii, *Theory of Laminar Film Condensation* (Springer, New York, NY, 1991)
72. G.F. Hewitt, G.L. Shires, T.R. Bott, *Process Heat Transfer* (CRC Press, Boca Raton, FL, 1994)
73. D.R. Webb, Condensation of vapour mixtures, *Heat Exchanger Design Handbook Update* (Begell House Publishers, New York, 1995)
74. D.Y. Shang, L.C. Zhong, Extensive study on laminar free film condensation from vapor–gas mixture. Int. J. Heat Mass Transfer **51**, 4300–4314 (2008)

Part I
Theoretical Foundation

Chapter 2
Basic Conservation Equations for Laminar Convection

Abstract In this chapter, the basic conservation equations related to laminar fluid flow conservation equations are introduced. On this basis, the corresponding conservation equations of mass, momentum, and energy for steady laminar forced convection boundary layer are obtained.

2.1 Continuity Equation

The conceptual basis for the derivation of the continuity equation of fluid flow is the mass conservation law. The control volume for the derivation of continuity equation is shown in Fig. 2.1 in which the mass conservation principle is stated as

$$\dot{m}_{\text{increment}} = \dot{m}_{\text{in}} - \dot{m}_{\text{out}}. \tag{2.1}$$

where $\dot{m}_{\text{increment}}$ expresses the mass increment per unit time in the control volume, $\dot{m}_{\text{in}}$ represents the mass flowing into the control volume per unit time, and $\dot{m}_{\text{out}}$ is the mass flowing out of the control volume per unit time. The dot notation signifies a unit time.

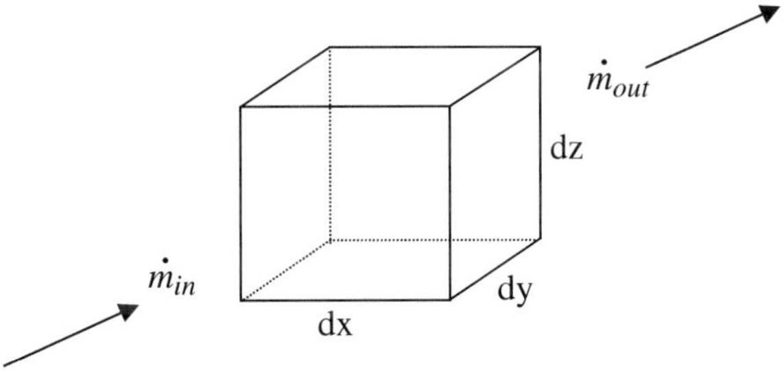

Fig. 2.1 Control volume for derivation of the continuity equations

D.Y. Shang, *Theory of Heat Transfer with Forced Convection Film Flows*, Heat and Mass Transfer, DOI 10.1007/978-3-642-12581-2_2,

In the control volume, the mass of fluid flow is given by $\rho \, \mathrm{d}x \, \mathrm{d}y \, \mathrm{d}z$, and the mass increment per unit time in the control volume can be expressed as

$$\dot{m}_{\text{increment}} = \frac{\partial \rho}{\partial \tau} \mathrm{d}x \, \mathrm{d}y \, \mathrm{d}z. \tag{2.2}$$

The mass flowing per unit time into the control volume in the x direction is given by $\rho w_x \mathrm{d}y \, \mathrm{d}z$. The mass flowing out of the control volume in a unit time in the x direction is given by $\left[\rho w_x + \frac{\partial(\rho w_x)}{\partial x}\mathrm{d}x\right] \mathrm{d}y \, \mathrm{d}z$. Thus, the mass increment per unit time in the x direction in the control volume is given by $\frac{\partial(\rho w_x)}{\partial x}\mathrm{d}x \, \mathrm{d}y \, \mathrm{d}z$. Similarly, the mass increments in the control volume in the y and z directions per unit time are given by $\frac{\partial(\rho w_y)}{\partial y}\mathrm{d}y \, \mathrm{d}x \, \mathrm{d}z$ and $\frac{\partial(\rho w_z)}{\partial z}\mathrm{d}z \, \mathrm{d}x \, \mathrm{d}y$, respectively. We thus obtain

$$\dot{m}_{\text{out}} - \dot{m}_{\text{in}} = \left[\frac{\partial(\rho w_x)}{\partial x} + \frac{\partial(\rho w_y)}{\partial y} + \frac{\partial(\rho w_z)}{\partial z}\right] \mathrm{d}x \, \mathrm{d}y \, \mathrm{d}z. \tag{2.3}$$

With (2.2) and (2.3), (2.1) becomes in Cartesian coordinates:

$$\frac{\partial \rho}{\partial \tau} + \frac{\partial(\rho w_x)}{\partial x} + \frac{\partial(\rho w_y)}{\partial y} + \frac{\partial(\rho w_z)}{\partial z} = 0, \tag{2.4}$$

or in the vector notation

$$\frac{\partial \rho}{\partial \tau} + \nabla \cdot (\rho \vec{W}) = 0, \tag{2.5}$$

where $\vec{W} = i w_x + j w_y + k w_z$ is the fluid velocity.

For steady state, the vector and Cartesian forms of the continuity equation are given by

$$\frac{\partial}{\partial x}(\rho w_x) + \frac{\partial}{\partial y}(\rho w_y) + \frac{\partial}{\partial z}(\rho w_z) = 0, \tag{2.6}$$

and

$$\frac{\partial}{\partial x}(\rho w_x) + \frac{\partial}{\partial y}(\rho w_y) = 0, \tag{2.7}$$

respectively for three- and two- dimensional continuity equations.

2.2 Momentum Equation (Navier–Stokes Equations)

The control volume for derivation of the momentum equation of fluid flow is shown in Fig. 2.2. Take an enclosed surface A that includes the control volume. Meanwhile, take $\vec{F}_{\mathrm{m}}$ as mass force per unit mass fluid, $\vec{F}_{\mathrm{m,v}}$ as the total mass force in the control volume, $\vec{\tau}_n$ as surface force per unit mass fluid flow at per area of surface, $\vec{\tau}_{n,A}$ as surface force in the control volume, $\dot{G}_{\mathrm{increment}}$ as momentum increment of the per unit mass fluid flow at unit time, and $G_{\mathrm{increment}}$ as momentum increment of the fluid flow per unit time in the volume. According to the momentum law, the momentum increment of the fluid flow per unit time in the control volume equals the sum of total mass force and surface forced in the same volume, as

$$G_{\mathrm{increment}} = \vec{F}_{\mathrm{m,v}} + \vec{\tau}_{n,A}. \tag{2.8}$$

$\vec{F}_{\mathrm{m,v}}$, $\vec{\tau}_{n,A}$, and $G_{\mathrm{increment}}$ in the control volume are expressed as, respectively,

$$\vec{F}_{\mathrm{m,v}} = \int_V \rho \vec{F}_{\mathrm{m}} \, \mathrm{d}V, \tag{2.9}$$

$$\vec{\tau}_{n,A} = \int_A \vec{\tau}_n \, dA, \tag{2.10}$$

$$G_{\mathrm{increment}} = \frac{D}{D\tau} \int_V \rho \vec{W} \, \mathrm{d}V. \tag{2.11}$$

According to tensor calculation, the right side of (2.10) is expressed as

$$\int_A \vec{\tau}_n \mathrm{d}A = \int_V \nabla \cdot [\tau] \mathrm{d}V, \tag{2.12}$$

where $\nabla \cdot [\tau]$ is divergence of the shear force tensor.

With (2.9), (2.10), (2.11), and (2.12), (2.8) is rewritten as

$$\frac{D}{D\tau} \int_V \rho \vec{W} \, \mathrm{d}V = \int_V \rho \vec{F}_{\mathrm{m}} \, \mathrm{d}V + \int_V \nabla \cdot [\tau] \mathrm{d}V, \tag{2.13}$$

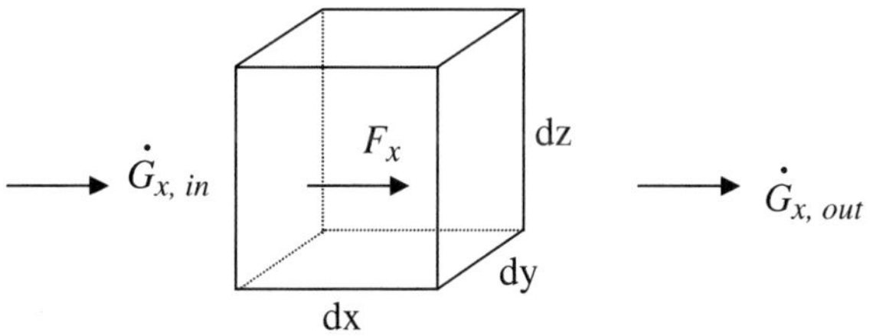

Fig. 2.2 Control volume for derivation of momentum equations

i.e.

$$\int_V \left\{ \frac{D(\rho \vec{W})}{D\tau} - \rho \vec{F} - \nabla \cdot [\tau] \right\} \mathrm{d}V = 0. \tag{2.14}$$

Therefore, we have

$$\frac{D(\rho \vec{W})}{D\tau} = \rho \vec{F} + \nabla \cdot [\tau]. \tag{2.15}$$

This is the Navier–Stokes equations of fluid flow. For Cartesian coordinates, (2.15) can be expressed as

$$\frac{D(\rho w_x)}{D\tau} = \frac{\partial \tau_{xx}}{\partial x} + \frac{\partial \tau_{yx}}{\partial y} + \frac{\partial \tau_{zx}}{\partial z} + \rho g_x, \tag{2.16}$$

$$\frac{D(\rho w_y)}{D\tau} = \frac{\partial \tau_{xy}}{\partial x} + \frac{\partial \tau_{yy}}{\partial y} + \frac{\partial \tau_{zy}}{\partial z} + \rho g_y, \tag{2.17}$$

$$\frac{D(\rho w_z)}{D\tau} = \frac{\partial \tau_{xz}}{\partial x} + \frac{\partial \tau_{yz}}{\partial y} + \frac{\partial \tau_{zz}}{\partial z} + \rho g_z, \tag{2.18}$$

where

$$\tau_{xx} = -\left[p + \frac{2}{3}\mu\left(\frac{\partial w_x}{\partial x} + \frac{\partial w_y}{\partial y} + \frac{\partial w_z}{\partial z}\right)\right] + 2\mu\frac{\partial w_x}{\partial x},$$

$$\tau_{yy} = -\left[p + \frac{2}{3}\mu\left(\frac{\partial w_x}{\partial x} + \frac{\partial w_y}{\partial y} + \frac{\partial w_z}{\partial z}\right)\right] + 2\mu\frac{\partial w_y}{\partial y},$$

$$\tau_{zz} = -\left[p + \frac{2}{3}\mu\left(\frac{\partial w_x}{\partial x} + \frac{\partial w_y}{\partial y} + \frac{\partial w_z}{\partial z}\right)\right] + 2\mu\frac{\partial w_z}{\partial z},$$

$$\tau_{xy} = \tau_{yx} = \mu\left(\frac{\partial w_y}{\partial x} + \frac{\partial w_x}{\partial y}\right),$$

$$\tau_{yz} = \tau_{zy} = \mu\left(\frac{\partial w_z}{\partial y} + \frac{\partial w_y}{\partial z}\right),$$

$$\tau_{zx} = \tau_{xz} = \mu\left(\frac{\partial w_x}{\partial z} + \frac{\partial w_z}{\partial x}\right),$$

g_x, g_y, and g_z are gravity accelerations in x, y, and z directions, respectively.

Then, (2.16) to (2.18) become

$$\begin{aligned}\frac{D(\rho w_x)}{D\tau} = &-\frac{\partial p}{\partial x} + 2\frac{\partial}{\partial x}\left(\mu\frac{\partial w_x}{\partial x}\right) + \frac{\partial}{\partial y}\left[\mu\left(\frac{\partial w_x}{\partial y} + \frac{\partial w_y}{\partial x}\right)\right] + \frac{\partial}{\partial z}\left[\mu\left(\frac{\partial w_x}{\partial z} + \frac{\partial w_z}{\partial x}\right)\right] \\ &- \frac{\partial}{\partial x}\left[\frac{2}{3}\mu\left(\left(\frac{\partial w_x}{\partial x} + \frac{\partial w_y}{\partial y} + \frac{\partial w_z}{\partial z}\right)\right]\right] + \rho g_x,\end{aligned} \tag{2.19}$$

$$\frac{D(\rho w_y)}{D\tau} = -\frac{\partial p}{\partial y} + \frac{\partial}{\partial x}\left[\mu\left(\frac{\partial w_x}{\partial y} + \frac{\partial w_y}{\partial x}\right)\right] + 2\frac{\partial}{\partial y}\left(\mu\frac{\partial w_y}{\partial y}\right) + \frac{\partial}{\partial z}\left[\mu\left(\frac{\partial w_y}{\partial z} + \frac{\partial w_z}{\partial y}\right)\right]$$
$$-\frac{\partial}{\partial y}\left[\frac{2}{3}\mu\left(\frac{\partial w_x}{\partial x} + \frac{\partial w_y}{\partial y} + \frac{\partial w_z}{\partial z}\right)\right] + \rho g_y, \tag{2.20}$$

$$\frac{D(\rho w_z)}{D\tau} = -\frac{\partial p}{\partial z} + \frac{\partial}{\partial x}\left[\mu\left(\frac{\partial w_x}{\partial z} + \frac{\partial w_z}{\partial x}\right)\right] + \frac{\partial}{\partial y}\left[\mu\left(\frac{\partial w_y}{\partial z} + \frac{\partial w_z}{\partial y}\right)\right] + 2\frac{\partial}{\partial z}\left[\mu\frac{\partial w_z}{\partial z}\right)$$
$$-\frac{\partial}{\partial z}\left[\frac{2}{3}\mu\left(\frac{\partial w_x}{\partial x} + \frac{\partial w_y}{\partial y} + \frac{\partial w_z}{\partial z}\right)\right] + \rho g_z \tag{2.21}$$

For steady state, the momentum equations (2.19) – (2.21) are given as follows, respectively,

$$\rho\left(\frac{\partial w_x}{\partial x}w_x + \frac{\partial w_x}{\partial y}w_y + \frac{\partial w_x}{\partial z}w_z\right) + w_x\left(w_x\frac{\partial \rho}{\partial x} + w_y\frac{\partial \rho}{\partial y} + w_z\frac{\partial \rho}{\partial z}\right) =$$
$$-\frac{\partial p}{\partial x} + 2\frac{\partial}{\partial x}\left(\mu\frac{\partial w_x}{\partial x}\right) + \frac{\partial}{\partial y}\left[\mu\left(\frac{\partial w_x}{\partial y} + \frac{\partial w_y}{\partial x}\right)\right] + \frac{\partial}{\partial z}\left[\mu\left(\frac{\partial w_x}{\partial z} + \frac{\partial w_z}{\partial x}\right)\right]$$
$$-\frac{\partial}{\partial x}\left[\frac{2}{3}\mu\left(\frac{\partial w_x}{\partial x} + \frac{\partial w_y}{\partial y} + \frac{\partial w_z}{\partial z}\right)\right] + \rho g_x, \tag{2.22}$$

$$\rho\left(\frac{\partial w_y}{\partial x}w_x + \frac{\partial w_y}{\partial y}w_y + \frac{\partial w_y}{\partial z}w_z\right) + w_y\left(w_x\frac{\partial \rho}{\partial x} + w_y\frac{\partial \rho}{\partial y} + w_z\frac{\partial \rho}{\partial z}\right) =$$
$$-\frac{\partial p}{\partial y} + \frac{\partial}{\partial x}\left[\mu\left(\frac{\partial w_x}{\partial y} + \frac{\partial w_y}{\partial x}\right)\right] + 2\frac{\partial}{\partial y}\left(\mu\frac{\partial w_y}{\partial y}\right) + \frac{\partial}{\partial z}\left[\mu\left(\frac{\partial w_y}{\partial z} + \frac{\partial w_z}{\partial y}\right)\right]$$
$$-\frac{\partial}{\partial y}\left[\frac{2}{3}\mu\left(\frac{\partial w_x}{\partial x} + \frac{\partial w_y}{\partial y} + \frac{\partial w_z}{\partial z}\right)\right] + \rho g_y, \tag{2.23}$$

$$\rho\left(\frac{\partial w_z}{\partial x}w_x + \frac{\partial w_z}{\partial y}w_y + \frac{\partial w_z}{\partial z}w_z\right) + w_z\left(w_x\frac{\partial \rho}{\partial x} + w_y\frac{\partial \rho}{\partial y} + w_z\frac{\partial \rho}{\partial z}\right) =$$
$$-\frac{\partial p}{\partial z} + \frac{\partial}{\partial x}\left[\mu\left(\frac{\partial w_x}{\partial z} + \frac{\partial w_z}{\partial x}\right)\right] + \frac{\partial}{\partial y}\left[\mu\left(\frac{\partial w_y}{\partial z} + \frac{\partial w_z}{\partial y}\right)\right] + 2\frac{\partial}{\partial z}\left(\mu\frac{\partial w_z}{\partial z}\right)$$
$$-\frac{\partial}{\partial z}\left[\frac{2}{3}\mu\left(\frac{\partial w_x}{\partial x} + \frac{\partial w_y}{\partial y} + \frac{\partial w_z}{\partial z}\right)\right] + \rho g_z. \tag{2.24}$$

Let us compare term $\rho\left(\frac{\partial w_x}{\partial x}w_x + \frac{\partial w_x}{\partial y}w_y + \frac{\partial w_x}{\partial z}w_z\right)$ with term $w_x\left(w_x\frac{\partial \rho}{\partial x} + w_y\frac{\partial \rho}{\partial y} + w_z\frac{\partial \rho}{\partial z}\right)$ in (2.22). In general, derivatives $\frac{\partial w_x}{\partial x}$, $\frac{\partial w_x}{\partial y}$, and $\frac{\partial w_x}{\partial z}$ are much larger than the derivatives $\frac{\partial \rho_x}{\partial x}$, $\frac{\partial \rho_x}{\partial y}$, and $\frac{\partial \rho_x}{\partial z}$, respectively. In this case, the term $w_x\left(w_x\frac{\partial \rho}{\partial x} + w_y\frac{\partial \rho}{\partial y} + w_z\frac{\partial \rho}{\partial z}\right)$ is omitted, and (2.22) is rewritten as generally

$$\rho\left(\frac{\partial w_x}{\partial x}w_x+\frac{\partial w_x}{\partial y}w_y+\frac{\partial w_x}{\partial z}w_z\right)=-\frac{\partial p}{\partial x}+2\frac{\partial}{\partial x}\left(\mu\frac{\partial w_x}{\partial x}\right)$$
$$+\frac{\partial}{\partial y}\left[\mu\left(\frac{\partial w_x}{\partial y}+\frac{\partial w_y}{\partial x}\right)\right]+\frac{\partial}{\partial z}\left[\mu\left(\frac{\partial w_x}{\partial z}+\frac{\partial w_z}{\partial x}\right)\right]$$
$$-\frac{\partial}{\partial x}\left[\frac{2}{3}\mu\left(\frac{\partial w_x}{\partial x}+\frac{\partial w_y}{\partial y}+\frac{\partial w_z}{\partial z}\right)\right]+\rho g_x. \quad (2.25)$$

Similarly, in general, (2.23) and (2.24) are rewritten as respectively

$$\rho\left(\frac{\partial w_y}{\partial x}w_x+\frac{\partial w_y}{\partial y}w_y+\frac{\partial w_y}{\partial z}w_z\right)=-\frac{\partial p}{\partial y}+\frac{\partial}{\partial x}\left[\mu\left(\frac{\partial w_x}{\partial y}+\frac{\partial w_y}{\partial x}\right)\right]$$
$$+2\frac{\partial}{\partial y}\left(\mu\frac{\partial w_y}{\partial y}\right)+\frac{\partial}{\partial z}\left[\mu\left(\frac{\partial w_y}{\partial z}+\frac{\partial w_z}{\partial y}\right)\right]$$
$$-\frac{\partial}{\partial y}\left[\frac{2}{3}\mu\left(\frac{\partial w_x}{\partial x}+\frac{\partial w_y}{\partial y}+\frac{\partial w_z}{\partial z}\right)\right]+\rho g_y, \quad (2.26)$$

$$\rho\left(\frac{\partial w_z}{\partial x}w_x+\frac{\partial w_z}{\partial y}w_y+\frac{\partial w_z}{\partial z}w_z\right)=-\frac{\partial p}{\partial z}+\frac{\partial}{\partial x}\left[\mu\left(\frac{\partial w_x}{\partial z}+\frac{\partial w_z}{\partial x}\right)\right]$$
$$+\frac{\partial}{\partial y}\left[\mu\left(\frac{\partial w_y}{\partial z}+\frac{\partial w_z}{\partial y}\right)\right]+2\frac{\partial}{\partial z}\left(\mu\frac{\partial w_z}{\partial z}\right)$$
$$-\frac{\partial}{\partial z}\left[\frac{2}{3}\mu\left(\frac{\partial w_x}{\partial x}+\frac{\partial w_y}{\partial y}+\frac{\partial w_z}{\partial z}\right)\right]+\rho g_z \quad (2.27)$$

2.3 Energy Equation

The control volume for derivation of the energy equation of fluid flow is shown in Fig. 2.3. Take an enclosed surface A that includes the control volume. According to the first law of thermodynamics, we have the following equation:

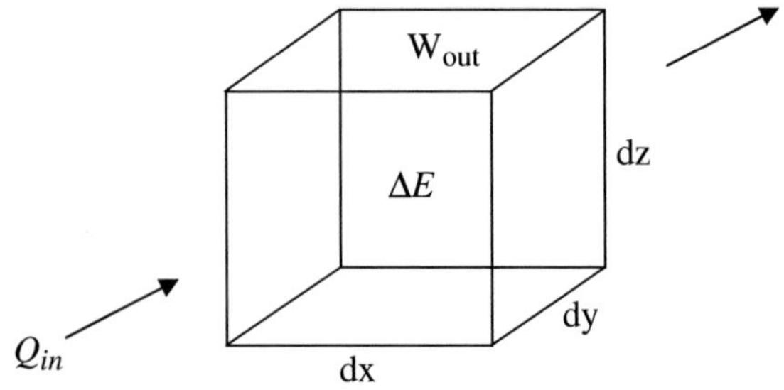

Fig. 2.3 Control volume for derivation of the energy equations of fluid flow

$$\Delta \dot{E} = \dot{Q} + \dot{W}_{\text{out}}, \tag{2.28}$$

where $\Delta \dot{E}$ is energy increment in the system per unit time, $\dot{Q}$ is heat increment in the system per unit time, and $\dot{W}_{\text{out}}$ denotes work done by the mass force and surface force on the system per unit time.

The energy increment per unit time in the system is described as

$$\Delta \dot{E} = \frac{D}{D\tau} \int_V \rho \left(e + \frac{W^2}{2} \right) \mathrm{d}V, \tag{2.29}$$

where τ denotes time, $\frac{W^2}{2}$ is the fluid kinetic energy per unit mass, W is fluid velocity, and the symbol e represents the internal energy per unit mass.

The work done by the mass force and surface force on the system per unit time is expressed as

$$\dot{W}_{\text{out}} = \int_V \rho \vec{F} \cdot \vec{W} \, \mathrm{d}V + \int_A \vec{\tau_n} \cdot \vec{W} \, \mathrm{d}A, \tag{2.30}$$

where $\vec{F}$ is the mass force per unit mass and $\vec{\tau_n}$ is surface force acting on unit area.

The heat increment entering into the system per unit time through thermal conduction is described by using Fourier's law as follows:

$$\dot{Q} = \int_A \lambda \frac{\partial t}{\partial n} \mathrm{d}A, \tag{2.31}$$

where n is normal line of the surface, and here the heat conduction is considered only.

With (2.29) to (2.31), (2.28) is rewritten as

$$\frac{D}{D\tau} \int_V \rho \left(e + \frac{W^2}{2} \right) \mathrm{d}V = \int_V \rho \vec{F} \cdot \vec{W} \, \mathrm{d}V + \int_A \vec{\tau_n} \cdot \vec{W} \, \mathrm{d}A + \int_A \lambda \frac{\partial t}{\partial n} \mathrm{d}A, \tag{2.32}$$

where

$$\frac{D}{D\tau} \int_V \rho \left(e + \frac{W^2}{2} \right) \mathrm{d}V = \int_V \frac{D}{D\tau} \left[\rho \left(e + \frac{W^2}{2} \right) \right] \mathrm{d}V, \tag{2.33}$$

$$\int_A \vec{\tau_n} \cdot \vec{W} \, \mathrm{d}A = \int_A \vec{n} [\tau] \cdot \vec{W} \, \mathrm{d}A = \int_A \vec{n} ([\tau] \cdot \vec{W}) \mathrm{d}A = \int_v \nabla \cdot ([\tau] \cdot \vec{W}) \mathrm{d}V, \tag{2.34}$$

$$\int_A \lambda \frac{\partial t}{\partial n} \mathrm{d}A = \int_V \nabla \cdot (\lambda \nabla t) \mathrm{d}V. \tag{2.35}$$

With (2.33), (2.34), and (2.35), (2.32) is rewritten as

$$\int_V \frac{D}{D\tau}\left[\rho\left(e+\frac{W^2}{2}\right)\right]\mathrm{d}V = \int_V \rho\,\vec{F}\cdot\vec{W}\,\mathrm{d}V + \int_v \nabla\cdot([\tau]\cdot\vec{W})\mathrm{d}V + \int_V \nabla\cdot(\lambda\nabla t)\mathrm{d}V. \tag{2.36}$$

Then

$$\frac{D}{D\tau}\left[\rho\left(e+\frac{W^2}{2}\right)\right] = \rho\,\vec{F}\cdot\vec{W} + \nabla\cdot([\tau]\cdot\vec{W}) + \nabla\cdot(\lambda\nabla t), \tag{2.37}$$

where $[\tau]$ denotes tensor of shear force.

Equation (2.37) is the energy equation.

Through tensor and vector analysis, (2.37) can be further derived into the following form:

$$\frac{D(\rho e)}{D\tau} = [\tau]\cdot[\varepsilon] + \nabla\cdot(\lambda\nabla t). \tag{2.38}$$

Equation (2.38) is another form of the energy equation. Here, $[\tau]\cdot[\varepsilon]$ is the scalar quantity product of force tensor $[\tau]$ and deformation rate tensor $[\varepsilon]$, and represents the work done by fluid deformation surface force. The physical significance of (2.38) is that the internal energy increment of fluid with unit volume during the unit time equals the sum of the work done by deformation surface force of fluid with unit volume and the heat entering the system.

With the general Newtonian law, the tensor of shear force is expressed as

$$[\tau] = 2\mu[\varepsilon] - \left(p + \frac{2}{3}\mu\nabla\cdot\vec{W}\right)[I], \tag{2.39}$$

where $[I]$ is unit tensor.

According to (2.39) the following equation can be obtained:

$$[\tau]\cdot[\varepsilon] = -p\nabla\cdot\vec{W} - \frac{2}{3}\mu(\nabla\cdot\vec{W})^2 + 2\mu[\varepsilon]^2. \tag{2.40}$$

Then, (2.38) can be rewritten as

$$\frac{D(\rho e)}{D\tau} = -p\nabla\cdot\vec{W} + \Phi + \nabla\cdot(\lambda\nabla t), \tag{2.41}$$

where $\Phi = -\frac{2}{3}\mu(\nabla\cdot\vec{W})^2 + 2\mu[\varepsilon]^2$ is viscous thermal dissipation, which is further described as

$$\Phi = \mu\left\{2\left(\frac{\partial w_x}{\partial x}\right)^2 + 2\left(\frac{\partial w_y}{\partial y}\right)^2 + 2\left(\frac{\partial w_z}{\partial z}\right)^2 + \left(\frac{\partial w_x}{\partial y} + \frac{\partial w_y}{\partial x}\right)^2 + \left(\frac{\partial w_y}{\partial z} + \frac{\partial w_z}{\partial y}\right)^2 + \left(\frac{\partial w_z}{\partial x} + \frac{\partial w_x}{\partial z}\right)^2 - \frac{2}{3}\left[\mathrm{div}(\vec{W})\right]^2\right\}, \tag{2.42}$$

where

$$\mathrm{div}(\vec{W}) = \frac{\partial w_x}{\partial x} + \frac{\partial w_y}{\partial y} + \frac{\partial w_z}{\partial z}.$$

According to continuity equation (2.5), we have

$$\nabla \cdot \vec{W} = -\frac{1}{\rho}\frac{D\rho}{D\tau} = \rho\frac{D}{D\tau}\left(\frac{1}{\rho}\right).$$

With the above equation, (2.41) is changed into the following form:

$$\left[\frac{D(\rho e)}{D\tau} + p\rho\frac{D}{D\tau}\left(\frac{1}{\rho}\right)\right] = \Phi + \nabla \cdot (\lambda \nabla t). \tag{2.43}$$

According to thermodynamics equation of fluid, we have

$$\frac{D(\rho h)}{D\tau} = \frac{D(\rho e)}{D\tau} + p\rho\frac{D}{D\tau}\left(\frac{1}{\rho}\right) + \frac{Dp}{D\tau}. \tag{2.44}$$

Then, Equation (2.43) can be expressed as the following enthalpy form:

$$\frac{D(\rho h)}{D\tau} = \frac{Dp}{D\tau} + \Phi + \nabla \cdot (\lambda \nabla t), \tag{2.45}$$

or

$$\frac{D(\rho c_p \cdot t)}{D\tau} = \frac{Dp}{D\tau} + \Phi + \nabla \cdot (\lambda \nabla t), \tag{2.46}$$

where $h = c_p \cdot t$, while c_p is specific heat.

In Cartesian form, the energy (2.46) can be rewritten as

$$\begin{aligned} &\frac{\partial(\rho c_p \cdot t)}{\partial \tau} + w_x\frac{\partial(\rho c_p \cdot t)}{\partial x} + w_y\frac{\partial(\rho c_p \cdot t)}{\partial y} + w_z\frac{\partial(\rho c_p \cdot t)}{\partial z} \\ &= \frac{DP}{D\tau} + \frac{\partial}{\partial x}\left(\lambda\frac{\partial t}{\partial x}\right) + \frac{\partial}{\partial y}\left(\lambda\frac{\partial t}{\partial y}\right) + \frac{\partial}{\partial z}\left(\lambda\frac{\partial t}{\partial z}\right) + \Phi. \end{aligned} \tag{2.47}$$

For steady state, the three-dimensional Cartesian form of the energy equation (2.47) is changed into

$$\begin{aligned} w_x\frac{\partial(\rho c_p t)}{\partial x} + w_y\frac{\partial(\rho c_p t)}{\partial y} + w_z\frac{\partial(\rho c_p t)}{\partial z} &= \frac{\partial}{\partial x}\left(\lambda\frac{\partial t}{\partial x}\right) + \frac{\partial}{\partial y}\left(\lambda\frac{\partial t}{\partial y}\right) \\ &\quad + \frac{\partial}{\partial z}\left(\lambda\frac{\partial t}{\partial z}\right) + \Phi. \end{aligned} \tag{2.48}$$

Above equation is usually approximately rewritten as

$$\rho\left[w_x\frac{\partial(c_p\cdot t)}{\partial x}+w_y\frac{\partial(c_p\cdot t)}{\partial y}+w_z\frac{\partial(c_p\cdot t)}{\partial z}\right]=\frac{\partial}{\partial x}\left(\lambda\frac{\partial t}{\partial x}\right)+\frac{\partial}{\partial y}\left(\lambda\frac{\partial t}{\partial y}\right)+\frac{\partial}{\partial z}\left(\lambda\frac{\partial t}{\partial z}\right)+\Phi, \quad (2.49)$$

where the temperature-dependence of density is ignored.

For steady state, the two-dimensional Cartesian form of the energy equation (2.48) is changed into

$$w_x\frac{\partial(\rho c_p t)}{\partial x}+w_y\frac{\partial(\rho c_p t)}{\partial y}=\frac{\partial}{\partial x}\left(\lambda\frac{\partial t}{\partial x}\right)+\frac{\partial}{\partial y}\left(\lambda\frac{\partial t}{\partial y}\right)+\Phi. \quad (2.50)$$

Above equation is usually approximately rewritten as

$$\rho\left[w_x\frac{\partial(c_p t)}{\partial x}+w_y\frac{\partial(c_p t)}{\partial y}\right]=\frac{\partial}{\partial x}\left(\lambda\frac{\partial t}{\partial x}t\right)+\frac{\partial}{\partial y}\left(\lambda\frac{\partial t}{\partial y}\right)+\Phi. \quad (2.51)$$

With the two-dimensional form, the viscous thermal dissipation is

$$\Phi=\mu\left\{2\left(\frac{\partial w_x}{\partial x}\right)^2+2\left(\frac{\partial w_y}{\partial y}\right)^2+\left(\frac{\partial w_x}{\partial y}+\frac{\partial w_y}{\partial x}\right)^2-\frac{2}{3}\left[div(\vec{W})\right]^2\right\}, \quad (2.52)$$

where

$$div(\vec{W})=\frac{\partial w_x}{\partial x}+\frac{\partial w_y}{\partial y}.$$

2.4 Governing Partial Differential Equations of Laminar Forced Convection Boundary Layers with Consideration of Variable Physical Properties

2.4.1 Principle of the Quantitative Grade Analysis

The general governing partial differential equations can be transformed to the related boundary layer equations. In fact, the only difference between the governing equations of free and forced convection is that buoyant force is only considered in the related momentum equation of free convection. Before the quantitative grade analysis, it is necessary to define its analytical standard. A normal quantitative grade is regarded as $\{1\}$, i.e. unit quantity grade, a very small quantitative grade is regarded as $\{\delta\}$, even very small quantitative grade is regarded as $\{\delta^2\}$, and so on. Then, $\frac{\{1\}}{\{1\}} = \{1\}$, $\frac{\{\delta\}}{\{\delta\}} = \{1\}$, $\frac{\{1\}}{\{\delta\}} = \{\delta^{-1}\}$, $\frac{\{1\}}{\{\delta^2\}} = \{\delta^{-2}\}$.

Furthermore, according to the theory of laminar forced boundary layer, the quantities of the velocity component w_x and the coordinate x can be regarded as unity, i.e. $\{w_x\} = \{1\}$ and $\{x\} = \{1\}$. The quantities of the velocity component w_y and the coordinate y should be regarded as δ, i.e. $\{w_y\} = \{\delta\}$ and $\{y\} = \{\delta\}$. However, the quantity of temperature t can be regarded as $\{t\} = \{1\}$, the quantity grade of the pressure gradient $\frac{\partial p}{\partial x}$ can be regarded as unity, i.e. $\left\{\frac{\partial p}{\partial x}\right\} = \{1\}$, but the quantity grade of the pressure gradient $\frac{\partial p}{\partial y}$ is only regarded as small quantity grade, i.e. $\left\{\frac{\partial p}{\partial y}\right\} = \{\delta\}$.

According to the quantitative grade of the general fluid, the quantitative grade of density ρ, thermal conductivity λ, absolute viscosity μ, and specific heat c_p are $\{\rho\} = \{1\}$, $\{\lambda\} = \{\delta^2\}$, $\{\mu\} = \{\delta^2\}$, and $\{c_p\} = \{1\}$. In addition, for gravity acceleration, we have $\{g_x\} = \{0\}$, and $\{g_y\} = \{1\}$ for forced convection.

2.4.2 Continuity Equation

Based on the (2.7), the steady-state two-dimensional continuity equation is given by

$$\frac{\partial}{\partial x}(\rho w_x) + \frac{\partial}{\partial y}(\rho w_y) = 0 \tag{2.53}$$

where variable fluid density is considered.

According to the above principle for the quantity analysis, now, we take the steady-state two-dimensional continuity equation (2.53) as example to do the quantity analysis as follows:

$$\begin{aligned} &\frac{\partial}{\partial x}(\rho w_x) + \frac{\partial}{\partial y}(\rho w_y) = 0 \\ &\frac{\{1\}}{\{1\}} \quad + \quad \frac{\{\delta\}}{\{\delta\}} \quad = 0 \\ \text{i.e.} \quad &\{1\} \quad + \quad \{1\} \quad = 0 \end{aligned} \tag{2.53a}$$

The above ratios of quantity grade related to the terms of (2.53) shows that both of the two terms of (2.53) should be kept, and (2.53) can be regarded as the continuity equation of the two-dimensionless steady-state forced convection with laminar two-dimensional boundary layer.

2.4.3 Momentum Equations (Navier–Stokes Equations)

The momentum equations (2.25) and (2.26) are rewritten as for steady two-dimensional convection

$$\rho\left(w_x\frac{\partial w_x}{\partial x}+w_y\frac{\partial w_x}{\partial y}\right)=-\frac{\partial p}{\partial x}+2\frac{\partial}{\partial x}\left(\mu\frac{\partial w_x}{\partial x}\right)+\frac{\partial}{\partial y}\left[\mu\left(\frac{\partial w_x}{\partial y}+\frac{\partial w_y}{\partial x}\right)\right]$$
$$-\frac{\partial}{\partial x}\left[\frac{2}{3}\mu\left(\frac{\partial w_x}{\partial x}+\frac{\partial w_y}{\partial y}\right)\right]+\rho g_x, \tag{2.54}$$

$$\rho\left(w_x\frac{\partial w_y}{\partial x}+w_y\frac{\partial w_y}{\partial y}\right)=-\frac{\partial p}{\partial y}+\frac{\partial}{\partial x}\left[\mu\left(\frac{\partial w_x}{\partial y}+\frac{\partial w_y}{\partial x}\right)\right]+2\frac{\partial}{\partial y}\left(\mu\frac{\partial w_y}{\partial y}\right)$$
$$-\frac{\partial}{\partial y}\left[\frac{2}{3}\mu\left(\frac{\partial w_x}{\partial x}+\frac{\partial w_y}{\partial y}\right)\right]+\rho g_y. \tag{2.55}$$

The quantity grades of the terms of (2.54) and (2.55) are expressed as follows, respectively:

$$\rho\left(w_x\frac{\partial w_x}{\partial x}+w_y\frac{\partial w_x}{\partial y}\right)=-\frac{\partial p}{\partial x}+2\frac{\partial}{\partial x}\left(\mu\frac{\partial w_x}{\partial x}\right)+\frac{\partial}{\partial y}\left[\mu\left(\frac{\partial w_x}{\partial y}+\frac{\partial w_y}{\partial x}\right)\right]$$
$$-\frac{\partial}{\partial x}\left[\frac{2}{3}\mu\left(\frac{\partial w_x}{\partial x}+\frac{\partial w_y}{\partial y}\right)\right]+\rho g_x$$

$$\{1\}(\{1\}\frac{\{1\}}{\{1\}}+\{\delta\}\frac{\{1\}}{\{\delta\}})=\{1\}+\frac{\{1\}}{\{1\}}\{\delta^2\}\frac{\{1\}}{\{1\}}+\frac{\{1\}}{\{\delta\}}\{\delta^2\}(\frac{\{1\}}{\{\delta\}}+\frac{\{\delta\}}{\{1\}})$$
$$-\frac{\{1\}}{\{1\}}\delta^2(\frac{\{1\}}{\{1\}}+\frac{\{\delta\}}{\{\delta\}})+\{1\}\{0\} \tag{2.54a}$$

$$\rho\left(w_x\frac{\partial w_y}{\partial x}+w_y\frac{\partial w_y}{\partial y}\right)=-\frac{\partial p}{\partial y}+\frac{\partial}{\partial x}\left[\mu\left(\frac{\partial w_x}{\partial y}+\frac{\partial w_y}{\partial x}\right)\right]+2\frac{\partial}{\partial y}\left(\mu\frac{\partial w_y}{\partial y}\right)$$
$$-\frac{\partial}{\partial y}\left[\frac{2}{3}\mu\left(\frac{\partial w_x}{\partial x}+\frac{\partial w_y}{\partial y}\right)\right]+\rho g_y$$

$$\{1\}(\{\delta\}+\{\delta\})=\{\delta\}+(\{\delta\}+\{\delta^3\})+\{\delta\}-(\{\delta\}(\{1\}+\{1\}))+\{1\}\{0\} \tag{2.55a}$$

Here, for forced convection, the volume forces ρg_x and ρg_y are regarded as zero.

Then, the quantity grades of (2.54a) is further simplified as follows, respectively:

$$\rho\left(w_x\frac{\partial w_x}{\partial x}+w_y\frac{\partial w_x}{\partial y}\right)=-\frac{\partial p}{\partial x}+2\frac{\partial}{\partial x}\left(\mu\frac{\partial w_x}{\partial x}\right)+\frac{\partial}{\partial y}\left[\mu\left(\frac{\partial w_x}{\partial y}+\frac{\partial w_y}{\partial x}\right)\right]$$
$$-\frac{\partial}{\partial x}\left[\frac{2}{3}\mu\left(\frac{\partial w_x}{\partial x}+\frac{\partial w_y}{\partial y}\right)\right]$$

$$\{1\}(\{1\}+\{1\}) = \{1\}+\{\delta^2\}+\{1\}+\{\delta^2\}-(\{\delta^2\}+\{\delta^2\}) \tag{2.54b}$$

Observing the quantity grades in (2.54a) it is found that the terms $2\frac{\partial}{\partial x}\left(\mu\frac{\partial w_x}{\partial x}\right)$, $\frac{\partial w_y}{\partial x}$ in term $\frac{\partial}{\partial y}\left[\mu\left(\frac{\partial w_x}{\partial y}+\frac{\partial w_y}{\partial x}\right)\right]$, $\frac{\partial}{\partial x}\left[\frac{2}{3}\mu\left(\frac{\partial w_x}{\partial x}+\frac{\partial w_y}{\partial y}\right)\right]$ are very small and can be ignored. Then, (2.54a) is simplified as follows:

$$\rho\left(w_x\frac{\partial w_x}{\partial x}+w_y\frac{\partial w_x}{\partial y}\right) = -\frac{\partial p}{\partial x}+\frac{\partial}{\partial y}\left[\mu\left(\frac{\partial w_x}{\partial y}\right)\right]. \tag{2.56}$$

Comparing the quantity grades of (2.55a) with those of (2.54b), it is found that the quantity grades of (2.55a) are very small. Then, (2.55) can be ignored, and only (2.56) is taken as the momentum equation of steady-state forced convection with two-dimensional boundary layer.

2.4.4 Energy Equations

The quantity grades of the terms of (2.51) for laminar two-dimensional energy equation is expressed as

$$\begin{aligned}&\rho\left[w_x\frac{\partial(c_p t)}{\partial x}+w_y\frac{\partial(c_p t)}{\partial y}\right] = \frac{\partial}{\partial x}\left(\lambda\frac{\partial t}{\partial x}\right)+\frac{\partial}{\partial y}\left(\lambda\frac{\partial t}{\partial y}\right)+\Phi\\ &\{1\}\left[\{1\}\cdot\frac{\{1\}}{\{1\}}+\{\delta\}\cdot\frac{\{1\}}{\{\delta\}}\right] = \frac{\{1\}}{\{1\}}\left(\delta^2\frac{\{1\}}{\{1\}}\right)+\frac{\{1\}}{\{\delta\}}\delta^2\frac{\{1\}}{\{\delta\}}\end{aligned} \tag{2.51a}$$

The quantity grade of (2.51a) is simplified to

$$\begin{aligned}&\rho\left[w_x\frac{\partial(c_p t)}{\partial x}+w_y\frac{\partial(c_p t)}{\partial y}\right] = \frac{\partial}{\partial x}\left(\lambda\frac{\partial t}{\partial x}\right)+\frac{\partial}{\partial y}\left(\lambda\frac{\partial t}{\partial y}\right)+\Phi\\ &\{1\}[\{1\}+\{1\}\cdot] = \{1\}(\delta^2\}+\{1\})\end{aligned} \tag{2.51b}$$

With the quantity grade analysis from (2.51b), it is seen that the term $\frac{\partial}{\partial x}\left(\lambda\frac{\partial t}{\partial x}\right)$ is very small compared with other terms, and then can be omitted from the equation. Then, the energy equation for steady-state forced convection with two-dimensional laminar boundary layer is simplified to

$$\rho\left[w_x\frac{\partial(c_p t)}{\partial x}+w_y\frac{\partial(c_p t)}{\partial y}\right] = \frac{\partial}{\partial y}\left(\lambda\frac{\partial t}{\partial y}\right)+\Phi. \tag{2.57}$$

Now, we analyze the viscous thermal dissipation Φ. According to (2.52), Φ is the following equation for the laminar stead state convection:

$$\Phi = \mu\left\{2\left(\frac{\partial w_x}{\partial x}\right)^2 + 2\left(\frac{\partial w_y}{\partial y}\right)^2 + \left(\frac{\partial w_x}{\partial y} + \frac{\partial w_y}{\partial x}\right)^2 - \frac{2}{3}\left(\frac{\partial w_x}{\partial x} + \frac{\partial w_y}{\partial y}\right)^2\right\}. \tag{2.52a}$$

With quantity grade analysis, the right side of (2.52a) is expressed as

$$\Phi = \mu\left\{2\left(\frac{\partial w_x}{\partial x}\right)^2 + 2\left(\frac{\partial w_y}{\partial y}\right)^2 + \left(\frac{\partial w_x}{\partial y} + \frac{\partial w_y}{\partial x}\right)^2 - \frac{2}{3}\left(\frac{\partial w_x}{\partial x} + \frac{\partial w_y}{\partial y}\right)^2\right\}$$

$$\{\delta^2\}\left(\left(\frac{\{1\}}{\{1\}}\right)^2 + \left(\frac{\{\delta\}}{\{\delta\}}\right)^2 + \left(\frac{\{1\}}{\{\delta\}} + \frac{\{\delta\}}{\{1\}}\right)^2 - \left(\frac{\{1\}}{\{1\}} + \frac{\{\delta\}}{\{\delta\}}\right)^2\right)$$

The quantity grade of the above equation is equivalent to

$$\{\delta^2\}\left(\{1\} + \{1\} + \left(\frac{\{1\}}{\{\delta^2\}} + \{\delta^2\}\right) - (\{1\} + \{1\})^2\right) \tag{2.52b}$$

With the quantity grade comparison, it is found that only the term $\mu\left(\frac{\partial w_x}{\partial y}\right)^2$ may be kept. Then, for the stead state forced convection with two-dimensional boundary layer, the viscous thermal dissipation is simplified to

$$\Phi = \mu\left(\frac{\partial w_x}{\partial y}\right)^2. \tag{2.52c}$$

With (2.52c), (2.57) is changed to the following equation

$$\rho\left[w_x\frac{\partial(c_p t)}{\partial x} + w_y\frac{\partial(c_p t)}{\partial y}\right] = \frac{\partial}{\partial y}\left(\lambda\frac{\partial t}{\partial y}\right) + \mu\left(\frac{\partial w_x}{\partial y}\right)^2 \tag{2.58}$$

as the steady-state energy equation with laminar forced convection for two-dimensional boundary layer.

In a later chapter, I will investigate the effect of the viscous thermal dissipation on heat transfer of laminar forced convection.

Now the basic governing partial differential equations for description of mass, momentum, and energy conservation of two-dimensional laminar steady-state forced convection boundary layers are shown as follows with consideration viscous thermal dissipation and variable physical properties:

$$\frac{\partial}{\partial x}(\rho w_x) + \frac{\partial}{\partial y}(\rho w_y) = 0, \tag{2.53}$$

$$\rho\left(w_x\frac{\partial w_x}{\partial x} + w_y\frac{\partial w_x}{\partial y}\right) = -\frac{\partial p}{\partial x} + \frac{\partial}{\partial y}\left[\mu\left(\frac{\partial w_x}{\partial y}\right)\right], \tag{2.56}$$

$$\rho\left[w_x\frac{\partial(c_p t)}{\partial x} + w_y\frac{\partial(c_p t)}{\partial y}\right] = \frac{\partial}{\partial y}\left(\lambda\frac{\partial t}{\partial y}\right) + \mu\left(\frac{\partial w_x}{\partial y}\right)^2. \tag{2.58}$$

Suppose a bulk flow with the velocity $w_{x,\infty}$ beyond the boundary layer, (2.56) is simplified to the following form:

$$w_{x,\infty}\frac{\partial w_{x,\infty}}{\partial x} = -\frac{1}{\rho}\frac{\mathrm{d}p}{\mathrm{d}x},$$

i.e.

$$\frac{\mathrm{d}p}{\mathrm{d}x} = -\rho w_{x,\infty}\frac{\partial w_{x,\infty}}{\partial x}.$$

Then (2.56) is changed to

$$\rho\left(w_x\frac{\partial w_x}{\partial x} + w_y\frac{\partial w_x}{\partial y}\right) = \rho w_{x,\infty}\frac{\partial w_{x,\infty}}{\partial x} + \frac{\partial}{\partial y}\left(\mu\left(\frac{\partial w_x}{\partial y}\right)\right) \tag{2.56a}$$

Obviously, if the main stream velocity $w_{x,\infty}$ is constant, i.e. $\frac{\partial w_{x,\infty}}{\partial x} = 0$, (2.56a) is identical to the following equations:

$$\rho\left(w_x\frac{\partial w_x}{\partial x} + w_y\frac{\partial w_x}{\partial y}\right) = \frac{\partial}{\partial y}\left[\mu\left(\frac{\partial w_x}{\partial y}\right)\right]. \tag{2.56b}$$

For rigorous solutions of the governing equations, the fluid temperature-dependent properties, such as density ρ, absolute viscosity μ, specific heat c_p, and thermal conductivity λ will be considered in the successive chapters of this book.

The laminar forced convection with two-dimensional boundary layer belongs to two-point boundary value problem, which is the basis of three-point boundary value problem for film condensation.

Obviously, the basic governing partial differential equations for mass, momentum, and energy conservation of two-dimensional laminar steady-state forced convection boundary layers with considering viscous thermal dissipation but ignoring the variable physical properties are obtained as based on (2.53), (2.56), and (2.58):

$$\frac{\partial w_x}{\partial x} + \frac{\partial w_y}{\partial y} = 0 \tag{2.59}$$

$$w_x\frac{\partial w_x}{\partial x} + w_y\frac{\partial w_x}{\partial y} = -\frac{1}{\rho}\frac{\partial p}{\partial x} + \nu\frac{\partial^2 w_x}{\partial y^2} \tag{2.60}$$

with $-\frac{1}{\rho}\frac{\mathrm{d}p}{\mathrm{d}x} = w_{x,\infty}\frac{\partial w_{x,\infty}}{\partial x}$

$$\left[w_x\frac{\partial t}{\partial x} + w_y\frac{\partial t}{\partial y}\right] = \frac{\nu}{\Pr}\frac{\partial^2 t}{\partial^2} + \left(\nu/c_p\right)\left(\frac{\partial w_x}{\partial y}\right)^2. \tag{2.61}$$

Now, the three-dimensional basic conservation equations for laminar convection and Two-dimensional basic conservation equations for laminar forced convection boundary layer can be summarized in Tables 2.1 and 2.2, respectively.

2.5 Summary

Table 2.1 Three-dimensional basic conservation equations for laminar convection (with consideration of variable physical properties)

Mass equation

$$\frac{\partial}{\partial x}(\rho w_x) + \frac{\partial}{\partial y}(\rho w_y) + \frac{\partial}{\partial z}(\rho w_z) = 0 \tag{2.6}$$

Momentum equation

$$\rho\left(\frac{\partial w_x}{\partial x}w_x + \frac{\partial w_x}{\partial y}w_y + \frac{\partial w_x}{\partial z}w_z\right) = -\frac{\partial p}{\partial x} + 2\frac{\partial}{\partial x}\left(\mu\frac{\partial w_x}{\partial x}\right) + \frac{\partial}{\partial y}\left[\mu\left(\frac{\partial w_x}{\partial y} + \frac{\partial w_y}{\partial x}\right)\right]$$
$$+\frac{\partial}{\partial z}\left[\mu\left(\frac{\partial w_x}{\partial z} + \frac{\partial w_z}{\partial x}\right)\right] - \frac{\partial}{\partial x}\left[\frac{2}{3}\mu\left(\frac{\partial w_x}{\partial x} + \frac{\partial w_y}{\partial y} + \frac{\partial w_z}{\partial z}\right)\right] + \rho g_x \tag{2.25}$$

$$\rho\left(\frac{\partial w_y}{\partial x}w_x + \frac{\partial w_y}{\partial y}w_y + \frac{\partial w_y}{\partial z}w_z\right) = -\frac{\partial p}{\partial y} + \frac{\partial}{\partial x}[\mu]\left(\frac{\partial w_x}{\partial y} + \frac{\partial w_y}{\partial x}\right) + 2\frac{\partial}{\partial y}\left(\mu\frac{\partial w_y}{\partial y}\right)$$
$$+\frac{\partial}{\partial z}\left[\mu\left(\frac{\partial w_y}{\partial z} + \frac{\partial w_z}{\partial y}\right)\right] - \frac{\partial}{\partial y}\left[\frac{2}{3}\mu\left(\frac{\partial w_x}{\partial x} + \frac{\partial w_y}{\partial y} + \frac{\partial w_z}{\partial z}\right)\right] + \rho g_y \tag{2.26}$$

$$\rho\left(\frac{\partial w_z}{\partial x}w_x + \frac{\partial w_z}{\partial y}w_y + \frac{\partial w_z}{\partial z}w_z\right) = -\frac{\partial p}{\partial z} + \frac{\partial}{\partial x}\left[\mu\left(\frac{\partial w_x}{\partial z} + \frac{\partial w_z}{\partial x}\right)\right]$$
$$+\frac{\partial}{\partial y}\left[\mu\left(\frac{\partial w_y}{\partial z} + \frac{\partial w_z}{\partial y}\right)\right] + 2\frac{\partial}{\partial z}\left(\mu\frac{\partial w_z}{\partial z}\right) - \frac{\partial}{\partial z}\left[\frac{2}{3}\mu\left(\frac{\partial w_x}{\partial x} + \frac{\partial w_y}{\partial y} + \frac{\partial w_z}{\partial z}\right)\right] + \rho g_z \tag{2.27}$$

Energy equation

$$\rho\left[w_x\frac{\partial(c_p\cdot t)}{\partial x} + w_y\frac{\partial(c_p\cdot t)}{\partial y} + w_z\frac{\partial(c_p\cdot t)}{\partial z}\right] = \frac{\partial}{\partial x}\left(\lambda\frac{\partial t}{\partial x}\right) + \frac{\partial}{\partial y}\left(\lambda\frac{\partial t}{\partial y}\right)$$
$$+\frac{\partial}{\partial z}\left(\lambda\frac{\partial t}{\partial z}\right) + \Phi \tag{2.49}$$

$$\Phi = \mu\left\{2\left(\frac{\partial w_x}{\partial x}\right)^2 + 2\left(\frac{\partial w_y}{\partial y}\right)^2 + 2\left(\frac{\partial w_z}{\partial z}\right)^2 + \left(\frac{\partial w_x}{\partial y} + \frac{\partial w_y}{\partial x}\right)^2\right.$$
$$\left. + \left(\frac{\partial w_y}{\partial z} + \frac{\partial w_z}{\partial y}\right)^2 + \left(\frac{\partial w_z}{\partial x} + \frac{\partial w_x}{\partial z}\right)^2 - \frac{2}{3}\left[div(\vec{W})\right]^2\right\}, \tag{2.42}$$

Table 2.2 Two-dimensional basic conservation equations for laminar forced convection boundary layer

With consideration of variable physical properties

Mass equation

$$\frac{\partial}{\partial x}(\rho w_x) + \frac{\partial}{\partial y}(\rho w_y) = 0 \tag{2.7}$$

Momentum equation

$$\rho\left(w_x \frac{\partial w_x}{\partial x} + w_y \frac{\partial w_x}{\partial y}\right) = -\frac{\partial p}{\partial x} + \frac{\partial}{\partial y}\left[\mu\left(\frac{\partial w_x}{\partial y}\right)\right]$$

$$-\frac{\mathrm{d}p}{\mathrm{d}x} = \rho w_{x,\infty}\frac{\partial w_{x,\infty}}{\partial x} \tag{2.56}$$

Energy equation

$$\rho\left[w_x \frac{\partial (c_p t)}{\partial x} + w_y \frac{\partial (c_p t)}{\partial y}\right] = \frac{\partial}{\partial y}\left(\lambda \frac{\partial t}{\partial y}\right) + \mu\left(\frac{\partial w_x}{\partial y}\right)^2 \tag{2.58}$$

(for consideration of viscous thermal dissipation $\mu\left(\frac{\partial w_x}{\partial y}\right)^2$)

With ignoring variable physical properties

Mass equation

$$\frac{\partial w_x}{\partial x} + \frac{\partial w_y}{\partial y} = 0 \tag{2.59}$$

Momentum equation

$$w_x \frac{\partial w_x}{\partial x} + w_y \frac{\partial w_x}{\partial y} = -\frac{1}{\rho}\frac{\partial p}{\partial x} + \nu \frac{\partial^2 w_x}{\partial y^2}$$

$$-\frac{1}{\rho}\frac{\mathrm{d}p}{\mathrm{d}x} = w_{x,\infty}\frac{\partial w_{x,\infty}}{\partial x} \tag{2.60}$$

Energy equation

$$\left[w_x \frac{\partial t}{\partial x} + w_y \frac{\partial t}{\partial y}\right] = \frac{\nu}{\Pr}\frac{\partial^2 t}{\partial^2} + (\nu/c_p)\left(\frac{\partial w_x}{\partial y}\right)^2 \tag{2.61}$$

(for consideration of viscous thermal dissipation $\mu\left(\frac{\partial w_x}{\partial y}\right)^2$)

Chapter 3
Review of Falkner-Skan Type Transformation for Laminar Forced Convection Boundary Layer

Abstract In this chapter, the traditional Falkner–Skan-type transformation for laminar forced convection boundary layer is reviewed. The typical two-dimensional basic conservation equations for laminar forced convection boundary layer are taken as example for derivation of the related similarity variables for Falkner–Skan-type transformation. By means of the stream function and the procedure with the method for group theory, the similarity intermediate function variable $f(\eta)$ is induced. Then, the velocity components are transformed to the related functions with the similarity intermediate function variable $f(\eta)$. On this basis, partial differential momentum equation of the forced convection boundary layer is transformed to related ordinary equation. At last, the limitations of the Falkner–Skan-type transformation are analyzed in detail.

3.1 Introduction

For solution of the laminar boundary layer problem, Falkner and Skan [1] proposed their similarity method in 1931, Now, the widely applied similarity analysis and transformation for the laminar convection and film flows is based on the Falkner-Skan type transformation. So far, the Falkner-Skan type transformation has been collected in numerous publications, but only some of them such as [2–14] are listed here for saving space. So, before the presentation of my related new similarity analysis method in this book for laminar forced convection and film flows, it is necessary here to review briefly the Falkner-Skan type similarity method.

3.2 Basic Conservation Equations

Based on (2.59) to (2.61), the typical two-dimensional basic conservation equations for laminar forced convection boundary layer are shown below with ignoring variable physical properties and viscous thermal dissipation:

$$\frac{\partial}{\partial x}(w_x) + \frac{\partial}{\partial y}(w_y) = 0, \tag{3.1}$$

D.Y. Shang, *Theory of Heat Transfer with Forced Convection Film Flows*,
Heat and Mass Transfer, DOI 10.1007/978-3-642-12581-2_3,

$$w_x \frac{\partial w_x}{\partial x} + w_y \frac{\partial w_x}{\partial y} = -\frac{1}{\rho_f}\frac{\partial p}{\partial x} + \nu_f \frac{\partial^2 w_x}{\partial y^2}, \tag{3.2}$$

$$-\frac{1}{\rho_f}\frac{\mathrm{d}p}{\mathrm{d}x} = w_{x,\infty}\frac{\partial w_{x,\infty}}{\partial x},$$

$$w_x \frac{\partial t}{\partial x} + w_y \frac{\partial t}{\partial y} = \frac{\nu_f}{\mathrm{Pr}_f}\frac{\partial^2 t}{\partial y^2}, \tag{3.3}$$

with the boundary condition equations

$$\mathrm{y} = 0\colon\ \mathrm{w_x} = 0,\ \ \mathrm{w_y} = 0, \qquad t = t_w, \tag{3.4}$$

$$\mathrm{y} \to \infty: \qquad w_x = w_{x,\infty},\ t = t_\infty. \tag{3.5}$$

Here, w_x and w_y are velocity components of the fluids in x and y directions, respectively, $w_{x,\infty}$ is the constant mainstream velocity, t is temperature. While the subscript "f" is induced in the equations for inferring in particular to the constant physical properties with the average temperature of boundary layer, i.e. $t_f = \frac{t_w + t_\infty}{2}$ (the same below). Here, (3.1), (3.2), and (3.3) will be taken as a basis for introduction of the Falkner-Skan type transformation method.

3.3 Derivation Review of Similarity Variables of Falkner-Skan Transformation on Laminar Forced Convection

For introducing the Falkner-Skan type transformation, first the solution of the above equations is expressed as the following form:

$$\frac{w_x}{w_{x,\infty}} = \phi_1(x, y), \tag{3.6}$$

$$\frac{t - t_\infty}{t_w - t_\infty} = \phi_2(x, y). \tag{3.7}$$

In some special cases, (3.6) and (3.7) can be rewritten as

$$\frac{\mathrm{w_x}}{\mathrm{w_{x,\infty}}} = \phi_1(\eta), \tag{3.8}$$

$$\frac{t - t_\infty}{t_w - t_\infty} = \phi_2(\eta), \tag{3.9}$$

where η is dimensionless similarity variable, a function of x and y. Later, it will be shown that η is proportional to $\frac{y}{\delta(x)}$ where $\delta(x)$ is some function of x. Then, the number of independent coordinate variables can be reduced from two to one.

A prerequisite of the Falkner-Skan transformation is the derivation of the similarity variables. The typical method for this procedure is group theory, which was discussed at length by Hansen [15] and Na [16]. For application of this method, a stream function $\psi(x, y)$ is assumed to satisfy the equation (3.1), i..e

$$\mathrm{w_x} = \frac{\partial \psi}{\partial \mathrm{y}}, \qquad \mathrm{w_y} = -\frac{\partial \psi}{\partial \mathrm{x}} \tag{3.10}$$

Using (3.10), (3.2) is transformed into

$$\frac{\partial \psi}{\partial y}\frac{\partial^2 \psi}{\partial x \partial y} - \frac{\partial \psi}{\partial x}\frac{\partial^2 \psi}{\partial y^2} = w_{x,\infty}\frac{dw_{x,\infty}}{dx} + \nu\frac{\partial^3 \psi}{\partial y^3} \tag{3.11}$$

Subsequently, (3.11) is solved by using the group theory. For this purpose, it is important to introduce the following linear transformation:

$$x = A^{a_1}\overline{x}, \quad y = A^{a_2}\overline{y}, \quad \psi = A^{a_3}\overline{\psi}, \quad \mathrm{w}_{x,\infty} = A^{a_4}\overline{w_{x,\infty}}, \tag{3.12}$$

where a_1, a_2, a_3 and a_4 are constant, and A is the transformation parameter.

With (3.12), (3.11) can be rewritten as

$$A^{2a_3-a_1-2a_2}\left(\frac{\partial \overline{\psi}}{\partial \overline{y}}\frac{\partial^2 \overline{\psi}}{\partial \overline{x}\partial \overline{y}} - \frac{\partial \overline{\psi}}{\partial \overline{x}}\frac{\partial^2 \overline{\psi}}{\partial \overline{x}^2}\right) = A^{2a_4-a_1}\overline{w_{x,\infty}}\frac{d\overline{w_{x,\infty}}}{d\overline{x}} + \nu A^{a_3-3a_2}\frac{\partial^3 \overline{\psi}}{\partial \overline{y}^3}. \tag{3.13}$$

From (3.13), due to the constancy of the variables, we compare the exponents of A in each of the terms and obtain

$$2a_3 - a_1 - 2a_2 = 2a_4 - a_1 = a_3 - 3a_2. \tag{3.14}$$

The incomplete solution of (3.14) is

$$a_3 = a_1 - a_2, \quad a_4 = a_3 - a_2. \tag{3.15}$$

Defining a new variable $a = \dfrac{a_2}{a_1}$, (3.15) can be rewritten as

$$\frac{a_3}{a_1} = 1 - a, \quad \frac{a_4}{a_1} = 2\frac{a_3}{a_1} - 1 = 1 - 2a. \tag{3.16}$$

From (3.12) we get

$$A = \left(\frac{x}{\overline{x}}\right)^{\frac{1}{a_1}} = \left(\frac{y}{\overline{y}}\right)^{\frac{1}{a_2}} = \left(\frac{\psi}{\overline{\psi}}\right)^{\frac{1}{a_3}} = \left(\frac{w_{x,\infty}}{\overline{w_{x,\infty}}}\right)^{\frac{1}{a_4}}. \tag{3.17}$$

Using (3.16), (3.17) can be transformed as

$$\frac{y}{x^a} = \frac{\overline{y}}{\overline{x}^a}, \quad \frac{\psi}{x^{1-a}} = \frac{\overline{\psi}}{(\overline{x})^{1-a}}, \qquad \frac{w_{x,\infty}}{x^{1-2a}} = \frac{\overline{w_{x,\infty}}}{(\overline{x})^{1-2a}}. \tag{3.18}$$

These combined variables in (3.18) are called absolute variables in the linear transformation of (3.12). According to Morgan theorem [15], if the boundary conditions of the velocity field of (3.4) and (3.5) can also be transformed, these absolute variables are similarity variables. We set now

$$\eta = \frac{y}{x^a}, \qquad f(\eta) = \frac{\psi}{x^{1-a}}, \tag{3.19}$$

$$h(\eta) = \frac{w_{x,\infty}}{x^{1-2a}} = \text{constant} = \text{C}. \tag{3.20}$$

Suppose the main stream velocity $w_{x,\infty}$ is only function of x, then it is never inconstant function of η. Because variable y is induced in η, $h(\eta)$ must be constant.

The second part of (3.19) can be rewritten as

$$\psi = x^{1-a} f(\eta)\frac{x^a}{x^a} = x^{1-2a} x^a f(\eta).$$

With (3.20) this equation can be transformed into

$$\psi = \frac{w_{x,\infty}}{C} x^a f(\eta). \tag{3.21}$$

By using the first equation of (3.10), the first equation of (3.19), and (3.21), w_x is rewritten as

$$w_x = \frac{\partial \psi}{\partial y} = \frac{\partial \psi}{\partial \eta}\frac{\partial \eta}{\partial y} = \frac{w_{x,\infty}}{C}\frac{\partial f(\eta)}{\partial \eta},$$

or

$$\frac{w_x}{w_{x,\infty}} = \frac{f'(\eta)}{C}. \tag{3.22}$$

Now, we set $m = 1 - 2a$. From (3.20) we obtain

$$w_{x,\infty} = Cx^m. \tag{3.23}$$

Using (3.23) the first equation of (3.19) will be transformed to

$$\eta = \frac{y}{x^{(1-m)/2}} = \left(\frac{w_{x,\infty}}{Cx}\right)^{1/2} y = \left(\frac{w_{x,\infty}x}{C}\right)^{1/2}\frac{y}{x}.$$

Replace C with $2\nu_f$, the above equation is

$$\eta = \frac{y}{x}\left(\frac{w_{x,\infty}x}{2\nu_f}\right)^{1/2} = \frac{y}{x}\left(\frac{1}{2}\text{Re}_{x,f}\right)^{1/2} \tag{3.24}$$

Here, $\mathrm{Re}_{x,f}$ is local Reynolds number without consideration of variable physical properties, i.e.

$$\mathrm{Re}_{x,f} = \frac{w_{x,\infty} x}{\nu_f}. \tag{3.25}$$

With (3.23), (3.21) will be

$$\psi = \frac{w_{x,\infty} x^{(1-m)/2}}{C} f(\eta) = x^{(1+m)/2} f(\eta) = \left(\frac{w_{x,\infty} x}{C}\right)^{1/2} f(\eta), \tag{3.26}$$

where Ψ has dimensions $\frac{m^2}{s}$. Replacing C with $(2\nu_f)^{-1}$, the correct dimension of Ψ can be obtained,

$$\psi = (2 w_{x,\infty} \nu_f x)^{1/2} f(\eta). \tag{3.27}$$

Equations (3.24), (3.25), (3.26), and (3.27) are defined as the Faulkner-Scan transformations for laminar forced convection.

Using (3.27) the velocity component w_x can be derived from the (3.10) as follows:

$$\begin{aligned} w_x &= (2 w_{x,\infty} \nu_f x)^{1/2} f'(\eta) \frac{\partial \eta}{\partial y} \\ &= (2 w_{x,\infty} \nu_f x)^{1/2} f'(\eta) \left(\frac{w_{x,\infty}}{2 \nu_f x}\right)^{1/2}, \end{aligned}$$

i.e.

$$w_x = w_{x,\infty} f'(\eta). \tag{3.28}$$

Employing (3.27), the velocity component $\mathrm{w_y}$ can be derived from the (3.10) as follows:

$$\begin{aligned} \mathrm{w_y} &= -(2 w_{x,\infty} \nu_f x)^{1/2} f'(\eta) \frac{\partial \eta}{\partial x} - \frac{1}{2} (2 w_{x,\infty} \nu_f)^{1/2} x^{-1/2} \cdot f(\eta) \\ &= -(2 w_{x,\infty} \nu_f x)^{1/2} f'(\eta) \frac{\partial \eta}{\partial x} - \frac{1}{2} (2 w_{x,\infty} \nu_f x)^{-1/2} \cdot 2 w_{x,\infty} \nu_f \cdot f(\eta). \end{aligned}$$

With (3.24), we have

$$\frac{\partial \eta}{\partial x} = -\frac{1}{2} \left(\frac{1}{2} \frac{w_{x,\infty}}{\nu_f x^3}\right)^{1/2} y = -\frac{1}{2} x^{-1} \eta, \tag{3.29}$$

$$\frac{\partial \eta}{\partial y} = \frac{1}{x} \left(\frac{1}{2} \mathrm{Re}_{x,f}\right)^{1/2} = \left(\frac{1}{2} \frac{w_{x,\infty}}{\nu_f x}\right)^{1/2}.$$

Then,

$$w_y = -(2w_{x,\infty}\nu_f x)^{1/2} f'(\eta)\left(-\frac{1}{2}x^{-1}\eta\right) - \frac{1}{2}(2w_{x,\infty}\nu_f x)^{-1/2}\cdot 2w_{x,\infty}\nu_f \cdot f(\eta)$$

$$= \frac{1}{2}\left(\frac{2w_{x,\infty}\nu_f}{x}\right)^{1/2}\eta\cdot f'(\eta) - \frac{1}{2}(2w_{x,\infty}\nu_f x)^{-1/2}\cdot 2w_{x,\infty}\nu_f\cdot f(\eta)$$

$$= \frac{1}{2}w_{x,\infty}\left(\frac{2\nu_f}{w_{x,\infty}x}\right)^{1/2}\eta\cdot f'(\eta) - \frac{1}{2}(2w_{x,\infty}\nu_f x)^{-1/2}\cdot 2w_{x,\infty}\nu_f\cdot f(\eta)$$

$$= \frac{1}{2}w_{x,\infty}\left(\frac{w_{x,\infty}x}{2\nu_f}\right)^{-1/2}\eta\cdot f'(\eta) - \frac{1}{2}\left(\frac{w_{x,\infty}x}{2\nu_f}\right)^{-1/2}\cdot w_{x,\infty}\cdot f(\eta)$$

Then

$$w_y = \frac{1}{2}w_{x,\infty}\left(\frac{1}{2}\mathrm{Re}_{x,f}\right)^{-1/2}[\eta\cdot f'(\eta) - f(\eta)] \tag{3.30}$$

Equation (3.9) can be rewritten as

$$\theta(\eta) = \frac{t - t_\infty}{t_w - t_\infty}. \tag{3.31}$$

Now the expressions with the dimensionless similarity analysis and transformation derived by the Falkner-Skan type transformation are summarized as

$$\eta = \frac{y}{x}\left(\frac{1}{2}\mathrm{Re}_{x,f}\right)^{1/2}, \tag{3.24}$$

$$\mathrm{Re}_{x,f} = \frac{w_{x,\infty}x}{\nu_f}, \tag{3.25}$$

$$w_x = w_{x,\infty}f'(\eta), \tag{3.28}$$

$$w_y = \frac{1}{2}w_{x,\infty}\left(\frac{1}{2}\mathrm{Re}_{x,f}\right)^{-1/2}[\eta\cdot f'(\eta) - f(\eta)], \tag{3.30}$$

$$\theta(\eta) = \frac{t - t_\infty}{t_w - t_\infty}. \tag{3.31}$$

3.4 Example of Similarity Transformation with Falkner-Skan Transformation

Now, the governing partial differential equations (3.2) and (3.3) can be transformed into the ordinary equations by (3.24), (3.25), (3.28), (3.30), and (3.31) with the similarity variables η, $\mathrm{Re}_{x,f}$, $f(\eta)$, and $\theta(\eta)$ obtained by the Falkner-Skan type transformation method. Since the stream function already satisfies (3.1), it is not necessary to transform (3.1).

From (3.24) and (3.28), we have

$$\frac{\partial w_x}{\partial x} = w_{x,\infty} f''(\eta) \frac{\partial \eta}{\partial x} = -\frac{1}{2} x^{-1} \eta w_{x,\infty} f''(\eta), \tag{3.32}$$

$$\frac{\partial w_x}{\partial y} = w_{x,\infty} f''(\eta) \frac{\partial \eta}{\partial y} = w_{x,\infty} f''(\eta) \left(\frac{1}{2} \frac{w_{x,\infty}}{\nu_f x} \right)^{1/2}, \tag{3.33}$$

$$\frac{\partial^2 w_x}{\partial y^2} = w_{x,\infty} f'''(\eta) \left(\frac{1}{2} \frac{w_{x,\infty}}{\nu_f x} \right). \tag{3.34}$$

On these bases, if the main stream velocity $w_{x,\infty}$ is constant, i.e. $\frac{\mathrm{d}w_{x,\infty}}{\mathrm{d}x} = 0$, (3.2) is transformed into

$$\begin{aligned} &- w_{x,\infty} f'(\eta) \frac{1}{2} x^{-1} \eta w_{x,\infty} f''(\eta) + \left(\frac{w_{x,\infty} \nu_f}{2x} \right)^{1/2} \\ &\quad [\eta f'(\eta) - f(\eta)] w_{x,\infty} f''(\eta) \left(\frac{1}{2} \frac{w_{x,\infty}}{\nu_f x} \right)^{1/2} \\ &= \nu_f \cdot w_{x,\infty} f'''(\eta) \left(\frac{1}{2} \frac{w_{x,\infty}}{\nu_f x} \right). \end{aligned}$$

The above equation is simplified to

$$f'''(\eta) + f(\eta) f''(\eta) = 0. \tag{3.35}$$

This is corresponding to the well-known Blasius's equation [17]. In fact, the original equation derived by Blasius is $2f'''(\eta) + f(\eta) f''(\eta) = 0$, which is corresponding to his supposition of the steam function $\psi = (w_{x,\infty} \nu_f x)^{1/2} f(\eta)$. While, for (3.27), the stream function ψ should be taken here as $\psi = (2 w_{x,\infty} \nu_f x)^{1/2} f(\eta)$.

In addition, the temperature derivatives in energy equation (3.3) can be derived as

$$\frac{\partial t}{\partial x} = (t_w - t_\infty) \theta'(\eta) \frac{\partial \eta}{\partial x} = -\frac{1}{2} x^{-1} \eta (t_w - t_\infty) \theta'(\eta), \tag{3.36}$$

$$\frac{\partial t}{\partial y} = (t_w - t_\infty) \theta'(\eta) \frac{\partial \eta}{\partial y} = (t_w - t_\infty) \theta'(\eta) \left(\frac{1}{2} \frac{w_{x,\infty}}{\nu_f x} \right)^{1/2}, \tag{3.37}$$

$$\frac{\partial^2 t}{\partial y^2} = (t_w - t_\infty) \theta''(\eta) \left(\frac{1}{2} \frac{w_{x,\infty}}{\nu_f x} \right). \tag{3.38}$$

Then, with (3.28), (3.30), (3.36) (3.37) and (3.38), (3.3) is transformed below

$$- w_{x,\infty} f'(\eta) \frac{1}{2} x^{-1} \eta (t_w - t_\infty) \theta'(\eta) + \left(\frac{w_{x,\infty} \nu_f}{2x} \right)^{1/2}$$

$$[\eta \cdot f'(\eta) - f(\eta)](t_w - t_\infty)\theta'(\eta)\left(\frac{1}{2}\frac{w_{x,\infty}}{\nu_f x}\right)^{1/2}$$
$$= \frac{\nu_f}{\Pr_f}(t_w - t_\infty)\theta''(\eta)\left(\frac{1}{2}\frac{w_{x,\infty}}{\nu_f x}\right)$$

Simplifying this equation, we obtain

$$-f'(\eta)\theta'(\eta)\eta + [\eta \cdot f'(\eta) - f(\eta)]\theta'(\eta) = \frac{1}{\Pr_f}\theta''(\eta),$$

i.e.

$$\theta''(\eta) + \Pr_f f(\eta)\theta'(\eta) = 0. \tag{3.39}$$

Now, the similarity governing ordinary equations for laminar forced convection transformed by Falkner-Skan transformation are collected below:

$$f'''(\eta) + f(\eta)f''(\eta) = 0, \tag{3.35}$$

$$\theta''(\eta) + \Pr_f f(\eta)\theta'(\eta) = 0. \tag{3.39}$$

The boundary conditions for equations (3.4) and (3.5) are now transformed into

$$\eta = 0: \quad f(\eta) = f'(\eta) = 0 \quad \theta(\eta) = 1 \tag{3.40}$$

$$\eta \to \infty: \quad f(\eta) = 1, \quad \theta(\eta) = 0 \tag{3.41}$$

3.5 Summary

Now it is time to summary the related basic conservation equations for laminar forced convection boundary layer, the similarity variables based on the Falkner-Skan type transformation, and the related transformed dimensionless similarity equations in Table 3.1.

3.6 Limitations of the Falkner-Skan Type Transformation

From the similarity governing equations on laminar forced convection boundary layer transformed by using Falkner-Skan type transformation, it is not difficult to find some disadvantages of this traditional similarity method.:

a. By means of group theory, derivation on a suitable stream function and obtaining the appropriate dimensional variables related to two-dimensional velocity components are complicated processes, which have caused a restriction for applica-

Table 3.1 Basic conservation equations for laminar forced convection boundary layer (without consideration of variable physical properties), the similarity variables based on the Falkner–Skan-type transformation, and the similarity conservation equations transformed by the Falkner-Skan transformation

Two-Dimensional Basic Conservation Equations for Laminar Forced Convection Boundary Layer

with ignoring variable physical properties

Mass equation

$$\frac{\partial w_x}{\partial x} + \frac{\partial w_y}{\partial y} = 0$$

Momentum equation

$$w_x \frac{\partial w_x}{\partial x} + w_y \frac{\partial w_x}{\partial y} = \nu_f \frac{\partial^2 w_x}{\partial y^2}$$

Energy equation

$$\left[w_x \frac{\partial t}{\partial x} + w_y \frac{\partial t}{\partial y} \right] = \frac{\nu_f}{\mathrm{Pr}_f} \frac{\partial^2 t}{\partial y^2}$$

$$\left(\text{with ignoring viscous thermal dissipation} \mu \left(\frac{\partial w_x}{\partial y} \right)^2 \right)$$

Boundary conditions

$$y = 0: \quad w_x = 0, \; w_y = 0, \quad t = t_w$$

$$y \to \infty: \quad w_x = w_{x,\infty}, \quad t = t_\infty$$

Equations with dimensionless similarity variables

Local Reynolds number $\mathrm{Re}_{x,f}$

$$\mathrm{Re}_{x,f} = \frac{w_{x,\infty} x}{\nu_f}$$

Coordinate variable η

$$\eta = \frac{y}{x} \left(\frac{1}{2} \mathrm{Re}_{x,f} \right)^{1/2}$$

Dimensionless temperature θ

$$\theta = \frac{t - t_\infty}{t_w - t_\infty}$$

Velocity component w_x

$$w_x = w_{x,\infty} \cdot f'(\eta)$$

Velocity component w_y

$$w_y = \frac{1}{2} w_{x,\infty} \left(\frac{1}{2} Re_{x,f} \right)^{-1/2} [\eta \cdot f'(\eta) - f(\eta)]$$

Table 3.1 (continued)

Transformed governing ordinary differential equations	
Mass equation	-
Momentum equation	$f'''(\eta) + f(\eta) \cdot f''(\eta) = 0$
Energy equation	$\theta''(\eta) + \mathrm{Pr}_f \cdot f(\eta) \cdot \theta'(\eta) = 0$
Boundary conditions	$\eta = 0 : f(\eta) = f'(\eta) = 0, \theta(\eta) = 1$ $\eta \to \infty : f'(\eta) = 1, \theta(\eta) = 0$

tion of the Falkner-Skan type transformation on extensive study of laminar forced convection boundary layer and two-phase film flows.

b. By using the Falkner-Skan type similarity transformation, the dimensionless function $f(\eta)$ and its derivatives become the main dimensionless similarity variables on momentum field, and the velocity components w_x and w_y are algebraic expressions with the function $f(\eta)$ and its derivatives. Then, inconvenience is caused for investigation of the velocity field of laminar forced convection. Such inconvenience even causes difficult investigation of two-phase forced film condensation.
c. With the Falkner-Skan type transformation, a difficulty is encountered for similarity transformation of the governing equations for the laminar forced convection with consideration of variable thermophysical properties. It is because that derivation for obtaining an appropriate stream function expression as well as the expressions of the two-dimensional velocity components is difficult work, in particular consideration of variable physical properties. Since rigorous consideration of variable physical properties is closely related to the reliability of investigation of laminar forced convection and two-phase film flows, this difficulty has hindered the research development of laminar forced convection and its two-phase film flows

To resolve the above problems caused by the traditional Falkner-Skan type transformation, in the following chapters, a related novel similarity analysis method will be presented for extensive investigation of laminar forced convection and two-phase film flows.

Exercises

1. In which special cases, (3.4) and (3.5) can be rewritten as $\dfrac{w_x}{w_{x,\infty}} = \phi_1(\eta)$ and $\dfrac{t - t_\infty}{t_w - t_\infty} = \phi_2(\eta)$ for similarity analysis and transformation?

2. Suppose that the dimensionless coordinate variable η in (3.24) is replaced by the form $\eta = \left(\frac{w_{x,\infty}}{\nu_f x}\right)^{1/2} y = \frac{y}{x}(\mathrm{Re}_{x,f})^{1/2}$, and the forms of other similarity variables are kept, please try to transform similarly the governing partial differential equations (3.1), (3.2), and (3.3) and the boundary condition equations (3.4) and (3.5) to the related ordinary differential equations.

References

1. V.M. Falkner, S.W. Skan, Some approximate solutions of the boundary layer equations. Phil. Mag. **12**, 865 (1931)
2. E.M. Sparrow and J.L. Gregg, The variable fluid property problem in free convection, Trans. ASME **80**, 879–886 (1958)
3. B. Gebhart, Natural convection flow, instability, and transition. J. Heat Transfer **91**, 293–309 (1969)
4. E.R.G. Eckert, R.M. Drake, *Analysis of Heat and Mass Transfer* (McGraw-Hill, New York, NY, 1972)
5. H. Schlichting, *Boundary Layer Theory.* trans. by J. Kestin (McGraw-Hill, New York, NY, 1979) pp. 316–317
6. D.D. Gray, A. Giogini, The validity of the Boussinesq approximation for liquids and gases. Int. J. Heat Mass Transfer **19** 545–551 (1977)
7. A.M. Clausing, S.N. Kempka, The influences of property variations on natural convection from vertical surfaces. J. Heat Transfer **103**, 609–612 (1981)
8. T. Cebeci, P. Bradshaw, *Physical and Computational Aspects of Convective Heat Transfer* (Springer, New York, NY 1984)
9. T. Fujii, *Theory of Laminar Film Condensation* (Springer, New York, NY, 1991)
10. Louis C. Burmeister, *Convective Heat Transfer,* 2nd edn, (Wiley, New York, NY, 1993)
11. S. Kakaç, Y. Yenner, *Convective Heat Transfer*, 2nd edn, (CRC Press, 1995)
12. A. Bejan, *Convection Heat Transfer,* 2nd edn, (Wiley, New York, NY, 1995)
13. I. Pop, D.B. Ingham, *Convective Heat Transfer – Mathematical and Computational Modelling of Viscous Fluids and Porous Media* (Elsevier, Amsterdam 2001)
14. T. Cebeci, *Convective Heat Transfer*, 2nd edn (Springer, Heidelberg 2002)
15. A.G. Hansen, *Similarity Analysis of Boundary Value Problems in Engineering* (Prentice Hall, Englewood Cliffs, NJ, 1964)
16. T.Y. Na, *Computational Methods in Engineering Boundary Value Problems* (Academic, New York, NY, 1979)
17. Blasius, H, Grenzschichten in Flussigkeiten mit kleiner Reibung. Z. Math. Phys. **56**, 1–37 (1908)

Chapter 4
A New Similarity Analysis Method for Laminar Forced Convection Boundary Layer

Abstract A new similarity analysis method with a new set of dimensionless similarity variables is provided for complete similarity transformation of the governing partial differential equations of laminar forced convection and two-phase film flows. First, the derivation of the Reynolds number together with the Nusselt number and Prandtl number is reviewed by means of Buckingham π-theorem and dimension analysis, where the Reynolds number is taken as the one of the new set of dimensionless analysis variables. Then the essential work focuses on derivation of equations for the dimensionless velocity components and the dimensionless coordinate variable, by means of a detailed analysis of quantity grade of the governing conservation partial differential equations of laminar forced convection. On this basis, the new similarity analysis method is produced for complete similarity transformation of the conservation partial differential equations of laminar forced convection and its film flows. Owing to dimensionless velocity components devoted in this chapter, the new similarity analysis method has obvious advantages compared with the Falkner–Skan transformation. These advantages are the following: (i) more convenient for consideration and treatment of the variable physical properties, (ii) more convenient for analysis and investigation of the two-dimensional velocity field, and (iii) more convenient for satisfaction of the interfacial mass transfer matching conditions in the numerical calculation and for rigorous investigation of mass transfer for two-phase film flows with three-point boundary problem. These advantages will be found from the successive chapters.

4.1 Introduction

In Chap. 3, we have briefly reviewed the Falkner-Skan type transformation. Up to now, Falkner-Skan type transformation is a main approach applied for similarity analysis and transformation of the governing partial differential equations of laminar forced convection and its two-phase film flows. With this method a stream function need to be induced, and then the dimensionless variable function $f(\eta)$ and its derivatives are included in the transformed similarity mathematical model. It can be found that the disadvantages of the method are: (i) difficult for analysis

D.Y. Shang, *Theory of Heat Transfer with Forced Convection Film Flows*,
Heat and Mass Transfer, DOI 10.1007/978-3-642-12581-2_4,

and investigation of the two-dimensional velocity field, (ii) difficult for satisfaction of the interfacial mass transfer matching conditions in the numerical calculation and difficult for rigorous investigation of mass transfer for two-phase film flows with three-point boundary problem, and (iii) difficult for consideration and treatment of the variable physical properties.

In [1], a new similarity analysis method was proposed on extensive investigation for heat transfer of laminar free convection, and collected in [2]. Since then the series related studies have already proved that this new similarity method is better alternative to the traditional Falkner-Skan type transformation for extensive studies on heat transfer of laminar free convection [3, 4], two-phase film flow boiling and condensation [5–7], and falling film flow of non-Newtonian power-law fluids [8–10]. In this chapter, a novel similarity analysis method will be reported for extensive investigation of laminar forced convection and two-phase film condensation, for resolving the difficulties caused by the Falkner-Skan type transformation method. First, a system of dimensionless similarity variables, such as Reynolds number, dimensionless coordinate variable, and dimensionless velocity components, is derived and determined through the analysis with the typical basis conservation equations on laminar forced convection. In derivation of the dimensionless similarity variables, it is never necessary to induce the stream function ψ, intermediate variable $f(\eta)$, and its derivatives. In this way, the above difficult points caused by the Falkner-Skan type transformation can be overcome, which can be found in the successive chapters of this book by using the present new similarity analysis method for extensive investigation of fluid forced convection and the two-phase forced film condensation.

4.2 Typical Basis Conservation Equations of Laminar Forced Convection

For presentation of the present new similarity analysis method, (3.1), (3.2), and (3.3) are taken as the governing partial differential equations for laminar forced film convection boundary layer without consideration of variable physical properties and viscous thermal dissipation:

$$\frac{\partial}{\partial \mathrm{x}}(\mathrm{w_x}) + \frac{\partial}{\partial \mathrm{y}}(\mathrm{w_y}) = 0, \tag{3.1}$$

$$w_x \frac{\partial w_x}{\partial x} + w_y \frac{\partial w_x}{\partial y} = \nu_f \frac{\partial^2 w_x}{\partial y^2}, \tag{3.2}$$

$$w_x \frac{\partial t}{\partial x} + w_y \frac{\partial t}{\partial y} = \frac{\nu_f}{\mathrm{Pr}_f} \frac{\partial^2 t}{\partial y^2}, \tag{3.3}$$

with the boundary condition equations

$$\mathrm{y} = 0: \mathrm{w_x} = 0, \quad w_y = 0, \quad t = t_w \tag{3.4}$$

$$\mathrm{y} \to \infty: \quad w_x = w_{x,\infty}(\text{constant}), \quad t = t_\infty. \tag{3.5}$$

Here, the subscript "f" demonstrates that the value of the related physical property is taken at the boundary layer average temperature $t_f = \frac{t_w + t_\infty}{2}$.

Equations (3.1), (3.2), and (3.3) with the constant properties are now taken as examples for presentation of the new similarity analysis method. First of all, it is necessary to briefly review the determination on the dimensionless similarity parameters for laminar forced convection.

4.3 Brief Review on Determination of Dimensionless Similarity Parameters (Number)

There are already three main dimensionless similarity parameters (number) for laminar forced convection, Nusselt number Nu, Reynolds number Re, and Prandtl number Pr. The Nusselt number is used to express heat transfer at a boundary (surface) within a fluid; the Reynolds number gives a measure of the ratio of inertial forces to viscous forces; and Pr is the ratio of momentum diffusivity (kinematic viscosity) and thermal diffusivity. Although only the Reynolds number is used for the similarity analysis and transformation of the governing partial differential equations of laminar forced convection and film flows, the review for determination on all these dimensionless similarity parameters will be presented here for fluid laminar forced convection. The popular method for creating the dimensionless similarity parameters is based on Buckingham π-theorem in dimensional analysis [11–21]. Here, the π-theorem with dimension analysis is applied to derive the dimensionless similarity parameters on laminar forced convection boundary layer.

With π-theorem, it is known that if a dependent physical phenomenon (variable) F_0 is completely determined by a set of n independent variables, in which r basic dimensions are included, the suitable dependent principle F_0 will be completely determined by $n - r$ dimensionless similarity parameters. According to π-theorem, the derivation on the dimensionless similarity parameters is done successively as following procedures.

4.3.1 *Select Whole Physical Independent Variables Dominating the Physical Phenomenon*

Observing governing equations (3.1), (3.2), and (3.3), it is seen that the whole physical independent variables dominating the physical phenomenon of laminar forced convection boundary layer can be obtained and expressed as the following form:

$$F_0(\lambda_f, w_{x,\infty}, x, \mu_f, \alpha_x, \rho_f, c_{p_f}) = 0. \tag{4.1}$$

Here, x is taken as the local length coordinate in the plate surface, the physical properties λ, μ, ρ, and c_p are thermal conductivity, absolute viscosity, density, and specific heat of the fluid, respectively, and α is heat transfer coefficient. While $w_{x,\infty}$

is the mainstream velocity. The above seven physical variables (i.e. $n = 7$) are whole independent physical variables, i.e. the decisive variables of fluid forced convection heat transfer.

4.3.2 Select Basic Dimension System

For investigation of the problem such as forced convection heat transfer, the following five physical dimensions are taken as basic dimensions: time [s], length [m], mass [kg], temperature [K], and quantity of heat [J]. While the dimensions of the above independent physical variables λ, $w_{x,\infty}$, x, μ, α, ρ, and c_p can be described by these basic dimensions, i.e. $\left[\frac{J}{m \cdot s \cdot K}\right]$, $\left[\frac{m}{s}\right]$, $[m]$, $\left[\frac{kg}{m \cdot s}\right]$, $\left[\frac{J}{m^2 \cdot s \cdot K}\right]$, $\left[\frac{kg}{m^3}\right]$, and $\left[\frac{J}{kg \cdot K}\right]$, respectively. Here the basic physical dimensions [J] and [K] exist as combined form, and then can be combined as an independent basic dimension $\left[\frac{J}{K}\right]$. In this case, the above five basic dimensions are changed to the following four basic dimensions [s], [m], [kg], and [J/K] (i.e. $r = 4$).

Here the number of the related dimensionless similarity physical parameters should be $n - r = 7 - 4 = 3$. According to the Buckingham's π-theorem, the dimensional analysis thus yields the result

$$F_0 = f(\pi_1, \pi_2, \pi_3), \tag{4.2}$$

where π_1, π_2, and π_3 are the dimensionless similarity parameters dependent on fluid laminar forced convection, and F_0 is the suitable dimensionless physical phenomenon (variable).

4.3.3 Determine the Dimensionless Similarity Parameters π_1, π_2, and π_3

Before determination of the dimensionless similarity physical parameters of the physical phenomenon, we should select the related physical variables. According to the experience, each physical parameter is constituted by r + 1 = 4 + 1 = 5 physical variables. Then, the dimensionless similarity parameters π_1, π_2, and π_3 can be expressed as the following equations, respectively:

$$\pi_1 = \lambda_f{}^{a1} \times w_{x,\infty}{}^{b1} \times x^{c1} \times \mu_f{}^{d1} \times \alpha_x = 0, \tag{4.3}$$

$$\pi_2 = \lambda_f{}^{a2} \times w_{x,\infty}{}^{b2} \times x^{c2} \times \mu_f{}^{d2} \times \rho_f = 0, \tag{4.4}$$

$$\pi_3 = \lambda_f{}^{a3} \times w_{x,\infty}{}^{b3} \times x^{c3} \times \mu_f{}^{d3} \times c_{p_f} = 0. \tag{4.5}$$

By using dimensional analysis, the following dimensional equation is obtained for the dimensionless similarity parameter π_1:

$$\left[\frac{J}{s \cdot K \cdot m}\right]^{a1} \cdot \left[\frac{m}{s}\right]^{b1} \cdot [m]^{c1} \cdot \left[\frac{kg}{m \cdot s}\right]^{d1} \cdot \left[\frac{J}{s \cdot K \cdot m^2}\right] = 0.$$

Obviously, the indexes a1 to d1 are suitable to the following equations:

For dimension [kg] balance: d1 = 0
For dimension [m] balance: −a1 + b1 + c1 − d1 − 2 = 0
For dimension [s] balance: −a1 − b1 − d1 − 1 = 0
For dimension [J/K]: a1 + 1 = 0
The solutions are a1 = −1, b1 = 0, c1 = 1, d1 = 0

Then the dimensionless similarity parameter π_1 is

$$\pi_1 = \frac{\alpha_x x}{\lambda_f}.$$

It is the well-known local Nusselt number without consideration of thermophysical physical properties, i.e.

$$Nu_{x,f} = \frac{\alpha_x x}{\lambda_f}, \tag{4.6}$$

where λ_f is the average thermal conductivity at the boundary layer average temperature

$$t_f = \frac{t_w + t_\infty}{2}.$$

By using dimensional analysis, the following dimensional equation is obtained for the dimensionless similarity parameter π_2:

$$\left[\frac{J}{s \cdot K \cdot m}\right]^{a2} \cdot \left[\frac{m}{s}\right]^{b2} \cdot [m]^{c2} \cdot \left[\frac{kg}{m \cdot s}\right]^{d2} \cdot \left[\frac{kg}{m^3}\right] = 0.$$

Obviously, the indexes a2 to d2 are suitable to the following equations:

For dimension [kg] balance: d2 + 1 = 0
For dimension [m] balance: −a2 + b2 + c2 − d2 − 3 = 0
For dimension [s] balance: −a2 − b2 − d2 = 0
For dimension [J/K]: a2 = 0
The solutions are a2 = 0, b2 = 1, c2 = 1, d2 = −1.

Then, we can obtain a following dimensionless parameter:

$$\pi_2 = \frac{w_{x,\infty} x \rho_f}{\mu_f} \quad \text{or} \quad \pi_2 = \frac{w_{x,\infty} x}{\nu_f}.$$

This is the well-known local Reynolds number without consideration of variable thermophysical properties, i.e.

$$\mathrm{Re}_{x,f} = \frac{w_{x,\infty} x}{\nu_f}. \tag{4.7}$$

By using dimensional analysis, the following dimensional equation is obtained for the dimensionless similarity parameter π_3:

$$\left[\frac{J}{s \cdot K \cdot m}\right]^{a3} \cdot \left[\frac{m}{s}\right]^{b3} \cdot [m]^{c3} \cdot \left[\frac{kg}{m \cdot s}\right]^{d3} \cdot \left[\frac{J}{kg \cdot K}\right] = 0.$$

Obviously, the indexes a3 to d3 are suitable to the following equations:

For dimension [kg] balance: $d3 - 1 = 0$
For dimension [m] balance: $-a3 + b3 + c3 - d3 = 0$
For dimension [s] balance: $-a3 - b3 - d3 = 0$
For dimension [J/K]: $a3 + 1 = 0$
The solutions are $a3 = -1$, $b3 = 0$, $c3 = 0$, $d3 = 1$.

Then

$$\pi_3 = \frac{\mu_f \cdot c_{p_f}}{\lambda_f}.$$

This is the average Prandtl number Pr_f, i.e.

$$\mathrm{Pr}_f = \frac{\mu_f \cdot c_{p_f}}{\lambda_f}. \tag{4.8}$$

So far, we have reviewed the derivation for obtaining thee dimensionless similarity parameters, Nusselt number, Reynolds number, and Prandtl number, respectively, for fluid laminar forced convection. On this basis, in the following section, we further investigate the dimensionless similarity variables related to the velocity field, which is the focal point of this chapter.

4.4 Investigation of the Dimensionless Similarity Variables on the Velocity Field

Here, with the basic momentum conservation equation, an analysis method will be applied for investigation of the dimensionless similarity variables on the velocity field of laminar forced convection boundary layer.

4.4.1 Derivation of Dimensionless Coordinate Variable

For investigation of the dimensionless similarity variables on the velocity field, first it is necessary to determine the related dimensionless coordinate variable.

In the boundary layer for laminar forced convection, the velocity component w_x can be regarded to have a quantitative grade equivalent to that of the mainstream velocity $w_{x,\infty}$, i.e.

$$w_x \propto w_{x,\infty}. \tag{4.9}$$

Meanwhile, with quantity grade analysis, (3.1) can be approximately rewritten as the following form:

$$\frac{w_{x,\infty}}{x} + \frac{w_y}{y} \propto 0,$$

or

$$w_y \propto \frac{w_{x,\infty} \cdot y}{x}. \tag{4.10}$$

With (3.9), (3.2) can be approximately expressed as the following quantity grade:

$$w_{x,\infty}\frac{w_{x,\infty}}{x} + \frac{w_{x,\infty}y}{x} \cdot \frac{w_{x,\infty}}{y} \propto \nu_f \frac{w_{x,\infty}}{y^2}.$$

Then

$$y^2 \propto \frac{\nu_f x}{w_{x,\infty}},$$

or

$$y \propto \sqrt{\frac{\nu_f x}{w_{x,\infty}}}. \tag{4.11}$$

According to the quantity grade relationship in (4.11), we can set a dimensionless variable η in the following equation:

$$\eta = \frac{y}{K\sqrt{\frac{\nu_f x}{w_{x,\infty}}}}.$$

Here, K is constant. Since the dimensionless variable η is related to the coordinates x and y, η can be regarded as the dimensionless coordinate variable.

Take K as $\sqrt{2}$, and the above dimensionless coordinate variable η is expressed as follows by using the local Reynolds number of (4.7):

$$\eta = \frac{y}{x}\left(\frac{1}{2}\mathrm{Re}_{x,f}\right)^{1/2}. \tag{4.12}$$

4.4.2 Derivation for Dimensionless Velocity Components

After determination of the expression of the dimensionless coordinate variable η, the investigation focuses on creation of the appropriate forms of the dimensionless velocity components. For this purpose, first, set $W_x(\eta)$ as the dimensionless velocity component in x coordinate and suppose the velocity component w_x is directly proportional to $W_x(\eta)$, mainstream velocity $w_{x,\infty}$, and the pth power of the local Reynolds number, i.e. $(\mathrm{Re}_{x,f})^p$. Additionally, set $W_y(\eta)$ as the dimensionless velocity component in y coordinate and suppose the velocity component w_y is directly proportional to $W_y(\eta)$, the mainstream velocity $w_{x,\infty}$ and qth power of the local Reynolds number, i.e. $(\mathrm{Re}_{x,f})^q$. Then, we have the following supposed equations about the two-dimensional velocity components w_x and w_y, respectively:

$$w_x = w_{x,\infty}\left(\frac{1}{2}\mathrm{Re}_{x,f}\right)^p \cdot W_x(\eta), \tag{4.13}$$

$$w_y = w_{x,\infty}\left(\frac{1}{2}\mathrm{Re}_{x,f}\right)^q W_y(\eta). \tag{4.14}$$

From (4.12) with (4.7), we have

$$\begin{aligned}\frac{\partial\eta}{\partial x} &= -\frac{1}{2}x^{-3/2}y\left(\frac{1}{2}\frac{w_{x,\infty}}{\nu_f}\right)^{1/2},\\ &= -\frac{1}{2}x^{-2}y\left(\frac{1}{2}\frac{w_{x,\infty}\cdot x}{\nu_f}\right)^{1/2},\\ &= -\frac{1}{2}x^{-1}\eta.\end{aligned} \tag{4.15}$$

$$\frac{\partial\eta}{\partial y} = \frac{1}{x}\left(\frac{1}{2}\mathrm{Re}_x\right)^{1/2}. \tag{4.16}$$

The partial derivative $\frac{\partial w_x}{\partial x}$ is expressed as

$$\begin{aligned}\frac{\partial w_x}{\partial x} &= w_{x,\infty}\cdot p\left(\frac{1}{2}\mathrm{Re}_{x,f}\right)^{p-1}\left(\frac{1}{2}\frac{w_{x,\infty}}{\nu_f}\right)\cdot W_x(\eta)\\ &\quad + w_{x,\infty}\left(\frac{1}{2}\mathrm{Re}_{x,f}\right)^p\frac{\mathrm{d}W_x(\eta)}{\mathrm{d}\eta}\cdot\frac{\partial\eta}{\partial x},\\ &= x^{-1}w_{x,\infty}\cdot p\left(\frac{1}{2}\frac{w_{x,\infty}x}{\nu_f}\right)^{p-1}\left(\frac{1}{2}\frac{w_{x,\infty}x}{\nu_f}\right)\cdot W_x(\eta)\\ &\quad + w_{x,\infty}\left(\frac{1}{2}\mathrm{Re}_{x,f}\right)^p\frac{\mathrm{d}W_x(\eta)}{\mathrm{d}\eta}\cdot\frac{\partial\eta}{\partial x},\end{aligned}$$

$$= px^{p-1} w_{x,\infty} \left(\frac{1}{2}\frac{w_{x,\infty}}{\nu_f}\right)^p \cdot W_x(\eta) + w_{x,\infty}\left(\frac{1}{2}\mathrm{Re}_{x,f}\right)^p \frac{\mathrm{d}W_x(\eta)}{\mathrm{d}\eta}\cdot\frac{\partial\eta}{\partial x},$$

$$= px^{-1} w_{x,\infty} \left(\frac{1}{2}\mathrm{Re}_{x,f}\right)^p \cdot W_x(\eta) + w_{x,\infty}\left(\frac{1}{2}\mathrm{Re}_{x,f}\right)^p \frac{\mathrm{d}W_x(\eta)}{\mathrm{d}\eta}\cdot\frac{\partial\eta}{\partial x}.$$

With (4.15), the above equation becomes

$$\frac{\partial w_x}{\partial x} = px^{-1} w_{x,\infty} \left(\frac{1}{2}\mathrm{Re}_{x,f}\right)^p \cdot W_x(\eta) + w_{x,\infty}\left(\frac{1}{2}\mathrm{Re}_{x,f}\right)^p \cdot \frac{\mathrm{d}W_x(\eta)}{\mathrm{d}\eta}\left(-\frac{1}{2}x^{-1}\eta\right),$$

$$= x^{-1}\left(\frac{1}{2}\mathrm{Re}_{x,f}\right)^p \cdot w_{x,\infty}\left[p\cdot W_x(\eta) - \frac{1}{2}\eta\frac{\mathrm{d}W_x(\eta)}{\mathrm{d}\eta}\right]. \tag{4.17}$$

With (4.14) we have

$$\frac{\partial w_x}{\partial y} = \left(\frac{1}{2}\mathrm{Re}_{x,f}\right)^p w_{x,\infty}\frac{\mathrm{d}W_x(\eta)}{\mathrm{d}\eta}\cdot\frac{\partial\eta}{\partial y}.$$

With (4.16) the above equation becomes

$$\frac{\partial w_x}{\partial y} = \left(\frac{1}{2}\mathrm{Re}_{x,f}\right)^p w_{x,\infty}\frac{\mathrm{d}W_x(\eta)}{\mathrm{d}\eta}\cdot\frac{1}{x}\left(\frac{1}{2}\mathrm{Re}_{x,f}\right)^{1/2}, \tag{4.18}$$

$$= \frac{1}{x}\left(\frac{1}{2}\mathrm{Re}_{x,f}\right)^{p+1/2} w_{x,\infty}\frac{\mathrm{d}W_x(\eta)}{\mathrm{d}\eta}.$$

Also

$$\frac{\partial^2 w_x}{\partial y^2} = \frac{1}{x}\left(\frac{1}{2}\mathrm{Re}_{x,f}\right)^{p+1/2} w_{x,\infty}\frac{\mathrm{d}^2W_x(\eta)}{\mathrm{d}\eta^2}\cdot\frac{\partial\eta}{\partial y},$$

$$= \frac{1}{x}\left(\frac{1}{2}\mathrm{Re}_{x,f}\right)^{p+1/2} w_{x,\infty}\frac{\mathrm{d}^2W_x(\eta)}{\mathrm{d}\eta^2}\cdot\frac{1}{x}\left(\frac{1}{2}\mathrm{Re}_{x,f}\right)^{1/2}, \tag{4.19}$$

$$= \frac{1}{x^2}\left(\frac{1}{2}\mathrm{Re}_{x,f}\right)^{p+1} w_{x,\infty}\frac{\mathrm{d}^2W_x(\eta)}{\mathrm{d}\eta^2}.$$

Then (3.2) is transformed to

$$w_{x,\infty}\left(\frac{1}{2}\mathrm{Re}_{x,f}\right)^{p}\cdot W_x(\eta)\cdot x^{-1}\left(\frac{1}{2}\mathrm{Re}_{x,f}\right)^{p}w_{x,\infty}\left[p\cdot W_x(\eta)-\frac{1}{2}\eta\frac{\mathrm{d}W_x(\eta)}{\mathrm{d}\eta}\right]$$

$$+\,w_{x,\infty}\left(\frac{1}{2}\mathrm{Re}_{x,f}\right)^{q}\cdot W_y(\eta)\cdot\frac{1}{x}\left(\frac{1}{2}\mathrm{Re}_{x,f}\right)^{p+1/2}w_{x,\infty}\frac{\mathrm{d}W_x(\eta)}{\mathrm{d}\eta}$$

$$=\nu_f\frac{1}{x^2}\left(\frac{1}{2}\mathrm{Re}_{x,f}\right)^{p+1}w_{x,\infty}\frac{\mathrm{d}^2W_x(\eta)}{\mathrm{d}\eta^2}.$$

The above equation is rewritten as

$$w_{x,\infty}\left(\frac{1}{2}\mathrm{Re}_{x,f}\right)^{p}\cdot W_x(\eta)\cdot x^{-1}\left(\frac{1}{2}\mathrm{Re}_{x,f}\right)^{p}w_{x,\infty}\left[p\cdot W_x(\eta)-\frac{1}{2}\eta\frac{\mathrm{d}W_x(\eta)}{\mathrm{d}\eta}\right]$$

$$+\,w_{x,\infty}\left(\frac{1}{2}\mathrm{Re}_{x,f}\right)^{q}\cdot W_y(\eta)\frac{1}{x}\left(\frac{1}{2}\mathrm{Re}_{x,f}\right)^{p+1/2}w_{x,\infty}\frac{\mathrm{d}W_x(\eta)}{\mathrm{d}\eta}$$

$$=\frac{\nu_f}{w_{x,\infty}\cdot x}\cdot\frac{1}{x}\left(\frac{1}{2}\mathrm{Re}_{x,f}\right)^{p+1}\cdot w_{x,\infty}{}^2\cdot\frac{\mathrm{d}^2W_x(\eta)}{\mathrm{d}\eta^2}.$$

i.e.

$$w_{x,\infty}\left(\frac{1}{2}\mathrm{Re}_{x,f}\right)^{p}\cdot W_x(\eta)\cdot x^{-1}\left(\frac{1}{2}\mathrm{Re}_{x,f}\right)^{p}w_{x,\infty}\left[pW_x(\eta)-\frac{1}{2}\eta\frac{\mathrm{d}W_x(\eta)}{\mathrm{d}\eta}\right]$$

$$+\,w_{x,\infty}\left(\frac{1}{2}\mathrm{Re}_{x,f}\right)^{q}\cdot W_y(\eta)\cdot\frac{1}{x}\left(\frac{1}{2}\mathrm{Re}_{x,f}\right)^{p+1/2}w_{x,\infty}\frac{\mathrm{d}W_x(\eta)}{\mathrm{d}\eta}$$

$$=\frac{1}{2x}\left(\frac{1}{2}\mathrm{Re}_{x,f}\right)^{p}w_{x,\infty}{}^2\frac{\mathrm{d}^2W_x(\eta)}{\mathrm{d}\eta^2}.$$

The above equation is divided by $\frac{1}{x}\left(\frac{1}{2}\mathrm{Re}_{x,f}\right)^{2p}\cdot w_{x,\infty}{}^2$ and simplified to

$$W_x(\eta)\cdot\left[p\cdot W_x(\eta)-\frac{1}{2}\eta\frac{dW_x(\eta)}{d\eta}\right]+W_y(\eta)\cdot\left(\frac{1}{2}\mathrm{Re}_{x,f}\right)^{-p+q+1/2}\frac{dW_x(\eta)}{d\eta}$$

$$=\left(\frac{1}{2}\right)\left(\frac{1}{2}\mathrm{Re}_{x,f}\right)^{-p}\frac{d^2W_x(\eta)}{d\eta^2}\tag{4.20}$$

If (4.20) becomes a complete similarity dimensionless equations, obviously we should have

$$-p + q + \frac{1}{2} = 0. \tag{4.21}$$

$$-p = 0. \tag{4.22}$$

Then, $p = 0$ and $q = -\frac{1}{2}$.

Then the supposed equations (4.13) and (4.14) for the two-dimensional velocity components of laminar forced boundary layer are expressed as, respectively,

$$w_x = w_{x,\infty} \cdot W_x(\eta), \tag{4.23}$$

$$w_y = w_{x,\infty} \cdot \left(\frac{1}{2}\mathrm{Re}_{x,f}\right)^{-1/2} \cdot W_y(\eta), \tag{4.24}$$

or

$$W_x(\eta) = \frac{w_x}{w_{x,\infty}}, \tag{4.23a}$$

$$W_y(\eta) = \frac{w_y}{w_{x,\infty}}\left(\frac{1}{2}\mathrm{Re}_{x,f}\right)^{1/2}. \tag{4.24a}$$

These are expressions for the new similarity velocity components $W_x(\eta)$ and $W_y(\eta)$ derived by using the new similarity analysis method for laminar forced convection boundary layer.

The related expressions on the two-dimensional velocity components w_x and w_y based on the Falkner-Skan transformation are

$$w_x = w_{x,\infty} f'(\eta), \tag{3.28}$$

$$w_y = \frac{1}{2} w_{x,\infty} \left(\frac{1}{2}\mathrm{Re}_{x,f}\right)^{-1/2} [\eta \cdot f'(\eta) - f(\eta)]. \tag{3.30}$$

Comparing (3.28) and (3.30) with (4.23) and (4.24), respectively, we can obtain the following equations for relationship of the new similarity velocity components $W_x(\eta)$ and $W_y(\eta)$ with the dimensionless function $f(\eta)$:

$$W_x(\eta) = f'(\eta), \tag{4.25}$$

$$W_y(\eta) = \left(\frac{1}{2}\right)\left[\eta f'(\eta) - f(\eta)\right]. \tag{4.26}$$

From (4.23), (4.24), (4.25), and (4.26), the following points can be concluded from comparison of the dimensionless similarity variables $W_x(\eta)$ and $W_y(\eta)$ with the dimensionless function $f(\eta)$ on the momentum velocity field: The derived dimensionless similarity velocity components $W_x(\eta)$ and $W_y(\eta)$ based on the present new similarity analysis method are directly proportional to the related velocity components w_x and w_y, while the relationships of dimensionless function $f(\eta)$ based on the Falkner-Skan type transformation with the velocity components w_x and w_y are more complicated, and even complicated issue caused by (3.30), which expresses a more complicated relationship between $f(\eta)$ and w_y.

On contrary, the simpler relationships of the dimensionless similarity variables $W_x(\eta)$ and $W_y(\eta)$ with the related velocity component w_x and w_y lead to two following advantages for the dimensionless governing equations derived by the new similarity analysis method compared with those by the Falkner-Skan type transformation: (i) more convenient to consider and treat the variable thermophysical properties, and (ii) more convenient to treat and satisfy the interfacial matching conditions for further rigorous investigation of heat and mass transfer of laminar forced film condensation. In the following chapters, these advantages with the present new similarity analysis method can be found successively.

Additionally, for dimensionless similarity analysis and transformation of the governing energy partial differential equation of fluid laminar forced convection, (3.3), we take the well-known form of the dimensionless temperature variable $\theta(\eta)$ as

$$\theta(\eta) = \frac{t - t_\infty}{t_w - t_\infty}. \tag{4.27}$$

Up to now, the whole set of equations for the dimensionless similarity analysis variables, i.e. the local Reynolds number $\mathrm{Re}_{x,f}$, dimensionless similarity variables η, dimensionless velocity components $W_x(\eta)$ and $W_y(\eta)$, and dimensionless temperature θ, based on the new similarity analysis method can be collected below for a clear expression:

$$\mathrm{Re}_{x,f} = \frac{w_{x,\infty} x}{\nu_f}, \tag{4.7}$$

$$\eta = \frac{y}{x}\left(\frac{1}{2}\mathrm{Re}_{x,f}\right)^{1/2}, \tag{4.12}$$

$$w_x = w_{x,\infty} \cdot W_x(\eta), \tag{4.23}$$

$$w_y = w_{x,\infty} \cdot \left(\frac{1}{2}\mathrm{Re}_{x,f}\right)^{-1/2} \cdot W_y(\eta), \tag{4.24}$$

$$\theta(\eta) = \frac{t - t_\infty}{t_w - t_\infty}. \tag{4.27}$$

Comparing the above set of similarity dimensionless variable for the new similarity analysis method with those in (3.24), (3.25), (3.28), (3.30), and (3.31) for the Falkner-Skan type transformation, it is found that their differences are in the dimensionless similarity variables related to the velocity field. However, in the next section, we will find that these differences will lead to quite different similarity dimensionless equations for laminar forced convection boundary layer based on the present new similarity analysis method from those based on the Falkner-Skan type transformation.

4.5 Application Example of the New Similarity Analysis Method

In this section, the expressions (4.7), (4.12), (4.23), (4.24), and (4.27), respectively, for the new system of similarity variables $\mathrm{Re}_{x,f}$, η, W_x, W_y, and θ based on the new similarity analysis method will be applied for complete similarity transformation of the governing partial differential equations (3.1) (3.2) and (3.3) with their boundary condition equations (3.4) and (3.5) of laminar forced convection boundary layer for ignoring constant properties.

4.5.1 Similarity Transformation of (3.1)

With (4.12) and (4.23), we have

$$\frac{\partial w_x}{\partial x} = w_{x,\infty} \frac{\mathrm{d}W_x(\eta)}{\mathrm{d}\eta} \cdot \frac{\partial \eta}{\partial x}.$$

Where with (4.7), we have

$$\begin{aligned} \frac{\partial \eta}{\partial x} &= -\frac{1}{2} x^{-3/2} y \left(\frac{1}{2} \frac{w_{x,\infty}}{\nu_f} \right)^{1/2} \\ &= -\frac{1}{2} x^{-2} y \left(\frac{1}{2} \frac{w_{x,\infty} \cdot x}{\nu_f} \right)^{1/2} \\ &= -\frac{1}{2} x^{-1} \eta \end{aligned} \tag{4.28}$$

Then

$$\frac{\partial w_x}{\partial x} = -\frac{1}{2} x^{-1} \eta \cdot w_{x,\infty} \frac{\mathrm{d}W_x(\eta)}{\mathrm{d}\eta}. \tag{4.29}$$

With (4.24) we have

$$\frac{\partial w_y}{\partial y} = w_{x,\infty} \left(\frac{1}{2} \mathrm{Re}_{x,f} \right)^{-1/2} \frac{\mathrm{d}W_y(\eta)}{\mathrm{d}\eta} \cdot \frac{\partial \eta}{\partial y}.$$

Where according to (4.12)

$$\frac{\partial \eta}{\partial y} = x^{-1}\left(\frac{1}{2}\mathrm{Re}_{x,f}\right)^{1/2}. \tag{4.30}$$

Then

$$\begin{aligned}\frac{\partial w_y}{\partial y} &= w_{x,\infty}\left(\frac{1}{2}\mathrm{Re}_{x,f}\right)^{-1/2}\frac{\mathrm{d}W_y(\eta)}{\mathrm{d}\eta}x^{-1}\left(\frac{1}{2}\mathrm{Re}_{x,f}\right)^{1/2}. \\ &= x^{-1}w_{x,\infty}\frac{\mathrm{d}W_y(\eta)}{\mathrm{d}\eta}\end{aligned} \tag{4.31}$$

Then (3.1) is changed to

$$-\frac{1}{2}x^{-1}\eta \cdot w_{x,\infty}\frac{\mathrm{d}W_x(\eta)}{\mathrm{d}\eta} + x^{-1}w_{x,\infty}\frac{\mathrm{d}W_y(\eta)}{\mathrm{d}\eta} = 0.$$

The above equation is simplified to

$$-\eta\frac{\mathrm{d}W_x(\eta)}{\mathrm{d}\eta} + 2\frac{\mathrm{d}W_y(\eta)}{\mathrm{d}\eta} = 0. \tag{4.32}$$

4.5.2 Similarity Transformation of (3.2)

With (4.12) and (4.23) we have

$$\frac{\partial w_x}{\partial y} = w_{x,\infty}\frac{\mathrm{d}W_x(\eta)}{\mathrm{d}\eta}\cdot\frac{\partial \eta}{\partial y}.$$

With (4.12) we have

$$\frac{\partial \eta}{\partial y} = x^{-1}\left(\frac{1}{2}\mathrm{Re}_{x,f}\right)^{1/2}. \tag{4.33}$$

Then

$$\frac{\partial w_x}{\partial y} = x^{-1}\left(\frac{1}{2}\mathrm{Re}_{x,f}\right)^{1/2} w_{x,\infty}\frac{\mathrm{d}W_x(\eta)}{\mathrm{d}\eta}, \tag{4.34}$$

$$\begin{aligned}\frac{\partial^2 w_x}{\partial y^2} &= x^{-1}\left(\frac{1}{2}\mathrm{Re}_{x,f}\right)^{1/2} w_{x,\infty}\frac{\mathrm{d}^2W_x(\eta)}{\mathrm{d}\eta^2}\cdot\frac{\mathrm{d}\eta}{\mathrm{d}y} \\ &= x^{-1}\left(\frac{1}{2}\mathrm{Re}_{x,f}\right)^{1/2} w_{x,\infty}\frac{\mathrm{d}^2W_x(\eta)}{\mathrm{d}\eta^2}x^{-1}\left(\frac{1}{2}\mathrm{Re}_{x,f}\right)^{1/2}. \\ &= x^{-2}\left(\frac{1}{2}\mathrm{Re}_{x,f}\right)\cdot w_{x,\infty}\frac{\mathrm{d}^2W_x(\eta)}{\mathrm{d}\eta^2}\end{aligned} \tag{4.35}$$

Then (3.2) is changed to

$$-\frac{1}{2}w_{x,\infty}W_x(\eta)\cdot x^{-1}\eta\cdot w_{x,\infty}\frac{\mathrm{d}W_x(\eta)}{\mathrm{d}\eta}+w_{x,\infty}\left(\frac{1}{2}\mathrm{Re}_{x,f}\right)^{-1/2}$$

$$W_y(\eta)\cdot x^{-1}\left(\frac{1}{2}\mathrm{Re}_{x,f}\right)^{1/2}w_{x,\infty}\frac{\mathrm{d}W_x(\eta)}{\mathrm{d}\eta}$$

$$=\nu_f\cdot x^{-2}\left(\frac{1}{2}\mathrm{Re}_{x,f}\right)\cdot w_{x,\infty}\frac{\mathrm{d}^2W_x(\eta)}{\mathrm{d}\eta^2}.$$

The above equation is simplified to

$$-\frac{1}{2}w_{x,\infty}W_x(\eta)\cdot x^{-1}\eta\cdot w_{x,\infty}\frac{\mathrm{d}W_x(\eta)}{\mathrm{d}\eta}+w_{x,\infty}\cdot W_y(\eta)\cdot x^{-1}\cdot w_{x,\infty}\frac{\mathrm{d}W_x(\eta)}{\mathrm{d}\eta}$$

$$=\nu_f\cdot x^{-2}\left(\frac{1}{2}\frac{w_{x,\infty}x}{\nu_f}\right)\cdot w_{x,\infty}\frac{\mathrm{d}^2W_x(\eta)}{\mathrm{d}\eta^2}.$$

The above equation is divided by $\frac{1}{2}x^{-1}\cdot w_{x,\infty}{}^2$, and simplified to

$$[-\eta\cdot W_x(\eta)+2W_y(\eta)]\frac{\mathrm{d}W_x(\eta)}{\mathrm{d}\eta}=\frac{\mathrm{d}^2W_x(\eta)}{\mathrm{d}\eta^2}. \tag{4.36}$$

4.5.3 Similarity Transformation of (3.3)

With (4.27), we have

$$\frac{\partial t}{\partial x}=(t_w-t_\infty)\frac{\mathrm{d}\theta(\eta)}{\mathrm{d}\eta}\cdot\frac{\partial\eta}{\partial x}.$$

With (4.28), the above equation is expressed as

$$\frac{\partial t}{\partial x}=-\frac{1}{2}x^{-1}\eta(t_w-t_\infty)\frac{d\theta(\eta)}{d\eta} \tag{4.37}$$

With (4.27), we have

$$\frac{\partial t}{\partial y}=(t_w-t_\infty)\frac{d\theta(\eta)}{d\eta}\cdot\frac{\partial\eta}{\partial y}$$

With (4.33), the above equation is expressed as

$$\frac{\partial t}{\partial y} = (t_w - t_\infty)\frac{\mathrm{d}\theta(\eta)}{\mathrm{d}\eta}x^{-1}\left(\frac{1}{2}\mathrm{Re}_{x,f}\right)^{1/2}, \tag{4.38}$$

$$\frac{\partial^2 t}{\partial y^2} = (t_w - t_\infty)x^{-1}\left(\frac{1}{2}\mathrm{Re}_{x,f}\right)^{1/2}\frac{\mathrm{d}^2\theta(\eta)}{\mathrm{d}\eta^2}\cdot\frac{\partial\eta}{\partial y}, \tag{4.39}$$

$$= (t_w - t_\infty)x^{-1}\left(\frac{1}{2}\mathrm{Re}_{x,f}\right)^{1/2}\frac{\mathrm{d}^2\theta(\eta)}{\mathrm{d}\eta^2}\cdot x^{-1}\left(\frac{1}{2}\mathrm{Re}_{x,f}\right)^{1/2},$$

$$= \frac{1}{2}(t_w - t_\infty)x^{-2}\mathrm{Re}_{x,f}\frac{\mathrm{d}^2\theta(\eta)}{\mathrm{d}\eta^2}.$$

Then (3.3) is changed to

$$w_{x,\infty}W(\eta)_x\cdot\left[-\frac{1}{2}x^{-1}\eta\cdot(t_w - t_\infty)\frac{\mathrm{d}\theta(\eta)}{\mathrm{d}\eta}\right] + w_{x,\infty}\left(\frac{1}{2}\mathrm{Re}_{x,f}\right)^{-1/2}$$

$$W_y(\eta)\cdot(t_w - t_\infty)\frac{\mathrm{d}\theta(\eta)}{\mathrm{d}\eta}x^{-1}\left(\frac{1}{2}\mathrm{Re}_{x,f}\right)^{1/2}$$

$$= \frac{\nu_f}{\mathrm{Pr}_f}\frac{1}{2}(t_w - t_\infty)x^{-2}\mathrm{Re}_{x,f}\frac{\mathrm{d}^2\theta(\eta)}{\mathrm{d}\eta^2}.$$

i.e.

$$-\frac{1}{2}w_{x,\infty}W_x(\eta)\cdot x^{-1}\eta\cdot(t_w - t_\infty)\frac{\mathrm{d}\theta(\eta)}{\mathrm{d}\eta} + w_{x,\infty}W_y(\eta)\cdot(t_w - t_\infty)\frac{\mathrm{d}\theta(\eta)}{\mathrm{d}\eta}x^{-1}$$

$$= \frac{1}{2}\frac{\nu_f}{\mathrm{Pr}_f}(t_w - t_\infty)x^{-2}\mathrm{Re}_{x,f}\frac{\mathrm{d}^2\theta(\eta)}{\mathrm{d}\eta^2}.$$

With the definition of $\mathrm{Re}_{x,f}$ in (4.7), the above equation is further simplified to

$$-\frac{1}{2}w_{x,\infty}W_x(\eta)\cdot x^{-1}\eta\cdot(t_w - t_\infty)\frac{\mathrm{d}\theta(\eta)}{\mathrm{d}\eta} + w_{x,\infty}W_y(\eta)(t_w - t_\infty)\frac{\mathrm{d}\theta(\eta)}{\mathrm{d}\eta}x^{-1}$$

$$= \frac{1}{2}\frac{\nu_f}{\mathrm{Pr}_f}(t_w - t_\infty)x^{-2}\frac{w_{x,\infty}x}{\nu_f}\frac{\mathrm{d}^2\theta(\eta)}{\mathrm{d}\eta^2}.$$

The above equation is divided by $\frac{1}{2}x^{-1}w_{x,\infty}(t_w - t_\infty)$, and simplified to

$$-W_x(\eta)\cdot\eta\frac{\mathrm{d}\theta(\theta)}{\mathrm{d}\eta} + 2W_y(\eta)\cdot\frac{\mathrm{d}\theta(\eta)}{\mathrm{d}\eta} = \frac{1}{\mathrm{Pr}_f}\frac{\mathrm{d}^2\theta(\eta)}{\mathrm{d}\eta^2},$$

i.e.

$$[-\eta \cdot W_x(\eta) + 2W_y(\eta)]\frac{d\theta(\eta)}{d\eta} = \frac{1}{\mathrm{Pr}_f}\frac{d^2\theta(\eta)}{d\eta^2}. \tag{4.40}$$

Up to now, by using the new similarity analysis method, the governing partial differential equations (3.1) to (3.3) have been transformed respectively to the following governing dimensionless ordinary differential equations:

$$-\eta\frac{dW_x(\eta)}{d\eta} + 2\frac{dW_y(\eta)}{d\eta} = 0, \tag{4.32}$$

$$[-\eta \cdot W_x(\eta) + 2W_y(\eta)]\frac{dW_x(\eta)}{d\eta} = \frac{d^2W_x(\eta)}{d\eta^2}, \tag{4.36}$$

$$[-\eta \cdot W_x(\eta) + 2W_y(\eta)]\frac{d\theta}{d\eta} = \frac{1}{\mathrm{Pr}_f}\frac{d^2\theta(\eta)}{d\eta^2}. \tag{4.40}$$

Meanwhile, with the dimensionless similarity variables, the boundary condition equations (3.4) and (3.5) are transformed to the followings:

$$\eta = 0 : W_x(\eta) = 0, \quad W_y(\eta) = 0, \quad \theta(\eta) = 1 \tag{4.41}$$

$$\eta \to \infty : W_x(\eta) = 1, \quad \theta(\eta) = 0 \tag{4.42}$$

It is found that the above dimensionless ordinary differential equations transformed by the present new similarity analysis method are quite different from the related ones obtained by the Falkner-Skan type transformation in Chap. 3.

It follows that (4.7), (4.12), (4.23), (4.24), and (4.27) for the set of dimensionless similarity variables, i.e. Reynolds number, dimensionless coordinate variable, dimensionless velocity components, and dimensionless temperature, are available for complete similarity analysis and transformation of governing partial differential equations on laminar forced convection boundary layer. These dimensionless variables become whole system of the similarity analysis variables of the present new similarity analysis method. In the successive chapters, these dimensionless similarity variables will be spread to extensive investigations on laminar forced convection and even for two-phase film forced condensation with consideration of various physical factors, including temperature- and concentration-dependent physical properties.

4.6 Comparison of the Two Similarity Methods

Now it is interesting to compare the present similarity analysis method with the Falkner-Skan type transformation.

4.6.1 Different Derivation Process of the Dimensionless Similarity Variables on Momentum Field

The dimensionless similarity variables on momentum field corresponding to the Falkner-Skan type transformation were derived by the group theory. In the derivation, the stream function was induced, and dimensionless function $f(\eta)$ was taken as the intermediate variable related to two-dimensional velocity components. Such derivation process is corresponding to a more complicated procedure.

The dimensionless similarity variables on momentum field corresponding to the present similarity analysis method are derived by the analysis on the basic mass and momentum conservation equations of laminar forced convection boundary layer. First, the appropriate dimensionless models for the two-dimensional similarity velocity components are respectively supposed to be a product of the related velocity component, mainstream velocity, and a power of local Reynolds number. Then determination of the exponent of local Reynolds number becomes the key work. It is seen that such derivation processes of the appropriate relations are distinct and clear by using the present similarity analysis method.

4.6.2 Different Dimensionless Expressions on Momentum Field

By comparing the dimensionless similarity variables in (3.35) and (3.39) derived by the Falkner-Skan type transformation with those in (4.32), (4.36), and (4.40) obtained by the present similarity analysis method, it is found that their difference is in the expressions of the similarity momentum field. The latter dimensionless similarity variables on the momentum field are dimensionless velocity components, proportional directly to the mainstream velocity and related velocity component for x and y coordinates, and also a power of the Reynolds number. But the former dimensionless similarity variables on the momentum field have more complicated form with the velocity component w_y. Such more complicated form on the velocity component w_y related to the dimensionless similarity variable on momentum field based on the Falkner-Skan type transformation will hinder a series of extensive investigations of laminar forced convection and two-phase film forced condensation, such as for consideration of momentum field, variable physical properties, two-phase film flow interfacial matching conditions, and condensate mass transfer, and so on.

4.6.3 Different Similarity Analysis Governing Mathematical Models

The quite different dimensionless similarity variables on momentum fields based on the Falkner-Skan type transformation and the present similarity analysis method lead to quite different dimensionless similarity governing mathematical models. The similarity governing mathematical model based on the Falkner-Skan transformation causes a difficulty for clarifying the two-dimensional flow velocity field, for consideration of variable physical properties, and even for further investigating heat and mass transfer of two-phase film condensation. It will be found in the successive chapters that the dimensionless similarity governing mathematical model based on the present similarity analysis method can resolve the above shortages caused by the traditional Falkner-Skan transformation, can be suitable for broad respects of research on laminar forced convection and two-phase film condensation, such as that with consideration and treatment of concentration and temperature-dependent physical properties, for dealing with the interfacial physical matching conditions, and for deep investigation of heat and mass transfer.

4.7 Remarks

In this chapter, a new similarity analysis method is reported with a detailed derivation for similarity analysis and transformation of laminar forced convection and two-phase film flows. A system of dimensionless similarity variables are provided including the dimensionless coordinate variable η, local Reynolds number $\mathrm{Re}_{x,f}$, dimensionless velocity components $W_x(\eta)$ and $W_y(\eta)$, and dimensionless temperature variable $\theta(\eta)$ for complete similarity analysis and transformation of a system of governing partial differential equations of laminar forced convection. In particular, the expressions of the similarity variables, i.e. dimensionless velocity components devoted in this book, have special advantages over those of the Falkner–Skan-type transformation for resolving the challenges of extensive investigations of laminar forced convection and its film condensation. These advantages can be found in successive chapters.

First, the derivation of the Reynolds number together with the Nusselt number and Prandtl number is reviewed by means of π-theorem for dimension analysis, where the local Reynolds number is useful in determination of the dimensionless similarity variables. The next work focuses on derivation of the dimensionless similarity variables on the momentum field by means of a series of special analyses on the governing partial differential equations. These derived similarity variables on the momentum field are dimensionless coordinate variable η and dimensionless velocity components $W_x(\eta)$ and $W_y(\eta)$. On this basis, the system of the dimensionless similarity variables is proposed for the similarity analysis and transformation of the governing partial differential equations on laminar forced convection. Finally, by means of the system of provided equations with the dimensionless similarity variables, the typical partial differential equations on laminar forced convection are completely transformed to the related dimensionless differential equations. It verified that the provided system of the dimensionless similarity variables are available for the complete similarity analysis and transformation of the governing partial differential equations of the laminar forced convection, and the present new similarity analysis method is valid.

The biggest difference of the present similarity analysis method from the Falkner-Skan type transformation is in their dimensionless similarity variables on the momentum field. The similarity variable for momentum field based on the Falkner-Skan type transformation is a dimensionless similarity intermediate variable $f(\eta)$, and the velocity component w_y has a complicated algebraic relation to the intermediate variable $f(\eta)$ and its derivatives. However, based on the present similarity analysis method, the velocity components are directly proportional to the related dimensionless similarity velocity components. These different dimensionless similarity variables lead to quite different similarity governing mathematical models produced by the two different similarity analysis methods, and the following big inconveniences exist in the model transformed by the Falkner-Skan type transformation. They are (i) inconvenience for analysis and investigation of the two-dimensional velocity field, (ii) inconvenience for satisfaction of the interfacial physical matching conditions in calculation for two-phase film condensation, (iii) inconvenience for rigorous investigation of mass transfer for two-phase film flows problem with three-point boundary problem, and (iv) inconvenience for consideration of variable physical properties. It can be found in the successive chapters of this book that the above big inconveniences caused by the Falkner-Skan type transformation can be resolved well by the present similarity analysis method.

Exercises

1. Please explain why the following equations can be set up originally before derivation of the expressions on the dimensionless similarity velocity components $W_x(\eta)$ and $W_y(\eta)$:

$$w_x = w_{x,\infty}\left(\frac{1}{2}\mathrm{Re}_{x,f}\right)^p \cdot W_x(\eta),$$

$$w_y = w_{x,\infty}\left(\frac{1}{2}\mathrm{Re}_{x,f}\right)^q W_y(\eta).$$

2. Suppose that the dimensionless coordinate variable η in (4.17) is replaced by the form $\eta = \frac{y}{x}\left(\mathrm{Re}_{x,f}\right)^{1/2}$, and the forms of other dimensionless similarity variables are kept, please try to transform similarly the governing partial differential equations (4.1), (4.2), and (4.3) and the boundary condition equations (4.4) and (4.5) to the related ordinary differential equations.

References

1. D.Y. Shang, B.X. Wang, Effect of variable thermophysical properties on laminar free convection of gas. Int. J. Heat Mass Transfer **33**(7), 1387–1395 (1990)
2. D.Y. Shang, *Free Convection Film Flows and Heat Transfer* (Springer Berlin, Heidelberg, New York, NY, 2006)
3. D.Y. Shang, B.X. Wang, Effect of variable thermophysical properties on laminar free convection of polyatomic gas. Int. J. Heat Mass Transfer **34**(3), 749–755 (1991)

4. Shang, B.X. Wang, Y. Wang, Y. Quan, Study on liquid laminar free convection with consideration of variable thermophysical properties. Int. J. Heat Mass Transfer **36**(14), 3411–3419 (1993)
5. D.Y. Shang , B.X. Wang, L.C. Zhong, A study on laminar film boiling of Liquid along an isothermal vertical plate in a pool with consideration of variable thermophysical properties. Int. J. Heat Mass Transfer **37**(5), 819–828 (1994)
6. D.Y. Shang, T. Adamek, Study on laminar film condensation of saturated steam on a vertical flat plate for consideration of various physical factors including variable thermophysical properties. Wärme- und Stoffübertragung **30**, 89–100 (1994)
7. D.Y. Shang, B.X. Wang, An extended study on steady-state laminar film condensation of a superheated vapor on an isothermal vertical plate. Int. J. Heat Mass Transfer **40**(4), 931–941 (1997)
8. H.I. Andersson, D.Y. Shang, An extended study of hydrodynamics of gravity- driven film flow of power-law fluids. Fluid Dyn Res **22**, 345–357 (1998)
9. D.Y. Shang, H. Andersson, Heat transfer in gravity-driven film flow of power-law fluids. Int. J. Heat Mass Transfer **42**(11), 2085–2099 (1999)
10. D.Y. Shang, J. Gu, Analyses of pseudo-similarity and boundary layer thickness for non-Newtonian falling film flow. Heat Mass Transfer **41**(1), 44–50, (2004)
11. E. Buckingham, On physically similar systems; illustrations of the use of dimensional equations. Phys. Rev. **4**, 345–376 (1914)
12. E. Buckingham, The principle of similitude. Nature **96**, 396–397 (1915)
13. E. Buckingham, Model experiments and the forms of empirical equations. Trans. A.S.M.E **37**, 263–296 (1915)
14. S.G. Taylor, The formation of a blast wave by a very intense explosion I. theoretical discussion. Proc. Roy. Soc. A **201**, 159–174 (1950)
15. S.G. Taylor, The formation of a blast wave by a very intense explosion II. the atomic explosion of 1945. Proc. Roy. Soc. A **201**, 175–186 (1950)
16. H. Hanche-Olsen, Buckingham's pi-theorem, TMA4195 Mathematical modelling, NTNU (2004)
17. S J. Kline *Similitude and Approximation Theory* (Springer, New York, NY, 1986)
18. F. Y.M. Wan, *Mathematical Models and their Analysis* (Harper & Row Publishers, New York, NY, 1989)
19. G.W. Hart *Multidimensional Analysis: Algebras and Systems for Science and Engineering* (Springer, Heidelberg, 1995)
20. Vignaux, G.A., Dimensional analysis in data modelling, Victoria University of Wellington, 1991. Retrieved December 15 (2005)
21. Mike Sheppard, Systematic Search for Expressions of Dimensionless Constants using the NIST database of Physical Constants (2007)

Part II
Laminar Forced Convection

Chapter 5
Heat Transfer on Laminar Forced Convection with Ignoring Variable Physical Properties and Viscous Thermal Dissipation

Abstract The present new similarity analysis method is applied for similarity transformation of governing partial differential equations for laminar forced convection without consideration of variable physical properties and viscous thermal dissipation, and then the complete similarity mathematical model is obtained. In the transformed dimensionless equations, dimensionless similarity velocity components $W_x(\eta)$ and $W_y(\eta)$ describe clearly the forced convection momentum field. The rigorous numerical solutions on velocity and temperature fields, as well as heat transfer equations, are obtained, which are very well coincident to Blasius and Pohlhausen solutions. All these prove that the present new similarity analysis method is valid.

5.1 Introduction

Blasius [1] described the two-dimensional boundary layer of laminar forced convection on a semi-infinite plate parallel to a main constant flow $w_{x,\infty}$. He transformed the partial differential equations related to the momentum field to a single ordinary differential equation by transforming the coordinate system. Then, the Blasius transformation approach was generalised by Falkner-Skan type transformation [2] for outer flows of the form $= cx^m$ results, and then the Falkner-Skan type transformation has had widespread application and widely collected in numerous books, such as [3–11] for laminar forced convection with boundary layer analysis. However, the Falkner-Skan type transformation has some disadvantages, which will cause difficulties for consideration of variable physical properties, and for treatment of the two-phase film flow interfacial physical matching conditions. The latter advantage will further lead to a difficulty for reliable investigation of two-phase flow condensation heat and mass transfer. In order to overcome these disadvantages, the new similarity analysis method was presented in detail in Chap. 4. From this chapter onwards, this new similarity analysis method will be successively used in extensive investigation on the laminar forced convection and two-phase film condensation. First, in this chapter, a typical case on laminar forced convection will be touched for ignoring variable physical properties and viscous thermal dissipation.

D.Y. Shang, *Theory of Heat Transfer with Forced Convection Film Flows*,
Heat and Mass Transfer, DOI 10.1007/978-3-642-12581-2_5,

5.2 Basic Conservation Equations of Laminar Forced Convection

5.2.1 Governing Partial Differential Equations

In Fig. 5.1 the physical model and coordinate system of boundary layer with two-dimensional laminar forced convection are shown schematically. A flat plate is horizontally located in parallel fluid flow with its mainstream velocity $w_{x,\infty}$. The plate surface temperature is t_w and the fluid bulk temperature is t_∞. Then, a velocity boundary layer is produced near the plate. If t_w is not equal to t_∞, a temperature boundary layer will occur near the plate together with the related velocity boundary layer. Then, (3.1), (3.2), and (3.3) can be taken as follows for governing partial differential equations of laminar forced convection boundary layer without consideration of variable physical properties and viscous thermal dissipation:

$$\frac{\partial}{\partial \mathrm{x}}(\mathrm{w_x}) + \frac{\partial}{\partial \mathrm{y}}(\mathrm{w_y}) = 0, \tag{3.1}$$

$$w_x \frac{\partial w_x}{\partial x} + w_y \frac{\partial w_x}{\partial y} = \nu_f \frac{\partial^2 w_x}{\partial y^2}, \tag{3.2}$$

$$w_x \frac{\partial t}{\partial x} + w_y \frac{\partial t}{\partial y} = \frac{\nu_f}{\mathrm{Pr}_f} \frac{\partial^2 t}{\partial y^2}, \tag{3.3}$$

with the boundary condition equations

$$\mathrm{y} = 0: \ \mathrm{w_x} = 0, \ \ \mathrm{w_y} = 0, \ \ t = t_w, \tag{3.4}$$

$$\mathrm{y} \to \infty: \ \ w_x = w_{x,\infty} \text{ (constant)}, \ \ t = t_\infty, \tag{3.5}$$

where the subscript f of the physical property variables denotes that the related reference temperature is the average temperature $t_f = \dfrac{t_w + t_\infty}{2}$. Here in energy equation (4.3), the viscous thermal dissipation is neglected.

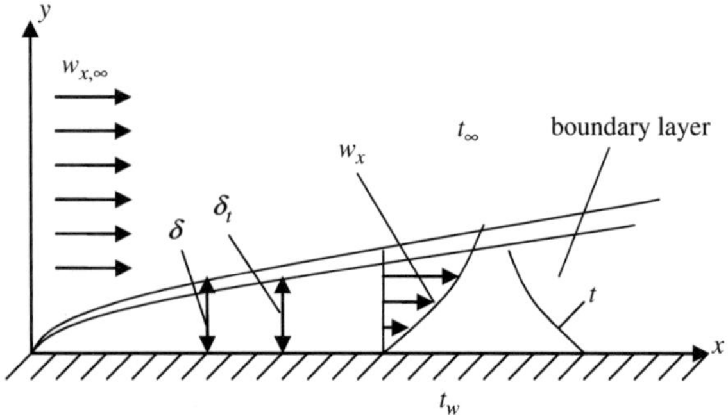

Fig. 5.1 Physical model and coordinate system of boundary layer for laminar forced convection on a horizontal flat plate

5.2.2 Similarity Transformation Variables

According to Chap. 4, the following equations with the new set of dimensionless similarity variables, such as the local Reynolds number $\mathrm{Re}_{x,f}$, dimensionless coordinate variable η, dimensionless velocity components $W_x(\eta)$ and $W_y(\eta)$, and dimensionless temperature $\theta(\eta)$ used for the similarity transformation of the above basic conservation equations of fluid laminar forced convection are

$$\mathrm{Re}_{x,f} = \frac{w_{x,\infty} x}{\nu_f}, \tag{4.7}$$

$$\eta = \frac{y}{x}\left(\frac{1}{2}\mathrm{Re}_{x,f}\right)^{1/2}, \tag{4.12}$$

$$w_x = w_{x,\infty} W_x(\eta), \tag{4.23}$$

$$w_y = w_{x,\infty}\left(\frac{1}{2}\mathrm{Re}_{x,f}\right)^{-1/2} W_y(\eta), \tag{4.24}$$

$$\theta(\eta) = \frac{t - t_\infty}{t_w - t_w}. \tag{4.27}$$

5.2.3 Governing Ordinary Differential Equations

According to the derivation in Chap. 4, with the above dimensionless similarity variables, the typical basic conservation equations (3.1) to (3.3) with the boundary conditions (3.4) and (3.5) are transformed equivalently to the following dimensionless governing differential equations:

$$-\eta\frac{\mathrm{d}W_x(\eta)}{\mathrm{d}\eta} + 2\frac{\mathrm{d}W_y(\eta)}{\mathrm{d}\eta} = 0, \tag{4.32}$$

$$[-\eta \cdot W_x(\eta) + 2W_y(\eta)]\frac{\mathrm{d}W_x(\eta)}{\mathrm{d}\eta} = \frac{\mathrm{d}^2 W_x(\eta)}{\mathrm{d}\eta^2}, \tag{4.36}$$

$$[-\eta \cdot W_x(\eta) + 2W_y(\eta)]\frac{\mathrm{d}\theta(\eta)}{\mathrm{d}\eta} = \frac{1}{\mathrm{Pr}_f}\frac{\mathrm{d}^2\theta(\eta)}{\mathrm{d}\eta^2}, \tag{4.40}$$

with the dimensional boundary condition equations

$$\eta = 0: \quad W_x(\eta) = 0, \quad W_y(\eta) = 0, \quad \theta(\eta) = 1, \tag{4.41}$$

$$\eta \to \infty: \quad W_x(\eta) = 1, \quad \theta(\eta) = 0. \tag{4.42}$$

Here, in the dimensionless governing differential equations (4.32), (4.36), and (4.40) and their boundary condition equations (4.41) and (4.42) transformed based on the present new similarity analysis method, the dimensionless velocity components

$W_x(\eta)$ and $W_y(\eta)$ are proportional to velocity components w_x and w_y, respectively. It is quite different from the corresponding traditional dimensionless governing equations transformed by the Falkner-Skan type approach, in which the velocity component w_x and w_y have to be represented indirectly by the algebraic form with the variable $f(\eta)$ and its derivatives, especially the form related to the velocity component w_y is more complicated with $f(\eta)$ and its derivatives. Obviously, the present new similarity analysis method is more practical and convenient than the Falkner-Skan transformation.

5.3 Numerical Results

5.3.1 Velocity Fields

The governing dimensionless differential equations (4.32), (4.36), and (4.40) with their boundary condition equations (4.41) and (4.42) are solved by a shooting method with fifth-order Runge-Kutta integration. For solving the nonlinear problem, a variable mesh approach is applied to the numerical calculation programs. It can be seen that for the present case with ignoring the variable physical properties, the solutions of the dimensionless velocity components $W_x(\eta)$ and $W_y(\eta)$ are independent on Prandtl number. The accurate numerical results of velocity field are obtained. The numerical results for the dimensionless velocity components $W_x(\eta)$ and $W_y(\eta)$ are plotted in Figs. 5.2 and 5.3, and the typical values of them are listed in Table 5.1, respectively, with the dimensionless coordinate variable η.

On the other hand, the numerical solutions $f'(\eta)$ and values $\frac{1}{2}[\eta \cdot f'(\eta) - f(\eta)]$ are obtained from the Blasius equation, (3.35), with the boundary condition equations (3.40) and (3.41), and also listed in Table 5.1.

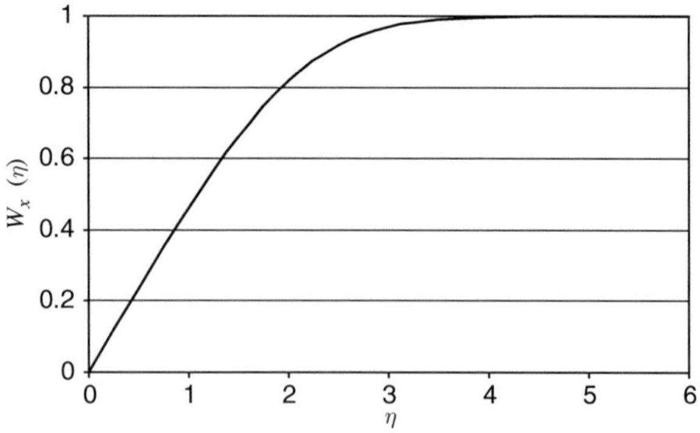

Fig. 5.2 Numerical results on the dimensionless velocity component $W_x(\eta)$ for laminar forced convection on a horizontal flat plate with ignoring the variable physical properties and viscous thermal dissipation

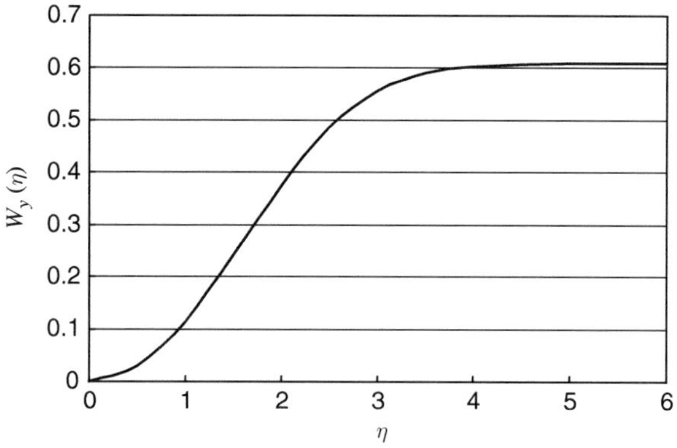

Fig. 5.3 Numerical results on dimensionless velocity component $W_y(\eta)$ for laminar forced convection on a horizontal flat plate with ignoring the variable physical properties and viscous thermal dissipation

It is seen that the numerical solutions $W_x(\eta)$ and $W_y(\eta)$ are very well identical to the solutions $f(\eta)'$ and $\frac{1}{2}[\eta \cdot f'(\eta) - f(\eta)]$, which are completely coincident to (4.25) and (4.26) for the relationship between the solutions based on the present new similarity analysis method and Kalkner-Skan type transformation. It means that the numerical results on velocity field based on the two kinds of similarity analysis method are equivalent. Clearly, it is more convenient to investigate the velocity fields with the solutions based on the present new similarity analysis method than those based on the Falkner-Skan type transformation.

5.3.2 Temperature Fields

The numerical solutions on dimensionless temperature $\theta(\eta)$ based on governing dimensionless differential equations (4.32), (4.36), and (4.40) and their boundary condition equations (4.41) and (4.42) are obtained. Some of the solutions $\theta(\eta)$ are plotted in Fig. 5.4 in a big range of average Prandtl number $\Pr_f$. It is found that increasing the average Prandtl number $\Pr_f$ will cause decrease of the thermal boundary layer thickness and increase of the temperature gradient. Meanwhile, with decreasing the average Prandtl number $\Pr_f$, the effect of the $\Pr_f$ variation on the temperature field will be more and more obvious.

5.4 Skin-Friction Coefficient

So far, the skin friction coefficient has been analyzed by means of Falkner-Skan type transformation for forced convection. In this section, we will analyze it under the present new similarity analysis method for laminar forced convection. The velocity

Table 5.1 Numerical solutions related to velocity field respectively based on the Falkner-Skan type transformation and the present new similarity analysis method for laminar forced convection on a horizontal flat plate without consideration of variable thermophysical properties and viscous thermal dissipation

	Solutions of (3.35) (transformed by Falkner-Skan pe transformation)			Solutions of (4.32) and (4.36) (transformed by the new similarity analysis method)	
η	$f(\eta)$	$f'(\eta)$	$\frac{1}{2}[\eta\cdot f'(\eta)-f(\eta)]$	$W_y(\eta)$	$W_x(\eta)$
0	0	0	0	0	0
0.2	0.009391	0.093905	0.004695	0.004695	0.093905
0.4	0.037549	0.187505	0.018727	0.018746	0.187605
0.6	0.084386	0.280505	0.041959	0.04198	0.280576
0.8	0.149575	0.371963	0.073998	0.073948	0.371964
1	0.232990	0.460632	0.113821	0.113821	0.460633
1.2	0.333657	0.545246	0.160319	0.160319	0.545247
1.4	0.450723	0.624386	0.211709	0.211709	0.624387
1.6	0.582956	0.696699	0.265881	0.265882	0.6967
1.8	0.728872	0.761057	0.320515	0.320516	0.761058
2	0.886796	0.816694	0.373296	0.373297	0.816695
2.2	1.054946	0.863303	0.42216	0.422161	0.863305
2.4	1.231527	0.901065	0.465515	0.465515	0.901066
2.6	1.414823	0.930601	0.50237	0.50237	0.930602
2.8	1.603282	0.952875	0.532384	0.532384	0.952876
3	1.795567	0.969054	0.555798	0.555798	0.969055
3.2	1.990580	0.980365	0.573294	0.573294	0.980366
3.4	2.187466	0.987970	0.585816	0.585816	0.987971
3.6	2.385589	0.992888	0.594404	0.594403	0.992889
3.8	2.584497	0.995944	0.600045	0.600045	0.995945
4	2.783885	0.997770	0.603598	0.603597	0.997771
4.2	2.983554	0.998818	0.605741	0.605741	0.998819
4.4	3.183381	0.999396	0.606981	0.606981	0.999397
4.6	3.383294	0.999703	0.60767	0.607669	0.999704
4.8	3.583252	0.999859	0.608036	0.608036	0.99986
5	3.783232	0.999936	0.608224	0.608222	0.999937
5.2	3.983224	0.999971	0.608313	0.608314	0.999973
5.4	4.183220	0.999988	0.608358	0.608357	0.999989
5.6	4.383218	0.999995	0.608377	0.608376	0.999996
5.8	4.583217	0.999998	0.608386	0.608385	0.999999
6	4.783217	0.999999	0.608389	0.608388	1

gradient at the wall is important characteristic of the solution, and the local skin-friction coefficient $C_{f,x}$ is a dimensionless measure of the shear stress at the wall, i.e.

$$C_{x,f}=\frac{\tau_{w,x}}{\frac{1}{2}\rho_f w_{x,\infty}^2}=\frac{\mu_f\left(\frac{\partial w_x}{\partial y}\right)_{y=0}}{\frac{1}{2}\rho_f w_{x,\infty}^2}. \tag{5.1}$$

With (4.23), we have

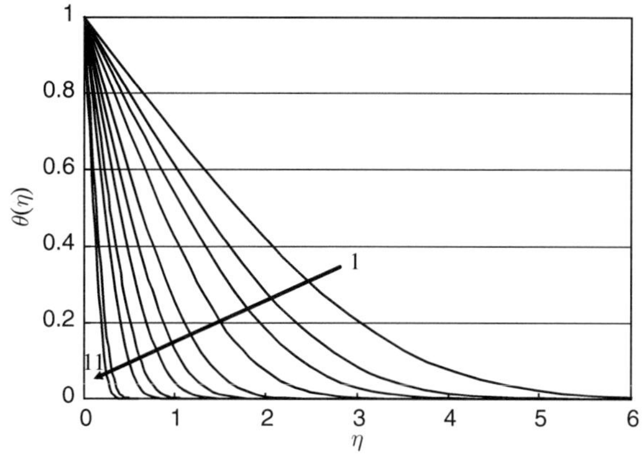

Fig. 5.4 Temperature profiles $\theta(\eta)$ of laminar forced convection on a horizontal flat plate with ignoring the variable thermophysical properties and viscous thermal dissipation
Note: Lines 1 to 11: the film average Prandtl number $\Pr_f = 0.3, 0.6, 1, 2, 5, 10, 20, 50, 100, 500$, and 1000

$$\left(\frac{\partial w_x}{\partial y}\right)_{y=0} = w_{x,\infty}\left(\frac{\mathrm{d}W_x(\eta)}{\mathrm{d}\eta}\right)_{\eta=0}\left(\frac{\partial \eta}{\partial y}\right)_{y=0}.$$

With (4.12), we have

$$\left(\frac{\partial \eta}{\partial y}\right)_{y=0} = x^{-1}\left(\frac{1}{2}\mathrm{Re}_{x,f}\right)^{1/2}.$$

Then

$$\left(\frac{\partial w_x}{\partial y}\right)_{y=0} = x^{-1}\left(\frac{1}{2}\mathrm{Re}_{x,f}\right)^{1/2} w_{x,\infty}\left(\frac{\mathrm{d}W_x(\eta)}{\mathrm{d}\eta}\right)_{\eta=0}.$$

Therefore, the local skin-friction coefficient $C_{f,x}$ is expressed as

$$\begin{aligned} C_{x,f} &= \frac{\mu_f x^{-1}\left(\frac{1}{2}\mathrm{Re}_{x,f}\right)^{1/2} w_{x,\infty}\left(\frac{\mathrm{d}W_x(\eta)}{\mathrm{d}\eta}\right)_{\eta=0}}{\frac{1}{2}\rho_f w_{x,\infty}{}^2} \\ &= \sqrt{2}\frac{\nu_f}{x w_{x,\infty}}(\mathrm{Re}_{x,f})^{1/2}\left(\frac{\mathrm{d}W_x(\eta)}{\mathrm{d}\eta}\right)_{\eta=0} \\ &= \sqrt{2}(\mathrm{Re}_{x,f})^{-1/2}\left(\frac{\mathrm{d}W_x(\eta)}{\mathrm{d}\eta}\right)_{\eta=0} \end{aligned} \quad (5.2)$$

Since the numerical solution $\left(\frac{\mathrm{d}W_x(\eta)}{\mathrm{d}\eta}\right)_{\eta=0} = 0.4696$, the well-known expression of local skin friction coefficient $C_{x,f}$ is obtained as

$$C_{x,f} == 0.664(\mathrm{Re}_{x,f})^{-1/2} \tag{5.3}$$

for laminar forced convection on a horizontal flat plate.

The average skin friction coefficient from $x = 0$ to x is described as

$$\overline{C}_{x,f} = \frac{1}{x}\int_0^x C_{x,f}\mathrm{d}x = 0.664\frac{1}{x}\int_0^x (\mathrm{Re}_{x,f})^{-1/2}\mathrm{d}x = 1.328\mathrm{Re}_{x,f}^{-1/2}. \tag{5.4}$$

5.5 Heat Transfer

5.5.1 Heat Transfer Analysis

The local heat transfer rate $(q_{x,f})_{Ec=0}$ at position x per unit area from the surface of the plate with ignoring the viscous thermal dissipation and variable physical properties can be calculated by Fourier's law as

$$(q_{x,f})_{\mathrm{Ec}=0} = -\lambda_f\left[\left(\frac{\partial t}{\partial y}\right)_{y=0}\right]_{f,\mathrm{Ec}=0}.$$

The subscript $\mathrm{Ec} = 0$ demonstrates that in the related investigation the viscous thermal dissipation is not considered, and the subscript f denotes the case that the boundary layer average temperature $t_f = \frac{t_w + t_\infty}{2}$ is taken as the reference temperature with ignoring the variable physical properties.

With (4.7), (4.12), and (4.27), we have

$$(q_{x,f})_{\mathrm{Ec}=0} = -\lambda_f(t_w - t_\infty)\left[\left(\frac{\mathrm{d}\theta(\eta)}{\mathrm{d}\eta}\right)_{\eta=0}\right]_{f,\mathrm{Ec}=0}\frac{\partial\eta}{\partial y},$$

$$(q_{x,f})_{\mathrm{Ec}=0} = \lambda_f(t_w - t_\infty)\left(\frac{1}{2}\mathrm{Re}_{x,f}\right)^{1/4} x^{-1}\left[-\left(\frac{\mathrm{d}\theta(\eta)}{\mathrm{d}\eta}\right)_{\eta=0}\right]_{f,\mathrm{Ec}=0}. \tag{5.5}$$

It should be indicated that in analysis and calculation of this chapter, the viscous thermal dissipation is not considered in the energy governing differential equation, and then the corresponding wall temperature gradient and local heat transfer rate are expressed as $\left[-\left(\frac{\mathrm{d}\theta(\eta)}{\mathrm{d}\eta}\right)_{\eta=0}\right]_{f,\mathrm{Ec}=0}$ and $(q_{x,f})_{\mathrm{Ec}=0}$. In next chapter, we will further find that the present case without consideration of the viscous thermal dissipation is corresponding to that for Eckert number Ec = 0.

The local heat transfer coefficient $[\alpha_{x,f}]_{\mathrm{Ec}=0}$ without consideration of the viscous thermal dissipation and variable physical properties, defined as $(q_{x,f})_{\mathrm{Ec}=0} = (\alpha_{x,f})_{\mathrm{Ec}=0}(t_w - t_\infty)$, will be given by

$$(\alpha_{x,f})_{\mathrm{Ec}=0} = \lambda_f \left(\frac{1}{2}\mathrm{Re}_{x,f}\right)^{1/2} x^{-1} \left[-\left(\frac{\mathrm{d}\theta(\eta)}{\mathrm{d}\eta}\right)_{\eta=0}\right]_{f,\mathrm{Ec}=0}. \quad (5.6)$$

Total heat transfer rate $(Q_{x,f})_{\mathrm{Ec}=0}$ from position $x = 0$ to x with width of b on the plate with ignoring the viscous thermal dissipation and variable physical properties is an integration

$$(Q_{x,f})_{\mathrm{Ec}=0} = \iint_A [q_{x,f}]_{\mathrm{Ec}=0}\ \mathrm{d}A = \int_0^x [q_{x,f}]_{\mathrm{Ec}=0}\ b\cdot \mathrm{d}x,$$

and hence

$$(Q_{x,f})_{\mathrm{Ec}=0} = \lambda_f b(t_w - t_\infty)\left[-\left(\frac{\mathrm{d}\theta(\eta)}{\mathrm{d}\eta}\right)_{\eta=0}\right]_{f,\mathrm{Ec}=0} \int_0^x \left(\frac{1}{2}\mathrm{Re}_{x,f}\right)^{1/2} x^{-1}\mathrm{d}x.$$

With (4.7) for definition of the local Reynolds number $\mathrm{Re}_{x,f}$, we obtain

$$(Q_{x,f})_{\mathrm{Ec}=0} = 2b\lambda_f(t_w - t_\infty)\left(\frac{1}{2}\mathrm{Re}_{x,f}\right)^{1/2}\left[-\left(\frac{\mathrm{d}\theta(\eta)}{\mathrm{d}\eta}\right)_{\eta=0}\right]_{f,\mathrm{Ec}=0}. \quad (5.7)$$

The average heat transfer rate $(\overline{Q}_{x,f})_{\mathrm{Ec}=0}$ from per unit plate area related to the plate area $A = b \times x$ with ignoring the viscous thermal dissipation and variable physical properties is

$$(\overline{Q}_{x,f})_{\mathrm{Ec}=0} = \frac{(Q_{x,f})_{\mathrm{Ec}=0}}{b \times x} = 2x^{-1}\lambda_f(t_w - t_\infty)\left(\frac{1}{2}\mathrm{Re}_{x,f}\right)^{1/2}\left[-\left(\frac{\mathrm{d}\theta(\eta)}{\mathrm{d}\eta}\right)_{\eta=0}\right]_{f,\mathrm{Ec}=0}. \quad (5.8)$$

The average heat transfer coefficient $([\overline{\alpha_{x,f}})_{\mathrm{Ec}=0}$ with ignoring the viscous thermal dissipation and variable physical properties defined as $(\overline{Q}_{x,f})_{\mathrm{Ec}=0} = (\overline{\alpha_{x,f}})_{\mathrm{Ec}=0}(t_w - t_\infty)$ is expressed as

$$(\overline{\alpha_{x,f}})_{\mathrm{Ec}=0} = 2\lambda_f \left(\frac{1}{2}\mathrm{Re}_{x,f}\right)^{1/2} x^{-1}\left[-\left(\frac{\mathrm{d}\theta(\eta)}{\mathrm{d}\eta}\right)_{\eta=0}\right]_{f,\mathrm{Ec}=0}. \quad (5.9)$$

The local Nusselt number without consideration of viscous thermal dissipation and variable physical properties defined by $(Nu_{x,f})_{\mathrm{Ec}=0} = \dfrac{[\alpha_{x,f}]_{\mathrm{Ec}=0}\cdot x}{\lambda_f}$ will be

$$(Nu_{x,f})_{\mathrm{Ec}=0} = \lambda_f \left(\frac{1}{2}\mathrm{Re}_{x,f}\right)^{1/2} x^{-1} \left[-\left(\frac{\mathrm{d}\theta(\eta)}{\mathrm{d}\eta}\right)_{\eta=0}\right]_{f,\mathrm{Ec}=0} \frac{x}{\lambda_f},$$

i.e.

$$(Nu_{x,f})_{\mathrm{Ec}=0} = \left(\frac{1}{2}\mathrm{Re}_{x,f}\right)^{1/2} \left[-\left(\frac{\mathrm{d}\theta(\eta)}{\mathrm{d}\eta}\right)_{\eta=0}\right]_{f,\mathrm{Ec}=0}. \tag{5.10}$$

The average Nusselt number without consideration of variable thermal dissipation and variable thermophysical properties is defined as $(\overline{Nu_{x,f}})_{\mathrm{Ec}=0} = \dfrac{(\overline{\alpha_{x,f}})_{\mathrm{Ec}=0}\, x}{\lambda_f}$, and hence

$$(\overline{Nu_{x,f}})_{\mathrm{Ec}=0} = 2\left(\frac{1}{2}\mathrm{Re}_{x,f}\right)^{1/2} \left[-\left(\frac{\mathrm{d}\theta(\eta)}{\mathrm{d}\eta}\right)_{\eta=0}\right]_{f,\mathrm{Ec}=0},$$

i.e.

$$(\overline{Nu_{x,f}})_{\mathrm{Ec}=0} = \sqrt{2}\mathrm{Re}_{x,f}{}^{1/2} \left[-\left(\frac{\mathrm{d}\theta(\eta)}{\mathrm{d}\eta}\right)_{\eta=0}\right]_{f,\mathrm{Ec}=0}. \tag{5.11}$$

5.5.2 Dimensionless Wall Temperature Gradient

In the heat transfer equations (5.5) to (5.11) obtained by heat transfer analysis, the dimensionless temperature gradient $\left[-\left(\frac{\mathrm{d}\theta(\eta)}{\mathrm{d}\eta}\right)_{\eta=0}\right]_{f,\mathrm{Ec}=0}$ is only one no-given variable for prediction of heat transfer. Then, evaluation of the dimensionless temperature gradient $\left[-\left(\frac{\mathrm{d}\theta(\eta)}{\mathrm{d}\eta}\right)_{\eta=0}\right]_{f,\mathrm{Ec}=0}$ becomes the important issue for the prediction of heat transfer results.

From Fig. 5.4, it is clear that the wall dimensionless temperature gradient $\left[-\left(\frac{\mathrm{d}\theta(\eta)}{\mathrm{d}\eta}\right)_{\eta=0}\right]_{f,\mathrm{Ec}=0}$ varies with Prandtl number Pr_f. A system of the numerical results of the wall temperature gradients $\left[-\left(\frac{\mathrm{d}\theta(\eta)}{\mathrm{d}\eta}\right)_{\eta=0}\right]_{f,\mathrm{Ec}=0}$ is obtained and plotted in Fig. 5.5, and some typical values are listed in Table 5.2. It is clear that decrease of average Prandtl number Pr_f will cause decrease of the wall temperature gradient $\left[-\left(\frac{\mathrm{d}\theta(\eta)}{\mathrm{d}\eta}\right)_{\eta=0}\right]_{f,\mathrm{Ec}=0}$ at an accelerative pace.

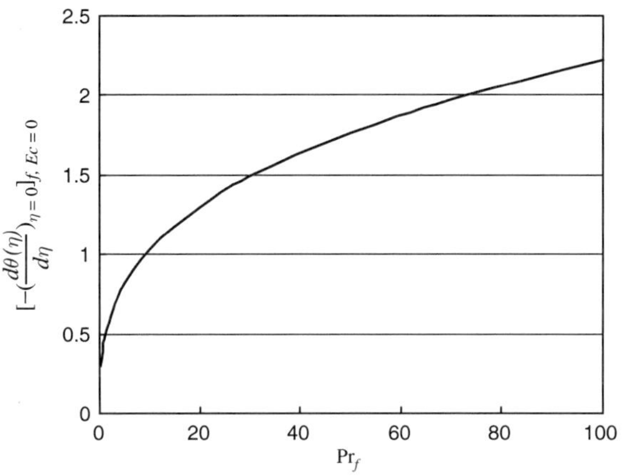

Fig. 5.5 Numerical solutions on wall temperature gradient $\left[-\left(\frac{\mathrm{d}\theta(\eta)}{\mathrm{d}\eta}\right)_{\eta=0}\right]_{f,\mathrm{Ec}=0}$ of laminar forced convection on a horizontal plate with ignoring the viscous thermal dissipation

Table 5.2 Numerical solutions and the predicted values for the wall temperature gradients $\left[-\left(\frac{\mathrm{d}\theta(\eta)}{\mathrm{d}\eta}\right)_{\eta=0}\right]_{f,\mathrm{Ec}=0}$ as well as their predicted deviations by using (5.12) for laminar forced convection on a horizontal flat plate with ignoring the viscous thermal dissipation and variable physical properties

	$\left[-\left(\frac{\mathrm{d}\theta(\eta)}{\mathrm{d}\eta}\right)_{\eta=0}\right]_{f,\mathrm{Ec}=0}$		
Pr_f	Numerical Solution	Predicted by (5.12)	Deviation predicted by (5.12)
0.3	0.30401	0.314366	3.34%
0.6	0.391718	0.396076	1.11%
1	0.469601	0.4696	0
2	0.597234	0.591659	−0.93%
3	0.686202	0.67728	−1.3%
5	0.815561	0.803005	−1.54%
10	1.029735	1.011723	−1.75%
20	1.298819	1.274691	−1.86%
30	1.487318	1.459156	−1.89%
50	1.763922	1.730021	−1.92%
70	1.97446	1.935356	−1.98%
100	2.222905	2.17969	−1.94%

By using a curve-matching approach, the numerical results of $\left[-\left(\frac{\mathrm{d}\theta(\eta)}{\mathrm{d}\eta}\right)_{\eta=0}\right]_{f,\mathrm{Ec}=0}$ for ignoring the thermal viscous dissipation and variable physical properties are here formulated to (5.12). Meanwhile, the values of $\left[-\left(\frac{\mathrm{d}\theta(\eta)}{\mathrm{d}\eta}\right)_{\eta=0}\right]_{f,\mathrm{Ec}=0}$ predicted by (5.12) are obtained for $0.3 \leq \mathrm{Pr}_f \leq 100$, and listed in Table 5.2 with their prediction deviations.

$$\left[-\left(\frac{\mathrm{d}\theta(\eta)}{\mathrm{d}\eta}\right)_{\eta=0}\right]_{f,\mathrm{Ec}=0} = 0.4696\mathrm{Pr}_f{}^{1/3}. \quad (0.3 \leq \mathrm{Pr}_f \leq 100) \tag{5.12}$$

In fact, (5.12) is coincident to the Pohlhausen equation [12] for prediction of the wall temperature gradient $\left[-\left(\frac{\mathrm{d}\theta(\eta)}{\mathrm{d}\eta}\right)_{\eta=0}\right]_{f,\mathrm{Ec}=0}$ for laminar forced convection on a horizontal flat plate, in which the viscous thermal dissipation and variable physical properties are ignored.

With (5.12), (5.5), (5.6), (5.7), (5.8), (5.9), (5.10), and (5.11) become available for simple and reliable prediction of heat transfer of laminar forced convection on horizontal flat plate, respectively, for ignoring the viscous thermal dissipation and variable thermophysical properties. For example, (5.7) and (5.11) will become the following ones, respectively:

$$[Q_{x,f}]_{\mathrm{Ec}=0} = 0.664b\lambda_f(t_w - t_\infty)\mathrm{Re}_{x,f}{}^{1/2}\mathrm{Pr}_f{}^{1/3}, \quad (0.3 \leq \mathrm{Pr}_f \leq 100) \tag{5.13}$$

$$[\overline{Nu}_{x,f}]_{\mathrm{Ec}=0} = 0.664\mathrm{Re}_{x,f}{}^{1/2}\mathrm{Pr}_f{}^{1/3}. \quad (0.3 \leq \mathrm{Pr}_f \leq 100) \tag{5.14}$$

Here, $[Q_{x,f}]_{\mathrm{Ec}=0}$ and $[\overline{Nu}_{x,f}]_{\mathrm{Ec}=0}$ denote the total heat transfer rate and average Nusselt number for laminar forced convection on a horizontal flat plate without consideration of viscous thermal dissipation and variable thermophysical properties. Since (5.12) is reliable, (5.13) and (5.14) also are reliable. (5.13) and (5.14) are equivalent to Pohlhausen equation [12] with ignoring the viscous thermal dissipation and variable physical properties.

The flow type of the laminar forced convection on a horizontal flat plate depends on Reynolds number, shape of the forward end, and the plate surface roughness. The critical Reynolds number from the laminar flow to the turbulent flow is generally located at $1 \times 10^5 < (\mathrm{Re}_{x,f})_{cr} < 3 \times 10^6$ for the forced convection boundary layer on a horizontal flat plate [11, 13, and 14]. However, $(\mathrm{Re}_{x,f})_{cr}$ can be taken as $(\mathrm{Re}_{x,f})_{cr} = 5 \times 10^5$ generally.

5.6 Summary

Now, it is time to summarize in Table 5.3 dimensional similarity variables, governing partial differential equations, similarity governing mathematical model, skin-friction coefficient, and related heat transfer equations based on the present similarity analysis method for laminar forced convection on a horizontal flat plate with ignoring variable physical properties and viscous thermal dissipation.

Table 5.3 Dimensional similarity variables, governing partial differential equations, similarity governing ordinary differential equations based on the present similarity analysis method, skin-friction coefficient, and heat transfer equations for laminar forced convection boundary layer on a horizontal flat plate with ignoring the variable physical properties and viscous thermal dissipation

---

Governing partial differential equations of laminar forced convection boundary layer with ignoring variable physical properties and viscous thermal dissipation

Mass equation

$$\frac{\partial w_x}{\partial x} + \frac{\partial w_y}{\partial y} = 0 \tag{3.1}$$

Momentum equation

$$w_x \frac{\partial w_x}{\partial x} + w_y \frac{\partial w_x}{\partial y} = \nu_f \frac{\partial^2 w_x}{\partial y^2} \tag{3.2}$$

Energy equation

$$w_x \frac{\partial t}{\partial x} + w_y \frac{\partial t}{\partial y} = \frac{\nu_f}{\mathrm{Pr}_f} \frac{\partial^2 t}{\partial^2} \tag{3.3}$$

Boundary conditions

$$\mathrm{y} = 0: \quad \mathrm{w_x} = 0, \quad \mathrm{w_y} = 0, \quad t = t_w$$

$$\mathrm{y} \to \infty: \quad w_x = w_{x,\infty}\ (\text{constant}), \quad t = t_\infty$$

Equations on dimensionless similarity variables

Local Reynolds number $\mathrm{Re}_{x,f}$

$$\mathrm{Re}_{x,f} = \frac{w_{x,\infty} x}{\nu_f}$$

Coordinate variable η

$$\eta = \frac{y}{x}\left(\frac{1}{2}\mathrm{Re}_{x,f}\right)^{1/2}$$

Temperature $\theta(\eta)$

$$\theta(\eta) = \frac{t - t_\infty}{t_w - t_\infty}$$

Velocity component w_x

$$w_x = w_{x,\infty} \cdot W_x(\eta)$$

Velocity component w_y

$$w_y = w_{x,\infty}\left(\frac{1}{2}\mathrm{Re}_x\right)^{-1/2} W_y(\eta)$$

Transformed governing ordinary differential equations

Mass equation

$$-\eta \cdot \frac{\mathrm{d}W_x(\eta)}{\mathrm{d}\eta} + 2\frac{\mathrm{d}W_y(\eta)}{\mathrm{d}\eta} = 0$$

Momentum equation

$$[-\eta \cdot W_x(\eta) + 2W_y(\eta)]\frac{\mathrm{d}W_x(\eta)}{\mathrm{d}\eta} = \frac{\mathrm{d}^2 W_x(\eta)}{\mathrm{d}\eta^2}$$

Table 5.3 (continued)

Energy equation	$[-\eta\cdot W_x(\eta)+2W_y(\eta)]\frac{\mathrm{d}\theta(\eta)}{\mathrm{d}\eta}=\frac{1}{\Pr_f}\frac{\mathrm{d}^2\theta(\eta)}{\mathrm{d}\eta^2}$
Boundary conditions	$\eta=0:\ W_x(\eta)=0,\ W_y(\eta)=0,\ \theta(\eta)=1$ $\eta\to\infty:\ W_x(\eta)=1,\ \theta(\eta)=0$
Skin-friction coefficient	
Local skin-friction coefficient $C_{f,x}$	$C_{x,f}==0.664{\rm Re}_{x,f}{}^{-1/2}$
Average skin-friction coefficient $\overline{C}_{x,f}$	$\overline{C}_{x,f}=1.328{\rm Re}_{x,f}{}^{-1/2}$
Equations related to heat transfer	
$\left[-\left(\frac{\mathrm{d}\theta(\eta)}{\mathrm{d}\eta}\right)_{\eta=0}\right]_{f,{\rm Ec}=0}$	$\left[-\left(\frac{\mathrm{d}\theta(\eta)}{\mathrm{d}\eta}\right)_{\eta=0}\right]_{f,{\rm Ec}=0}=0.4696\Pr_f{}^{1/3}\quad(0.3\le\Pr_f\le100)$
Local values $(Nu_{x,f})_{{\rm Ec}=0}=\frac{[\alpha_{x,f}]_{{\rm Ec}=0}\cdot x}{\lambda_f}$	$(q_{x,f})_{{\rm Ec}=0}=\lambda_f(t_w-t_\infty)\left(\frac{1}{2}{\rm Re}_{x,f}\right)^{1/4}x^{-1}\left[-\left(\frac{\mathrm{d}\theta(\eta)}{\mathrm{d}\eta}\right)_{\eta=0}\right]_{f,{\rm Ec}=0}$ $(Nu_{x,f})_{{\rm Ec}=0}=\left(\frac{1}{2}{\rm Re}_{x,f}\right)^{1/2}\left[-\left(\frac{\mathrm{d}\theta(\eta)}{\mathrm{d}\eta}\right)_{\eta=0}\right]_{f,{\rm Ec}=0}$
Average values $(\overline{Nu_{x,f}})_{{\rm Ec}=0}=\frac{(\overline{\alpha_{x,f}})_{{\rm Ec}=0}x}{\lambda_f}$	$(\overline{Q}_{x,f})_{{\rm Ec}=0}=2x^{-1}\lambda_f(t_w-t_\infty)\left(\frac{1}{2}{\rm Re}_{x,f}\right)^{1/2}\left[-\left(\frac{\mathrm{d}\theta(\eta)}{\mathrm{d}\eta}\right)_{\eta=0}\right]_{f,{\rm Ec}=0}$ $(\overline{Nu_{x,f}})_{{\rm Ec}=0}=\sqrt{2}{\rm Re}_{x,f}{}^{1/2}\left[-\left(\frac{\mathrm{d}\theta(\eta)}{\mathrm{d}\eta}\right)_{\eta=0}\right]_{f,{\rm Ec}=0}$
Total value	$(Q_{x,f})_{{\rm Ec}=0}=2b\lambda_f(t_w-t_\infty)\left(\frac{1}{2}{\rm Re}_{x,f}\right)^{1/2}\left[-\left(\frac{\mathrm{d}\theta(\eta)}{\mathrm{d}\eta}\right)_{\eta=0}\right]_{f,{\rm Ec}=0}$

5.7 Remarks

The new similarity analysis method is proposed in this chapter, which is used for treatment of complete similarity analysis and transformation of the governing partial differential equations of laminar forced convection. The difference of the dimensionless similarity variables for the present new similarity analysis method from those for the Falkner-Skan type transformation is that the former similarity variables on the momentum field are expressed by the dimensionless velocity components $W_x(\eta)$ and $W_y(\eta)$ and the latter similarity variables on the momentum field are expressed by a dimensionless intermediate variable $f(\eta)$. This difference leads to quite different similarity governing mathematical models produced by the two different similarity analysis methods. In this chapter, a study of a simple case for laminar forced convection with ignoring the viscous thermal dissipation and variable physical properties demonstrates that the convenience of the present new similarity analysis method for investigation of the velocity field with the simple and direct relations of the dimensionless similarity velocity components $W_x(\eta)$ and $W_y(\eta)$ to the velocity components w_x and w_y. This convenience brought by the present similarity analysis method will be further benefit for deep investigation of heat transfer, and will be more and more obvious for even complicated case of laminar forced convection and two-phase film condensation.

Rigorous numerical solutions on velocity and temperature fields as well as heat transfer equations are obtained, which are very well coincident to those based on Blasius and Pohlhausen solutions. All these prove that the present new similarity analysis method is valid.

The present new similarity analysis method for laminar forced convection is a spread of the similarity analysis method reported in the book [1] for laminar free convection and two-phase film flows, and can be widely used in a series of cases of laminar forced convection such as for laminar forced convection with consideration of temperature-independent physical properties and even for consideration of temperature and concentration-dependent physical properties, and so on.

5.8 Calculation Example

Example 1 A flat plate with $b = 1\,\text{m}$ in width and $x = 0.25\,\text{m}$ in length is horizontally located in water flowing horizontally with $w_{x,\infty} = 1.3\,\text{m/s}$. The water temperature is $t_\infty = 10°\text{C}$, and surface temperature of the plate is $t_w = 70\,°\text{C}$.

Questions: If ignoring the variable physical properties and viscous thermal dissipation, please

(1) judge if the forced convection is laminar;
(2) evaluate the dimensionless temperature gradient on the plate with ignoring the variable physical properties and viscous thermal dissipation;
(3) calculate the forced convection heat transfer on the plate without consideration of viscous thermal dissipation and variable physical properties;

(4) calculate the average skin-friction coefficient with ignoring the variable physical properties and viscous thermal dissipation.

Solution 1: The boundary temperature will be

$$t_f = \frac{t_w + t_\infty}{2} = \left(\frac{70+10}{2}\right) \text{°C} = 40\text{°C}$$

The related water properties at the average temperature t_f are listed in Table 5.4.

Table 5.4 Physical properties of water at the average temperature t_f

	t_w, °C	t_∞, °C	t_f, °C		
Temperature	70	10	40		
Physical properties			$\Pr_f$	λ_f, W/(m °C)	ν_f, m²/s
			4.32	0.63	0.6564×10^{-6}

The local Reynolds number with ignoring the variable physical properties will be

$$\text{Re}_{x,f} = \frac{w_{x,\infty} x}{\nu_f} = \frac{1.3 \times 0.25}{0.6564 \times 10^{-6}} = 495124.9 < 5 \times 10^6.$$

Then the flow can be regarded as laminar flow.

Solution 2: With (5.13), the dimensionless temperature gradient on the plate will be

$$\left[-\left(\frac{\mathrm{d}\theta}{\mathrm{d}\eta}\right)_{\eta=0}\right]_{f,\text{Ec}=0} = 0.4696 {\Pr_f}^{1/3} = 0.4696 \times 4.32^{1/3} = 0.764814$$

for ignoring the viscous thermal dissipation and variable physical properties.

Solution 3: With (5.7), the heat transfer rate of the plate will be

$$(Q_{x,f})_{\text{Ec}=0} = \sqrt{2} b \lambda_f (t_w - t_\infty) {\text{Re}_{x,f}}^{1/2} \left[-\left(\frac{\mathrm{d}\theta}{\mathrm{d}\eta}\right)_{\eta=0}\right]_{f,\text{Ec}=0}, \tag{5.7}$$

$$[Q_{x,f}]_{\text{Ec}=0} = \sqrt{2} \times 1 \times 0.63 \times (70-10) \times 495124.9^{1/2} \times 0.764814 = 28768.7\text{W}.$$

Solution 4: With (5.4), the average skin-friction coefficient with ignoring the variable physical properties will be

$$\overline{C}_{x,f} = 1.328\ {\text{Re}_{x,f}}^{-1/2} = 1.328 \times 495124.9^{-1/2} = 0.001887.$$

Example 2 Only the flow water is replaced by flow air, and all other conditions in example 1 are kept.

Questions:

(1) Please judge if the forced convection is laminar;
(2) Please evaluate the dimensionless temperature gradient on the plate with ignoring the variable physical properties and viscous thermal dissipation;
(3) Please calculate the forced convection heat transfer on the plate without consideration of viscous thermal dissipation and variable physical properties;
(4) Please calculate the average skin-friction coefficient with ignoring the variable physical properties and viscous thermal dissipation.

Solution 1: The related air properties at the average temperature t_f are listed in Table 6.5.

Table 5.5 Physical properties of air at the average temperature t_f

	t_w, °C	t_∞, °C	t_f, °C		
Temperature	70	10	40		
Physical properties			Pr_f	λ_f, W/(m °C)	ν_f, m^2/s
			0.7	0.02726	17.20×10^{-6}

The local Reynolds number will be

$$\mathrm{Re}_{x,f} = \frac{w_{x,\infty}x}{\nu_f} = \frac{1.3 \times 0.25}{17.20 \times 10^{-6}} = 18895.3 < 5 \times 10^5.$$

Then the flow can be regarded as laminar flow.

Solution 2: With (5.13), the dimensionless temperature gradient on the plate will be

$$\left[-\left(\frac{\mathrm{d}\theta}{\mathrm{d}\eta}\right)_{\eta=0}\right]_{f,\mathrm{Ec}=0} = 0.4696{\mathrm{Pr}_f}^{1/3} = 0.4696 \times 0.7^{1/3} = 0.41696$$

for ignoring the viscous thermal dissipation and variable physical properties.

Solution 3: With (5.7), the forced convection heat transfer result of air laminar forced convection with ignoring viscous thermal dissipation and variable physical properties will be

$$\begin{aligned}[Q_{x,f}]_{\mathrm{Ec}=0} &= \sqrt{2}b\lambda_f(t_w - t_\infty){\mathrm{Re}_{x,f}}^{1/2}\left[-\left(\frac{\mathrm{d}\theta}{\mathrm{d}\eta}\right)_{\eta=0}\right]_{f,\mathrm{Ec}=0}, \\ &= \sqrt{2} \times 1 \times 0.02726 \times (70 - 10) \times 18895.3^{1/2} \times 0.41696, \\ &= 132.6\,\mathrm{W}.\end{aligned}$$

Solution 4: With (5.4), the average skin-friction coefficient with ignoring variable physical properties is calculated as

$$\overline{C}_{x,f} = 1.328\ \mathrm{Re}_{x,f}{}^{-1/2} = 1.328 \times 18895.3^{-1/2} = 0.009661.$$

Exercise

1. Follow example 1. Only the wall temperature is changed to 90°C, and all other conditions are kept.
   Questions: If ignoring the variable physical properties and viscous thermal dissipation, please
   (1) judge if the forced convection is laminar;
   (2) evaluate the dimensionless temperature gradient on the plate by ignoring the variable physical properties and viscous thermal dissipation;
   (3) calculate the forced convection heat transfer on the plate by ignoring the variable physical properties and viscous thermal dissipation;
   (4) calculate the average skin-friction coefficient by ignoring the variable physical properties and viscous thermal dissipation.

References

1. H. Blasius, Grenzschichten in Flussigkeiten mit kleiner Reibung, Z. Math. Phys. **56**, 1–37 (1908)
2. V.M. Falkner, S.W. Skan, Some approximate solutions of the boundary layer Equations. Phil. Mag, **12**, 865 (1931)
3. E.R.G. Eckert, R.M. Drake, Analysis of heat and mass transfer. (McGraw-Hill, NewYork, NY, 1972)
4. T. Cebeci, P. Bradshaw, Physical and Computational aspects of convective heat transfer. (Springer, New York, NY, 1984)
5. T. Fujii, *Theory of Laminar Film Condensation*. (Springer, New York, NY, 1991)
6. B. Louis, etc., *Convective Heat Transfer*, 2nd edn, (John Wiley & sons, Inc., 1993)
7. S. Kakaç, Y. Yenner *Convective Heat Transfer*, 2nd edn, CRC Press, 1995)
8. A. Bejan, *Convection Heat Transfer*, 2nd edn, (Wiley, New York, NY, 1995)
9. I. Pop, D.B. Ingham, *Convective heat transfer – Mathematical and Computational Modelling of Viscous Fluids and Porous Media* (Elsevier, Amsterdam, 2001)
10. T. Cebeci, *Convective Heat Transfer*, 2nd edn, (Springer, Heidelberg, 2002)
11. H. Schlichting, *Boundary-Layer Theory* (Springer, Heidelberg, 2004)
12. E. Pohlhausen, Der Wärmeaustausch zwischen festen Körpern und Flüssigkeiten mit kleiner Reibung und kleiner Wärmeleitung. Z. Angew. Math. Mech. **1**, 115–121, (1921)
13. J.H. Lienhard, *A Heat Transfer Textbook*, (Prentice-Hall, Englewood Cliffs, 1981)
14. S.W. Churchill, A comprehensive correlating equation for forced convection from flat plates. AICHE J. **22**(2), 264–268, (1976)

Chapter 6
Heat Transfer of Laminar Forced Convection with Consideration of Viscous Thermal Dissipation

Abstract The present new similarity analysis method is used for extensive study on heat transfer of laminar forced convection with consideration of viscous thermal dissipation. The proposed expressions of the dimensionless similarity velocity components related to the momentum field are proportional to the related velocity components, and it is obvious that the similarity governing mathematical model based on the present new similarity analysis method is more clear and direct than that transformed by the Falkner-Skan type transformation. With the complete similarity governing mathematical model based on the new similarity analysis method, a system of numerical solutions is found out in a wide range of the average Prandtl number and Eckert number. With these numerical solutions, it is clear that: (i) Increasing the Eckert number causes decrease of the positive heat transfer (from the wall to flow), and increase of the negative heat transfer (from the flow to the wall); and (ii) Increase of the Prandtl number will cause increase the effect of the viscous thermal dissipation on heat transfer. The system of key numerical solutions on wall dimensionless temperature gradient is formulated with the Eckert number and average Prandtl number, and the formulated equations are so reliable that they are very well coincident to the related numerical results. Combined with the theoretical equations on heat transfer, the formulated equations on wall dimensionless temperature gradient can be used to predict heat transfer simply and reliably for consideration of viscous thermal dissipation and average Prandtl number. Additionally, deviation of heat transfer rate predicted by ignoring the viscous thermal dissipation is formulated for judgement of the validity for ignoring the viscous thermal dissipation. Furthermore, a system of the adiabatic Eckert numbers is formulated with variation of average Prandtl number for judgement of the heat flux direction. It is clear that the adiabatic Eckert number will increase at an accelerative pace with decrease of the average Prandtl number, especially in the region with lower Prandtl number.

6.1 Introduction

In treatment of energy equation of forced convection, the viscous thermal dissipation is usually neglected in a lot of previous theoretical analysis and calculation. Ignoring the viscous thermal dissipation is equivalent to regard Eckert number Ec [1] as zero

D.Y. Shang, *Theory of Heat Transfer with Forced Convection Film Flows*,
Heat and Mass Transfer, DOI 10.1007/978-3-642-12581-2_6,

in the dimensionless form of the energy equation. As we know, the Eckert number, defined as $\mathrm{Ec} = \frac{w^2}{c_p \cdot \Delta t}$, expresses the ratio of a flow's kinetic energy to the boundary layer enthalpy difference, and is used to characterize viscous thermal dissipation of convection, especially for forced convection. So far, numerous investigations have been done for the effect of the viscous thermal dissipation on the fluid flows and heat transfer, such as those of [1–9], and some investigation results were further collected in a series books, such as [10, 11]. Those studies showed that in the fluid flows with high Eckert number, generated heat due to the viscous thermal dissipation dominates the fluid temperature, and then the Eckert number can never be taken as zero in the investigation of convection heat transfer.

Therefore, in theoretical study and engineering practice, it is very important to clarify the effect of viscous thermal dissipation on convection heat transfer, in order to understand in which condition the Eckert number can be neglected without any obvious deviation produced in the prediction result on heat transfer. In this present work, laminar forced convection on a horizontal flat plate is taken as an example for extensive investigation of these issues.

6.2 Governing Partial Differential Equations of Laminar Forced Convection

6.2.1 Governing Partial Differential Equations

The physical model and co-ordinate system of boundary layer with two-dimensional laminar forced convection are shown schematically in Fig. 5.1. A flat plate is horizontally located in parallel fluid flow with its mainstream velocity $w_{x,\infty}$. The plate surface temperature is t_w and the fluid bulk temperature is t_∞. Then, a velocity boundary layer is produced near the plate. If t_w is not equal to t_∞, a temperature boundary layer will occur near the plate. We assume that the velocity boundary layer is laminar. Then the governing partial differential equations (2.53), (2.56), and (2.58) for laminar forced convection boundary layer are simplified as follows for consideration of the viscous thermal dissipation but with ignoring the variable physical properties:

$$\frac{\partial w_x}{\partial x} + \frac{\partial w_y}{\partial y} = 0, \tag{6.1}$$

$$w_x \frac{\partial w_x}{\partial x} + w_y \frac{\partial w_x}{\partial y} = \nu_f \frac{\partial^2 w_x}{\partial y^2}, \tag{6.2}$$

$$w_x \frac{\partial t}{\partial x} + w_y \frac{\partial t}{\partial y} = \frac{\nu_f}{\mathrm{Pr}_f} \frac{\partial^2 t}{\partial y^2} + \frac{\nu_f}{c_{p_f}} \left(\frac{\partial w_x}{\partial y} \right)^2, \tag{6.3}$$

with the boundary conditions

$$y = 0: \quad w_x = 0, \quad w_y = 0, \quad t = t_w, \tag{6.4}$$

$$y \to \infty: \quad w_x = w_{x,\infty} \text{ (constant)}, \quad t = t_\infty, \tag{6.5}$$

where $\frac{\nu_f}{c_{pf}}\left(\frac{\partial w_x}{\partial y}\right)^2$ in the energy equation (6.3) is viscous thermal dissipation term. It expresses the viscous friction heat. Here the subscript f of the physical property variables denotes that the related reference temperature is the average temperature $t_f = \frac{t_w + t_\infty}{2}$ of the boundary layer.

6.2.2 Similarity Variables

The equations of the system of the dimensionless similarity variables for the laminar forced convection boundary layer with ignoring variable physical properties but consideration of viscous thermal dissipation are same as those in Chap. 5 without consideration of viscous thermal dissipation:

$$\mathrm{Re}_{x,f} = \frac{w_{x,\infty} x}{\nu_f}, \tag{4.7}$$

$$\eta = \frac{y}{x}\left(\frac{1}{2}\mathrm{Re}_{x,f}\right)^{1/2}, \tag{4.12}$$

$$w_x = w_{x,\infty} W_x(\eta), \tag{4.23}$$

$$w_y = w_{x,\infty}\left(\frac{1}{2}\mathrm{Re}_{x,f}\right)^{-1/2} W_y(\eta), \tag{4.24}$$

$$\theta(\eta) = \frac{t - t_\infty}{t_w - t_\infty}, \tag{4.27}$$

where η is dimensionless coordinate variable, $\mathrm{Re}_{x,f}$ is local Reynolds number with consideration of constant physical properties. $W_x(\eta)$ and $W_y(\eta)$ are dimensionless velocity components in x and y coordinates, respectively, and θ denotes dimensionless temperature.

6.2.3 Governing Ordinary Differential Equations

According to the similarity transformation in Chap. 5, (6.1) and (6.2) are transformed to the following ordinary equations respectively by the above equations with the dimensionless similarity variables $\mathrm{Re}_{x,f}$, η, $W_x(\eta)$, and $W_y(\eta)$:

$$-\eta\frac{\mathrm{d}W_x(\eta)}{\mathrm{d}\eta}+2\frac{\mathrm{d}W_y(\eta)}{\mathrm{d}\eta}=0, \tag{6.6}$$

$$[-\eta W_x(\eta)+2W_y(\eta)]\frac{\mathrm{d}W_x(\eta)}{\mathrm{d}\eta}=\frac{\mathrm{d}^2 W_x(\eta)}{\mathrm{d}\eta^2}. \tag{6.7}$$

While the similarity transformation of (6.3) is shown as follows.:

With (4.27), we have

$$\frac{\partial t}{\partial x}=(t_w-t_\infty)\frac{\mathrm{d}\theta(\eta)}{\mathrm{d}\eta}\frac{\partial\eta}{\partial x}.$$

While, with (4.28) we have

$$\frac{\partial\eta}{\partial x}=-\frac{1}{2}x^{-1}\eta.$$

Then

$$\frac{\partial t}{\partial x}=-\frac{1}{2}x^{-1}\eta(t_w-t_\infty)\frac{d\theta(\eta)}{d\eta}. \tag{a}$$

With (4.27), we have

$$\frac{\partial t}{\partial y}=(t_w-t_\infty)\frac{\mathrm{d}\theta(\eta)}{\mathrm{d}\eta}\frac{\partial\eta}{\partial y}.$$

From (4.12) we have

$$\frac{\partial\eta}{\partial y}=x^{-1}\left(\frac{1}{2}\mathrm{Re}_{x,f}\right)^{1/2}.$$

Then

$$\frac{\partial t}{\partial y}=(t_w-t_\infty)\frac{\mathrm{d}\theta(\eta)}{\mathrm{d}\eta}x^{-1}\left(\frac{1}{2}\mathrm{Re}_{x,f}\right)^{1/2}. \tag{b}$$

Therefore,

$$\frac{\partial^2 t}{\partial y^2}=\frac{1}{2}(t_w-t_\infty)\frac{\mathrm{d}^2\theta(\eta)}{\mathrm{d}\eta^2}x^{-2}\mathrm{Re}_{x,f}. \tag{c}$$

With (4.23), (4.24), (a), (b) and (c) and (4.34), (6.3) is changed to

$$-\frac{1}{2}x^{-1}\eta w_{x,\infty}W_x(\eta)(t_w-t_\infty)\frac{\mathrm{d}\theta(\eta)}{\mathrm{d}\eta}$$

$$+w_{x,\infty}\left(\frac{1}{2}\mathrm{Re}_{x,f}\right)^{-1/2}W_y(\eta)(t_w-t_\infty)\frac{\mathrm{d}\theta(\eta)}{\mathrm{d}\eta}x^{-1}\left(\frac{1}{2}\mathrm{Re}_{x,f}\right)^{1/2}$$

$$=\frac{1}{2}\frac{\nu_f}{\mathrm{Pr}_f}(t_w-t_\infty)\frac{\mathrm{d}^2\theta(\eta)}{\mathrm{d}\eta^2}x^{-2}\,\mathrm{Re}_{x,f}+\frac{\nu_f}{c_{p_f}}\left[x^{-1}\left(\frac{1}{2}\mathrm{Re}_{x,f}\right)^{1/2}w_{x,\infty}\frac{\mathrm{d}W_x(\eta)}{\mathrm{d}\eta}\right]^2$$

With the definition of $\mathrm{Re}_{x,f}$, the above equation is further simplified to

$$-\frac{1}{2}x^{-1}\eta w_{x,\infty}W_x(\eta)(t_w-t_\infty)\frac{\mathrm{d}\theta(\eta)}{\mathrm{d}\eta}+w_{x,\infty}W_y(\eta)(t_w-t_\infty)\frac{\mathrm{d}\theta(\eta)}{\mathrm{d}\eta}x^{-1}$$

$$=\frac{1}{2}\frac{\nu_f}{\mathrm{Pr}_f}(t_w-t_\infty)\frac{\mathrm{d}^2\theta(\eta)}{\mathrm{d}\eta^2}x^{-2}\frac{w_{x,\infty}x}{\nu_f}+\frac{\nu_f}{c_{p_f}}x^{-2}\left(\frac{1}{2}\frac{w_{x,\infty}x}{\nu_f}\right)w_{x,\infty}{}^2\left(\frac{\mathrm{d}W_x(\eta)}{\mathrm{d}\eta}\right)^2.$$

With the definition of Reynolds number, the above equation is divided by $\frac{1}{2}x^{-1}$ $(t_w-t_\infty)w_{x,\infty}$, and simplified to

$$[-\eta\cdot W_x(\eta)+2\cdot W_y(\eta)]\frac{\mathrm{d}\theta(\eta)}{\mathrm{d}\eta}=\frac{1}{\mathrm{Pr}_f}\frac{\mathrm{d}^2\theta(\eta)}{\mathrm{d}\eta^2}+\mathrm{Ec}\left(\frac{\mathrm{d}W_x(\eta)}{\mathrm{d}\eta}\right)^2. \qquad (6.8)$$

The boundary condition equations (6.4) and (6.5) are transformed to the following dimensionless ones:

$$\eta=0:W_x(\eta)=0,\quad W_y(\eta)=0,\ \theta(\eta)=1, \qquad (6.9)$$

$$\eta\to\infty:\ W_x(\eta)=1,\ \theta(\eta)=0, \qquad (6.10)$$

where

$$\mathrm{Ec}=\frac{w_{x,\infty}{}^2}{c_{p_f}(t_w-t_\infty)} \qquad (6.11)$$

is the Eckert number for descript of the viscous thermal dissipation.

6.3 Numerical Results

The governing dimensionless equations (6.6), (6.7), and (6.8) with the boundary condition equations (6.9) and (6.10) are solved by a shooting method with fifth-order Runge-Kutta integration. For solving the nonlinear problem, a variable mesh approach is applied to the numerical calculation programs. A system of accurate numerical results of velocity and temperature fields is obtained as follows.

6.3.1 Velocity Field

The numerical results for the dimensionless velocity components $W_x(\eta)$ and $W_y(\eta)$ with ignoring the variable physical properties have been plotted in Figs. 5.2 and 5.3, and some typical solutions are listed in Table 5.1. Due to the assumption of independent-temperature physical properties, the dimensionless velocity field is independent on the temperature field.

6.3.2 Temperature Fields

A system of numerical results for dimensionless temperature fields and the wall dimensionless temperature gradient is obtained, and some of the temperature profiles are plotted in Figs. 6.1, 6.2, 6.3, 6.4, 6.5, 6.6, 6.7, and 6.8 respectively with variations of average Prandtl number $\mathrm{Pr_f}$ and Eckert number Ec. It is found here that the temperature field is not only related to the Prandtl number, but also varies with the Eckert number Ec. Increasing the average Prandtl number $\mathrm{Pr_f}$ causes increase of the effect of the Eckert number Ec on the temperature field. It is seen that with increasing the average Prandtl number, the temperatures θ of some portion of the boundary layer flow are larger than unity (see Figs. 6.5, 6.6, and 6.7), which means that the temperatures of this portion of the boundary layer flow are so high that they

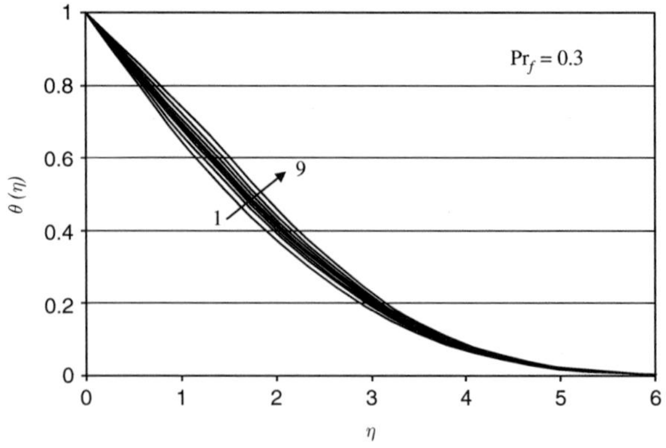

Fig. 6.1 Temperature profiles $\theta(\eta)$ of laminar forced convection on a horizontal flat plate for $\mathrm{Pr}_f = 0.3$ with consideration of viscous thermal dissipation but without consideration of variable physical properties (Lines 1 to 9: $\mathrm{Ec} = -1, -0.6, -0.3, -0.1, 0, 0.1, 0.3, 0.6$, and 1)

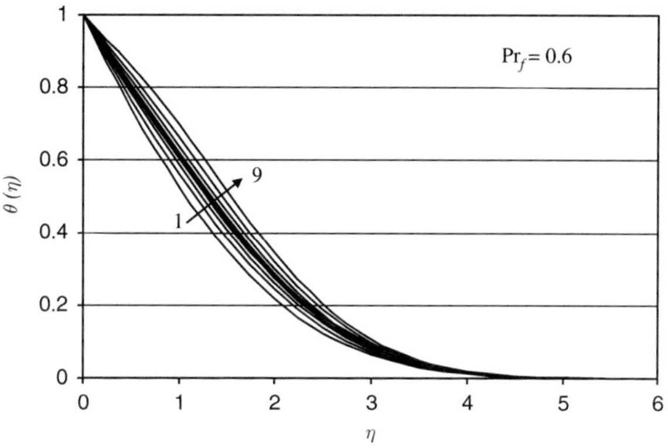

Fig. 6.2 Temperature profiles $\theta(\eta)$ of laminar forced convection on a horizontal flat plate for $\mathrm{Pr}_f = 0.6$ with consideration of viscous thermal dissipation but without consideration of variable physical properties (Lines 1 to 9: $\mathrm{Ec} = -1, -0.6, -0.3, -0.1, 0, 0.1, 0.3, 0.6$, and 1)

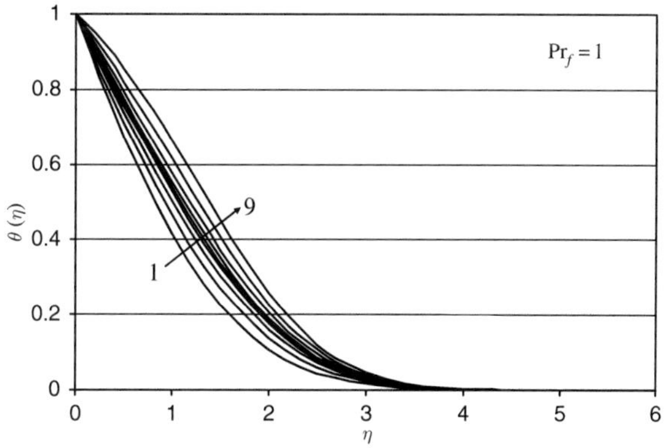

Fig. 6.3 Temperature profiles $\theta(\eta)$ of laminar forced convection on a horizontal flat plate for $\mathrm{Pr}_f = 1$ with consideration of viscous thermal dissipation but without consideration of variable physical properties (Lines 1 to 9: $\mathrm{Ec} = -1, -0.6, -0.3, -0.1, 0, 0.1, 0.3, 0.6$, and 1)

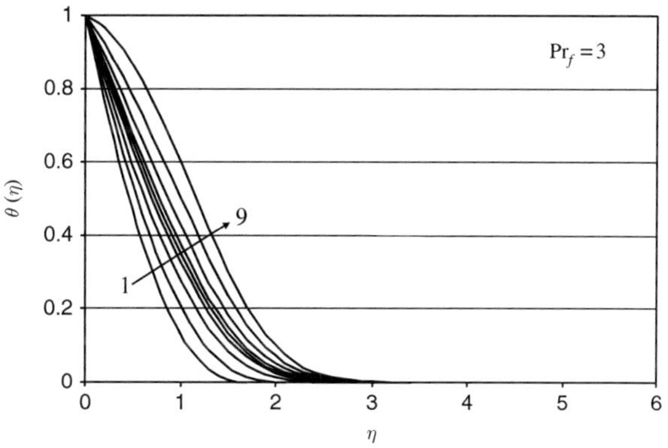

Fig. 6.4 Temperature profiles $\theta(\eta)$ of laminar forced convection on a horizontal flat plate for $\mathrm{Pr}_f = 3$ with consideration of viscous thermal dissipation but without consideration of variable physical properties (lines 1 to 9: $\mathrm{Ec} = -1, -0.6, -0.3, -0.1, 0, 0.1, 0.3, 0.6$, and 1)

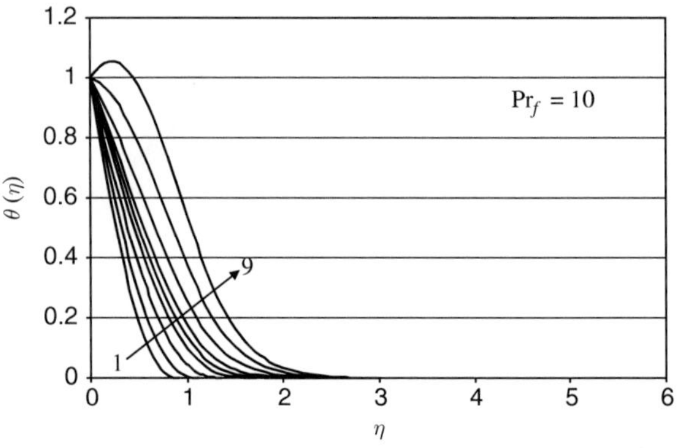

Fig. 6.5 Temperature profiles $\theta(\eta)$ of laminar forced convection on a horizontal flat plate for $\mathrm{Pr}_f = 10$ with consideration of viscous thermal dissipation but without consideration of variable physical properties (Lines 1 to 9: Ec = −1, −0.6, −0.3, −0.1, 0, 0.1, 0.3, 0.6, and 1)

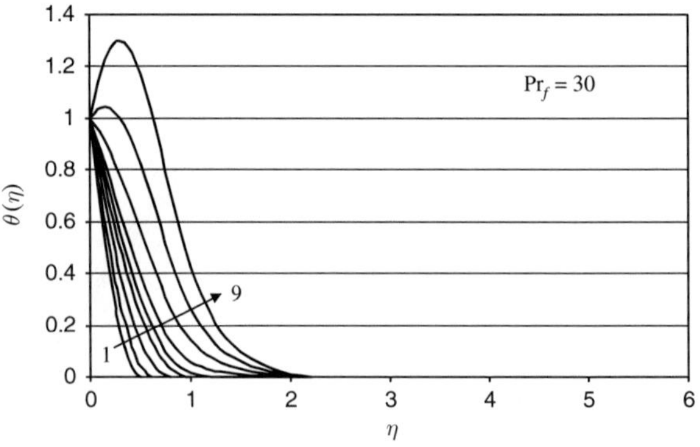

Fig. 6.6 Temperature profiles $\theta(\eta)$ of laminar forced convection on a horizontal flat plate for $\mathrm{Pr}_f = 30$ with consideration of viscous thermal dissipation but without consideration of variable physical properties (Lines 1 to 9: Ec = −1, −0.6, −0.3, −0.1, 0, 0.1, 0.3, 0.6, and 1)

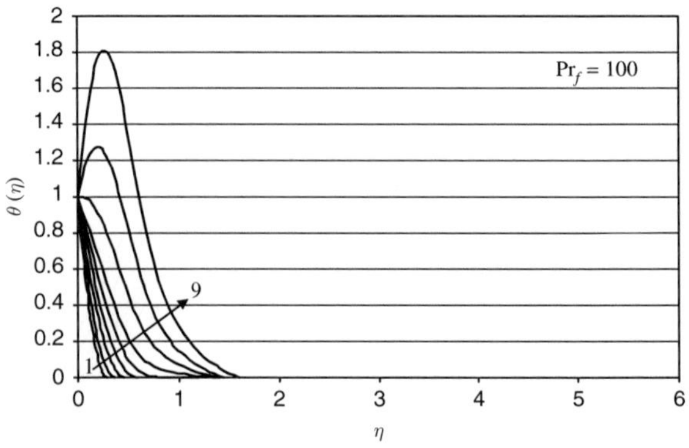

Fig. 6.7 Temperature profiles $\theta(\eta)$ of laminar forced convection on a horizontal flat plate for $\mathrm{Pr}_f = 100$ with consideration of viscous thermal dissipation but without consideration of variable physical properties (Lines 1 to 9: Ec = −1, −0.6, −0.3, −0.1, 0, 0.1, 0.3, 0.6, and 1)

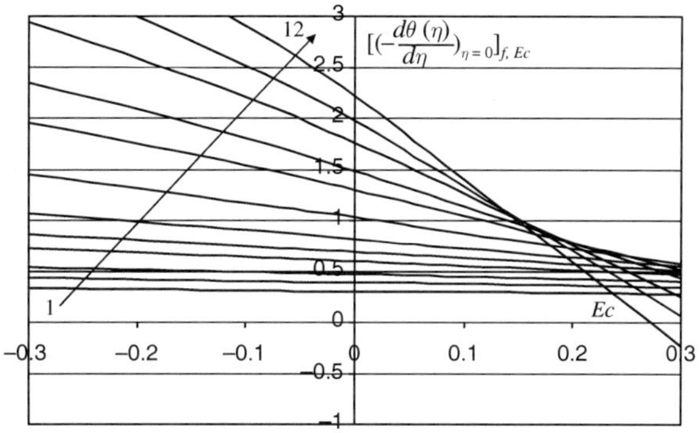

Fig. 6.8 Numerical solutions on dimensionless wall temperature gradient $\left[-\left(\frac{d\theta(\eta)}{d\eta}\right)_{\eta=0}\right]_{f,\text{Ec}}$ of laminar forced convection on a horizontal plate with ignoring variable physical properties but consideration of viscous thermal dissipation (Lines 1 to 12: for $\text{Pr}_f = 0.3$, 0.6, 1, 2, 3, 5, 10, 20, 30, 50, 70, and 100)

are larger than the boundary temperature due to the viscous thermal dissipation. This phenomenon is more and more obvious with increasing the average Prandtl number.

6.4 Heat Transfer Analysis

Consulting the heat transfer analysis in Chap. 5 for laminar forced convection boundary layer with ignoring the variable physical properties and viscous thermal dissipation, the related heat transfer analysis in this chapter with ignoring the variable physical properties but consideration of viscous thermal dissipation is as follows.

The local heat transfer rate $(q_{x,f})_{\text{Ec}}$ with ignoring the variable physical properties but consideration of viscous thermal dissipation at position x per unit area from the surface of the plate can be calculated and expressed as

$$(q_{x,f})_{\text{Ec}} = \lambda_f(t_w - t_\infty)\left(\frac{1}{2}\text{Re}_{x,f}\right)^{1/2} x^{-1}\left[-\left(\frac{d\theta(\eta)}{d\eta}\right)_{\eta=0}\right]_{f,\text{Ec}}, \tag{6.12}$$

where $\left[-\left(\frac{d\theta(\eta)}{d\eta}\right)_{\eta=0}\right]_{f,\text{Ec}}$ is wall dimensionless temperature gradient with ignoring the variable physical properties but consideration of viscous thermal dissipation. Obviously, according to (6.12), the positive value of heat flux $(q_{x,f})_{\text{Ec}}$ is corresponding to the direction from the wall to flow, and the negative value of heat flux $(q_{x,f})_{\text{Ec}=0}$ is corresponding to that from the flow to the wall.

The local heat transfer coefficient $(\alpha_{x,f})_{\text{Ec}}$, defined as $(q_{x,f})_{\text{Ec}} = (\alpha_{x,f})_{\text{Ec}}(t_w - t_\infty)$, with ignoring the variable physical properties but consideration of viscous thermal dissipation will be given by

$$(\alpha_{x,f})_{\mathrm{Ec}} = \lambda_f \left(\frac{1}{2}\mathrm{Re}_{x,f}\right)^{1/2} x^{-1} \left[-\left(\frac{\mathrm{d}\theta(\eta)}{\mathrm{d}\eta}\right)_{\eta=0}\right]_{f,\mathrm{Ec}}. \tag{6.13}$$

Total heat transfer rate $(Q_{x,f})_{\mathrm{Ec}}$ from position $x = 0$ to x with width of b on the plate with ignoring the variable physical properties but consideration of viscous thermal dissipation is

$$(Q_{x,f})_{\mathrm{Ec}} = \sqrt{2} b \lambda_f (t_w - t_\infty) \mathrm{Re}_{x,f}{}^{1/2} \left[-\left(\frac{\mathrm{d}\theta(\eta)}{\mathrm{d}\eta}\right)_{\eta=0}\right]_{f,\mathrm{Ec}}. \tag{6.14}$$

The average heat transfer rate $(\overline{Q}_{x,f})_{Ec}$ from per unit plate area related to the plate area $A = b \times x$ with ignoring the variable physical properties but consideration of viscous thermal dissipation is expressed as

$$(\overline{Q}_{x,f})_{\mathrm{Ec}} = \frac{(Q_{x,f})_{\mathrm{Ec}}}{b \times x} = \sqrt{2} x^{-1} \lambda_f (t_w - t_\infty) \mathrm{Re}_{x,f}{}^{1/2} \left[-\left(\frac{\mathrm{d}\theta(\eta)}{\mathrm{d}\eta}\right)_{\eta=0}\right]_{f,\mathrm{Ec}}. \tag{6.15}$$

The average heat transfer coefficient $(\overline{\alpha_{x,f}})_{\mathrm{Ec}}$ defined as $(\overline{Q}_{x,f})_{\mathrm{Ec}} (\overline{\alpha_{x,f}})_{\mathrm{Ec}} (t_w - t_\infty)$ with ignoring the variable physical properties but consideration of viscous thermal dissipation is expressed as

$$(\overline{\alpha_{x,f}})_{\mathrm{Ec}} = \sqrt{2} \lambda_f \mathrm{Re}_{x,f}{}^{1/2} x^{-1} \left[-\left(\frac{\mathrm{d}\theta(\eta)}{\mathrm{d}\eta}\right)_{\eta=0}\right]_{f,\mathrm{Ec}}. \tag{6.16}$$

The local Nusselt number with ignoring variable physical properties but consideration of viscous thermal dissipation, defined by $[Nu_{x,f}]_{\mathrm{Ec}} = \frac{[\alpha_{x,f}]_{\mathrm{Ec}} \cdot x}{\lambda_f}$, will be

$$[Nu_{x,f}]_{\mathrm{Ec}} = \left(\frac{1}{2}\mathrm{Re}_{x,f}\right)^{1/2} \left[-\left(\frac{\mathrm{d}\theta(\eta)}{\mathrm{d}\eta}\right)_{\eta=0}\right]_{f,\mathrm{Ec}}. \tag{6.17}$$

The average Nusselt number is defined as $(\overline{Nu_{x,f}})_{\mathrm{Ec}} = \frac{(\overline{\alpha_{x,f}})_{\mathrm{Ec}} \cdot x}{\lambda_f}$, with ignoring variable physical properties but consideration of viscous thermal dissipation, and hence

$$(\overline{Nu}_{x,f})_{\mathrm{Ec}} = \sqrt{2}\, \mathrm{Re}_{x,f}{}^{1/2} \left[-\left(\frac{\mathrm{d}\theta(\eta)}{\mathrm{d}\eta}\right)_{\eta=0}\right]_{f,\mathrm{Ec}}. \tag{6.18}$$

6.5 Formulated Equation of Dimensionless Wall Temperature Gradient

It is found that in the heat transfer (6.12), (6.13), (6.14), (6.15), (6.16), (6.17), and (6.18) obtained by heat transfer analysis, the dimensionless temperature gradient $\left[-\left(\frac{d\theta(\eta)}{d\eta}\right)_{\eta=0}\right]_{f,\mathrm{Ec}}$ is only one no-given variable for the prediction of heat transfer. In this section, we will further investigate its evaluation so that the heat transfer results can be predicted simply and reliably.

Figures 6.1, 6.2, 6.3, 6.4, 6.5, 6.6, 6.7, and 6.8 demonstrate that the temperature fields as well as the wall temperature gradient $\left[-\left(\frac{d\theta(\eta)}{d\eta}\right)_{\eta=0}\right]_{f,\mathrm{Ec}}$ for ignoring the variable physical properties but consideration of viscous thermal dissipation vary with Prandtl number Pr_f and Eckert number Ec. A system of the numerical results of the wall temperature gradients $\left[-\left(\frac{d\theta(\eta)}{d\eta}\right)_{\eta=0}\right]_{f,\mathrm{Ec}}$ are obtained and plotted in Fig. 6.8, and some of them are listed in Table 6.1. It is clear that increasing the Eckert number causes decrease of the wall temperature gradient $\left[-\left(\frac{d\theta(\eta)}{d\eta}\right)_{\eta=0}\right]_{f,\mathrm{Ec}}$. On the other hand, increase of Prandtl number $\mathrm{Pr_f}$ will cause increase in the effect of the viscous thermal dissipation on the dimensionless wall temperature gradient $\left[-\left(\frac{d\theta(\eta)}{d\eta}\right)_{\eta=0}\right]_{f,\mathrm{Ec}}$.

By using a curve-fitting approach, the dimensionless temperature gradient $\left[-\left(\frac{d\theta(\eta)}{d\eta}\right)_{\eta=0}\right]_{f,\mathrm{Ec}}$ with ignoring variable physical properties but consideration of viscous thermal dissipation is formulated based on the system of the rigorous numerical solutions, i.e.

$$\left[-\left(\frac{d\theta(\eta)}{d\eta}\right)_{\eta=0}\right]_{f,\mathrm{Ec}} = a\cdot Ec + \left[-\left(\frac{d\theta(\eta)}{d\eta}\right)_{\eta=0}\right]_{f,\mathrm{Ec}=0},$$

$$(-0.1 \le E_c \le 0.1) \text{ and } (0.3 \le \mathrm{Pr_f} \le 3) \tag{6.19}$$

$$\left[-\left(\frac{d\theta(\eta)}{d\eta}\right)_{\eta=0}\right]_{f,\mathrm{Ec}} = b\cdot \mathrm{Ec}^2 + c\cdot \mathrm{Ec} + \left[-\left(\frac{d\theta(\eta)}{d\eta}\right)_{\eta=0}\right]_{f,\mathrm{Ec}=0},$$

$$(-0.1 \le E_c \le 0.1) \text{ and } (3 \le \mathrm{Pr_f} \le 100) \tag{6.20}$$

Table 6.1 Numerical solutions and evaluated values on wall temperature gradient $\left[-\left(\frac{d\theta(\eta)}{d\eta}\right)_{\eta=0}\right]_{f,Ec}$ of laminar forced convection on a horizontal plate with ignoring variable physical properties but consideration of viscous thermal dissipation

		Ec						
		−0.1	−0.03	−0.01	0	0.01	0.03	0.1
Pr_f		$\left[-\left(\frac{d\theta(\eta)}{d\eta}\right)_{\eta=0}\right]_{f,Ec}$						
0.3	(1)	0.31224	0.306478	.304832	0.30401	0.303187	0.301542	0.29578
	(2)	0.322797	0.316895	0.315209	0.314366	0.31352	0.3118	0.30593
	(3)	0.0327	0.0329	0.0329	0.0329	0.0329	0.0330	0.0332
0.6	(1)	0.40685	0.396258	0.393231	0.39172	0.390204	0.387177	0.37658
	(2)	0.411049	0.400568	0.397573	0.396076	0.394579	0.391584	0.381103
	(3)	0.0102	0.01076	0.01092	0.011	0.0110	0.01125	0.0119
1	(1)	0.49308	0.476645	0.471949	0.46960	0.467252	0.462556	0.44612
	(2)	0.49291	0.476593	0.471931	0.4696	0.467269	0.462607	0.44629
	(3)	−0.00034	−0.00011	−3.81E-05	0	3.64E-05	0.00011	0.00038
2	(1)	0.63925	0.609839	0.601436	0.59723	0.593033	0.584630	0.55522
	(2)	0.633879	0.604325	0.595880	0.591659	0.587437	0.578993	0.549439
	(3)	−0.0085	−0.0091	−0.0093	−0.0094	−0.0095	−0.0097	−0.0105
3	(1)	0.74466	0.703738	0.692048	0.68620	0.680357	0.668666	0.62775
	(2)	0.736723	0.695115	0.683225	0.677280	0.671335	0.659445	0.61782384
	(3)	−0.0108	−0.0124	−0.01291	−0.0132	−0.0134	−0.01398	−0.0161
5	(1)	0.90349	0.842056	0.824401	0.81556	0.806717	.789030	0.72712
	(2)	0.889663	0.829055	0.8116931	0.8030047	0.794311	0.776909	0.71584454
	(3)	−0.0155	−0.0157	−0.0156	−0.0156	−0.0156	−0.0156	−0.0158
10	(1)	1.17655	1.074676	1.044835	1.02974	1.014506	.984024	0.87734
	(2)	1.159622	1.056645	1.026749	1.0117225	0.996643	0.966326	0.858559
	(3)	−0.0146	−0.01706	−0.0176	−0.0178	−0.0179	−0.0183	−0.0218
20	(1)	1.53638	1.372632	1.323778	1.29882	1.272997	1.221358	1.04062
	(2)	1.513667	1.34856	1.299520	1.2746905	1.249653	1.198956	1.014989
	(3)	−0.015	−0.01785	−0.0186	−0.0189	−0.0187	−0.01868	−0.0253
30	(1)	1.79972	1.584863	1.520512	1.48732	1.452399	1.382560	1.13813
	(2)	1.772507	1.55692	1.492103	1.4591563	1.425851	1.358166	1.109982
	(3)	−0.0154	−0.0179	−0.01906	−0.01936	−0.0186	−0.0179	−0.0254
50	(1)	2.20267	1.901544	1.812972	1.76392	1.713498	1.612651	1.25969
	(2)	2.168313	1.86832	1.776772	1.730021	1.68262	1.585869	1.226785
	(3)	−0.01584	−0.01778	−0.020374	−0.0195944	−0.01835	−0.01689	−0.026822
70	(1)	2.51602	2.145805	2.032571	1.97446	1.913859	1.783148	1.33413
	(2)	2.480924	2.108752	1.9940806	1.93535558	1.875704	1.753623	1.2971627
	(3)	−0.01415	−0.01757	−0.0193	−0.0202	−0.02035	−0.01685	−0.0285
100	(1)	2.90624	2.441331	2.299284	2.22291	2.142979	1.979899	1.40442
	(2)	2.867333	2.399785	2.2543694	2.17969012	2.103696	1.947766	1.360603
	(3)	−0.01357	−0.0173	−0.0199	−0.0198	−0.01867	−0.0165	−0.0322

Note: (1) numerical solution, (2) evaluated values by using (6.19) and (6.20), and (3) the evaluated deviations

where

$$a = 0.0138\text{Pr}_f^{\,2} - 0.2305\text{Pr}_f - 0.0164,$$

$$b = (-0.0007\text{Pr}_f + 0.002)\text{Pr}_f^{\,1.75}, \qquad (3 \le \text{Pr}_f \le 10)$$

$$b = 9\text{E-}05\text{Pr}_f^{\,2} - 0.08\text{Pr}_f + 0.5278, \qquad (10 \le \text{Pr}_f \le 100)$$

$$c = (-0.2027\text{Pr}_f - 0.0929)/\text{Pr}_f^{\,0.15}, \qquad (3 \le \text{Pr}_f \le 10)$$

$$c = (-0.3525Pr_f + 0.2819)/\text{Pr}_f^{\,1/3}, \qquad (10 \le \text{Pr}_f \le 100)$$

$$\left[-\left(\frac{\mathrm{d}\theta(\eta)}{\mathrm{d}\eta}\right)_{\eta=0}\right]_{f,\text{Ec}=0} = 0.4696\text{Pr}_f^{\,1/3}. \qquad (0.3 \le \text{Pr}_f \le 100) \tag{5.12}$$

Here, $\left[-\left(\frac{\mathrm{d}\theta(\eta)}{\mathrm{d}\eta}\right)_{\eta=0}\right]_{f,\text{Ec}=0}$ is the dimensionless temperature gradient with ignoring variable physical properties and viscous thermal dissipation for laminar forced convection on a horizontal flat plate, and (5.12) is identical to the Pohlhausen solution.

The results predicted by (6.19) and (6.20) are listed in Table 6.1 for comparing with the related numerical solutions, and the relative deviation caused by the prediction values are listed in Table 6.1 also. It is seen that the prediction values $\left[-\left(\frac{\mathrm{d}\theta(\eta)}{\mathrm{d}\eta}\right)_{\eta=0}\right]_{f,\text{Ec}}$ are very well coincident to the numerical solutions, and then (6.19) and (6.20) are reliable.

6.6 Heat Transfer Prediction Equation

With (6.19) and (6.20), (6.12), (6.13), (6.14), (6.15), (6.16), (6.17), and (6.18) become available for the prediction of heat transfer results on laminar forced convection on a horizontal flat plate with ignoring variable physical properties and consideration of viscous thermal dissipation. For example, (6.14) and (6.18) for plate heat transfer rate Q_x and average Nusselt number $\overline{Nu}_{x,f}$ become respectively

$$[Q_{x,f}]_{\text{Ec}} = \sqrt{2}b\lambda_f(t_w - t_\infty)\text{Re}_{x,f}^{\;1/2}\left\{a \cdot Ec + \left[-\left(\frac{\mathrm{d}\theta(\eta)}{\mathrm{d}\eta}\right)_{\eta=0}\right]_{\text{Ec}=0}\right\},$$

$$(-0.1 \le \text{Ec} \le 0.1) \text{ and } (0.3 \le \text{Pr}_f \le 3), \tag{6.21}$$

$$[Q_{x,f}]_{\text{Ec}} = \sqrt{2}b\lambda_f(t_w - t_\infty)\text{Re}_{x,f}^{\;1/2}\left\{b \cdot \text{Ec}^2 + c \cdot \text{Ec} + \left[-\left(\frac{\mathrm{d}\theta(\eta)}{\mathrm{d}\eta}\right)_{\eta=0}\right]_{\text{Ec}=0}\right\},$$

$$(-0.1 \le \text{Ec} \le 0.1) \text{ and } (3 \le \text{Pr}_f \le 100), \tag{6.22}$$

$$[\overline{Nu}_{x,f}]_{\mathrm{Ec}} = \sqrt{2}\mathrm{Re}_{x,f}{}^{1/2}\left\{a\cdot\mathrm{Ec} + \left[-\left(\frac{\mathrm{d}\theta(\eta)}{\mathrm{d}\eta}\right)_{\eta=0}\right]_{\mathrm{Ec}=0}\right\},$$

$$(-0.1 \le \mathrm{Ec} \le 0.1) \text{ and } (0.3 \le \mathrm{Pr_f} \le 3) \tag{6.23}$$

$$[\overline{Nu}_{x,f}]_{\mathrm{Ec}} = \sqrt{2}\mathrm{Re}_{x,f}{}^{1/2}\left\{b\cdot Ec^2 + c\cdot Ec + \left[\left(-\frac{\mathrm{d}\theta(\eta)}{\mathrm{d}\eta}\right)_{\eta=0}\right]_{\mathrm{Ec}=0}\right\},$$

$$(-0.1 \le \mathrm{Ec} \le 0.1) \text{ and } (3 \le \mathrm{Pr}_f \le 100), \tag{6.24}$$

where

$$a = 0.0138\mathrm{Pr_f}^2 - 0.2305\mathrm{Pr_f} - 0.0164,$$

$$b = (-0.0007\mathrm{Pr_f} + 0.002)\mathrm{Pr_f}^{1.75}, \quad (3 \le \mathrm{Pr_f} \le 10)$$

$$b = 9\text{E-}05\mathrm{Pr_f}^2 - 0.08\mathrm{Pr_f} + 0.5278, \quad (10 \le \mathrm{Pr} \le 100)$$

$$c = (-0.2027\mathrm{Pr_f} - 0.0929)/\mathrm{Pr_f}^{0.15}, \quad (3 \le \mathrm{Pr_f} \le 10)$$

$$c = (-0.3525Pr_{\mathrm{f}} + 0.2819)/\mathrm{Pr_f}^{1/3}, \quad (10 \le \mathrm{Pr_f} \le 100)$$

$$\left[\left(-\frac{\mathrm{d}\theta(\eta)}{\mathrm{d}\eta}\right)_{\eta=0}\right]_{f,\mathrm{Ec}=0} = 0.4696\mathrm{Pr_f}^{1/3}, \quad 0.3 \le \mathrm{Pr_f} \le 100). \tag{5.12}$$

The accuracies of (6.21), (6.22), (6.23), and (6.24) depend on those of (6.19) and (6.20), then are reliable for their practical prediction on heat transfer for laminar forced convection on a horizontal flat plate with ignoring variable physical properties but consideration of viscous thermal dissipation.

6.7 Heat Transfer Prediction Deviation Caused by Ignoring the Viscous Thermal Dissipation

If we take $E_{(\overline{Nu}_{x,f})_{\mathrm{Ec}=0}}$ as average Nusselt number prediction deviation caused by ignoring the viscous thermal dissipation, it can be expressed as $E_{(\overline{Nu}_{x,f})_{\mathrm{Ec}=0}} = \frac{(\overline{Nu}_{x,f})_{\mathrm{Ec}=0} - (\overline{Nu}_{x,f})_{\mathrm{Ec}}}{(\overline{Nu}_{x,f})_{\mathrm{Ec}=0}}$. Here, $(\overline{Nu}_{x,f})_{\mathrm{Ec}}$ denotes average Nusselt number for ignoring variable physical properties but consideration of viscous thermal dissipation, and $(\overline{Nu}_{x,f})_{\mathrm{Ec}=0}$ denotes average Nusselt number with ignoring variable physical properties and viscous thermal dissipation. With (5.11) and (6.18), $E_{(\overline{Nu}_{x,f})_{\mathrm{Ec}=0}}$ can be expressed as

$$E_{(\overline{Nu}_{x,f})_{\mathrm{Ec}=0}} = \frac{\sqrt{2}\mathrm{Re}_{x,f}{}^{1/2}\left[\left(-\frac{\mathrm{d}\theta(\eta)}{\mathrm{d}\eta}\right)_{\eta=0}\right]_{f,\mathrm{Ec}=0} - \sqrt{2}(\mathrm{Re}_{x,f})^{1/2}\left[\left(-\frac{\mathrm{d}\theta(\eta)}{\mathrm{d}\eta}\right)_{\eta=0}\right]_{f,\mathrm{Ec}}}{\sqrt{2}\mathrm{Re}_{x,f}{}^{1/2}\left[\left(-\frac{\mathrm{d}\theta(\eta)}{\mathrm{d}\eta}\right)_{\eta=0}\right]_{f,\mathrm{Ec}=0}},$$

i.e.

$$E_{(\overline{Nu}_{x,f})_{\mathrm{Ec}=0}} = \frac{\left[\left(-\frac{\mathrm{d}\theta(\eta)}{\mathrm{d}\eta}\right)_{\eta=0}\right]_{f,\mathrm{Ec}=0} - \left[\left(-\frac{\mathrm{d}\theta(\eta)}{\mathrm{d}\eta}\right)_{\eta=0}\right]_{f,\mathrm{Ec}}}{\left[\left(-\frac{\mathrm{d}\theta(\eta)}{\mathrm{d}\eta}\right)_{\eta=0}\right]_{f,\mathrm{Ec}=0}}. \tag{6.25}$$

Similarly,

$$E_{(Q_{x,f})_{\mathrm{Ec}=0}} = \frac{(Q_{x,f})_{\mathrm{Ec}=0} - (Q_{x,f})_{\mathrm{Ec}}}{(Q_{x,f})_{\mathrm{Ec}=0}} = \frac{\left[\left(-\frac{\mathrm{d}\theta(\eta)}{\mathrm{d}\eta}\right)_{\eta=0}\right]_{f,\mathrm{Ec}=0} - \left[\left(-\frac{\mathrm{d}\theta(\eta)}{\mathrm{d}\eta}\right)_{\eta=0}\right]_{f,\mathrm{Ec}}}{\left[\left(-\frac{\mathrm{d}\theta(\eta)}{\mathrm{d}\eta}\right)_{\eta=0}\right]_{f,\mathrm{Ec}=0}}, \tag{6.26}$$

where $E_{(Q_{x,f})_{\mathrm{Ec}=0}} = \frac{(Q_{x,f})_{\mathrm{Ec}=0}-(Q_{x,f})_{\mathrm{Ec}}}{(Q_{x,f})_{\mathrm{Ec}=0}}$ is the heat transfer prediction deviation caused by ignoring the viscous thermal dissipation.

On this basis, the numerical solutions on the heat transfer prediction deviation $E_{(Q_x)_{\mathrm{Ec}=0}}$ caused by ignoring the viscous thermal dissipation are found out, and plotted in Fig. 6.9. Then, it is clear that increasing the Eckert number causes increase of the positive deviation $E_{(Q_x)_{\mathrm{Ec}=0}}$, which demonstrates the effect of the viscous thermal dissipation on heat transfer. Meanwhile, increasing the Prandtl number Pr_f causes increase of the effect of the viscous thermal dissipation on the deviation $E_{(Q_x)_{\mathrm{Ec}=0}}$.

With (5.12), (6.19), and (6.20), (6.25) and (6.26) are changed to

$$E_{(Q_{x,f})_{\mathrm{Ec}=0}} = E_{(\overline{Nu}_{x,f})_{\mathrm{Ec}=0}} = -\frac{a\cdot\mathrm{Ec}}{0.4696\mathrm{Pr}_f{}^{1/3}}$$

$$(-0.1 \le \mathrm{Ec} \le 0.1) \quad \text{and} \quad (0.3 \le \mathrm{Pr_f} \le 3) \tag{6.27}$$

$$E_{(Q_{x,f})_{\mathrm{Ec}=0}} = E_{(\overline{Nu}_{x,f})_{\mathrm{Ec}=0}} = -\frac{b\cdot\mathrm{Ec}^2 + c\cdot\mathrm{Ec}}{0.4696\mathrm{Pr}_f{}^{1/3}}$$

$$(-0.1 \le \mathrm{Ec} \le 0.1) \quad \text{and} \quad (3 \le \mathrm{Pr_f} \le 100), \tag{6.28}$$

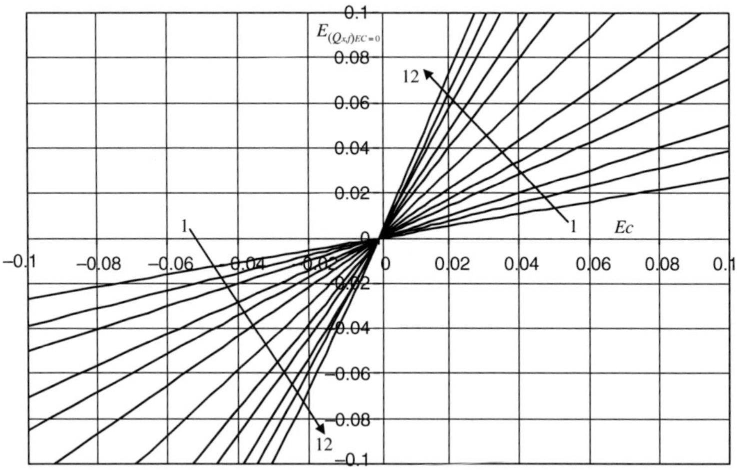

Fig. 6.9 Numerical solutions on $E_{(Q_{x,f})_{\mathrm{Ec}=0}}$ for laminar forced film convection on a horizontal flat plate with ignoring variable physical properties but with consideration of viscous thermal dissipation (Lines 1 to 12: for = 0.3. 0.6, 1, 2, 3, 5, 10, 20, 30, 50, 70, and 100)

where

$$a = 0.0138\mathrm{Pr_f}^2 - 0.2305\mathrm{Pr_f} - 0.0164$$

$$b = (-0.0007\mathrm{Pr_f} + 0.002)\mathrm{Pr_f}^{1.75} \qquad (3 \le \mathrm{Pr_f} \le 10)$$

$$b = 9\text{E-}05\mathrm{Pr_f}^2 - 0.08\mathrm{Pr_f} + 0.5278 \qquad (10 \le \mathrm{Pr_f} \le 100)$$

$$c = (-0.2027\mathrm{Pr_f} - 0.0929)/\mathrm{Pr_f}^{0.15} \qquad (3 \le \mathrm{Pr_f} \le 10)$$

$$c = (-0.3525Pr_f + 0.2819)\mathrm{Pr_f}^{1/3}, \qquad (10 \le \mathrm{Pr_f} \le 100)$$

$$\left[\left(-\frac{\mathrm{d}\theta(\eta)}{\mathrm{d}\eta}\right)_{\eta=0}\right]_{\mathrm{Ec}=0} = 0.4696\mathrm{Pr}_f{}^{1/3} \qquad (0.3 \le \mathrm{Pr}_f \le 100). \qquad (5.12)$$

Equations (6.27) and (6.28) are important for further investigation on laminar forced convection and film flows, because they are used to determine if it is not necessary to consider the viscous thermal dissipation without any obvious deviation.

6.8 Adiabatic Eckert Numbers

From the above analysis, it is clear that increasing Eckert number Ec causes the decrease of the positive heat transfer (from the plate to flow). When heat transfer decreases to zero, the related Eckert number Ec is so-called adiabatic Eckert number $\mathrm{Ec_a}$.

In fact, the adiabatic Eckert number Ec_a is so important that it is the judgement of the heat flux direction. The heat flux direction with $Ec < Ec_a$ is different from that with $Ec > Ec_a$. On the other hand, from Fig. 6.8, it is found that increasing the absolute deviation $|\Delta Ec| = |Ec - Ec_a|$ will cause increase of heat flux $|q_x|$ no matter heat flux direction.

A series of adiabatic Eckert numbers Ec_a are evaluated here numerically with average Prandtl number Pr_f, plotted in Fig. 6.10, and are listed in Table 6.2. It is seen that decreasing the Prandtl number will cause increase of the adiabatic Eckert number Ec_a at an accelerative pace, especially for lower Prandtl number region.

The numerical results of the adiabatic Eckert numbers Ec_a are formulated here as (6.29) by a curve-matching approach. The values of Ec_a predicted by (6.29) and its prediction deviations are obtained and listed in Table 6.2 also.

$$Ec_a = 2{Pr_f}^{[0.0135 \cdot \ln(Pr_f) - 0.4996]}. \qquad (0.3 \le Pr_f \le 100) \qquad (6.29)$$

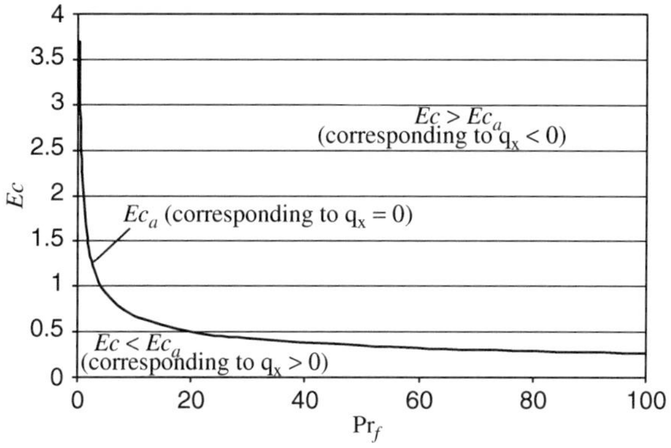

Fig. 6.10 Adiabatic Eckert number Ec_a with variation of average Prandtl number Pr_f for the laminar forced convection on the horizontal flat plate

Table 6.2 Adiabatic Eckert number Ec_a with variation of average Prandtl number Pr_f for the laminar forced convection on the horizontal flat plate

	Ec_a		
Pr_f	Numerical solutions	Predicted by (6.29)	Predicted deviations
0.3	3.699	3.72185	0.62%
0.6	2.589	2.590571	0.0607%
0.7 (air)	2.3933	2.394225	0.0386%
1 (water vapour)	2	2	0
2	1.4215	1.423811	0.16%
3	1.171	1.174185	0.27%
5	0.9221	0.926854	0.516%
10	0.6757	0.680009	0.638%
20	0.5027	0.50542	0.54%
30	0.4259	0.427449	0.36%
50	0.35015	0.348298	−0.53%
70	0.3077	0.305523	−0.707%
100	0.271345	0.26679	−1.68%

6.9 Summary

Now, it is time to summarize in Table 6.3 the equations on dimensional similarity variables, governing partial differential equations, governing similarity ordinary differential equations, and related heat transfer equations based on the present similarity analysis method for laminar forced convection on a horizontal flat plate with ignoring variable physical properties but consideration viscous thermal dissipation.

Table 6.3 Equations on dimensional similarity variables, governing partial differential equations, governing similarity ordinary differential equations, and heat transfer equations based on the present similarity analysis method for laminar forced convection boundary layer on a horizontal flat plate with ignoring the variable physical properties but consideration of viscous thermal dissipation

Governing partial differential equations of laminar forced convection boundary layer with ignoring variable physical properties but consideration of viscous thermal dissipation		
Mass equation	$\frac{\partial w_x}{\partial x}+\frac{\partial w_y}{\partial y}=0$	(3.1)
Momentum equation	$w_x\frac{\partial w_x}{\partial x}+w_y\frac{\partial w_x}{\partial y}=\nu_f\frac{\partial^2 w_x}{\partial y^2}$	(3.2)
Energy equation	$w_x\frac{\partial t}{\partial x}+w_y\frac{\partial t}{\partial y}=\frac{\nu_f}{\Pr_f}\frac{\partial^2 t}{\partial y^2}+\frac{\nu_f}{c_{p_f}}(\frac{\partial w_x}{\partial y})^2$	(3.3)
Boundary conditions	$y=0:\quad w_x=0,\ w_y=0,\quad t=t_w$	
	$y\to\infty:\quad w_x=w_{x,\infty}(\text{constant}),\quad t=t_\infty$	
Equations on dimensionless similarity variables		
Local Reynolds number $\mathrm{Re}_{x,f}$	$\mathrm{Re}_{x,f}=\frac{w_{x,\infty}x}{\nu_f}$	
Coordinate variable η	$\eta=\frac{y}{x}\left(\frac{1}{2}\mathrm{Re}_{x,f}\right)^{1/2}$	
Temperature θ	$\theta=\frac{t-t_\infty}{t_w-t_\infty}$	
Velocity component w_x	$w_x=w_{x,\infty}\cdot W_x(\eta)$	
Velocity component w_y	$w_y=w_{x,\infty}\left(\frac{1}{2}\mathrm{Re}_x\right)^{-1/2}W_y(\eta)$	
Transformed governing ordinary differential equations		
Mass equation	$-\eta\cdot\frac{\mathrm{d}W_x(\eta)}{\mathrm{d}\eta}+2\frac{\mathrm{d}W_y(\eta)}{\mathrm{d}\eta}=0$	
Momentum equation	$[-\eta\cdot W_x(\eta)+2W_y(\eta)]\frac{\mathrm{d}W_x(\eta)}{\mathrm{d}\eta}=\frac{\mathrm{d}^2W_x(\eta)}{\mathrm{d}\eta^2}$	
Energy equation	$[-\eta\cdot W_x(\eta)+2\cdot W_y(\eta)]\frac{\mathrm{d}\theta(\eta)}{\mathrm{d}\eta}=\frac{1}{\Pr_f}\frac{\mathrm{d}^2\theta(\eta)}{\mathrm{d}\eta^2}+\mathrm{Ec}\left(\frac{\mathrm{d}W_x(\eta)}{\mathrm{d}\eta}\right)^2$	
Boundary conditions	$\eta=0:\ W_x(\eta)=0,\ W_y(\eta)=0,\ \theta(\eta)=1$	
	$\eta\to\infty:\ W_x(\eta)=1,\ \theta(\eta)=0$	
Skin-friction coefficient		
Local skin-friction coefficient $C_{f,x}$	$C_{x,f}==0.664(\mathrm{Re}_{x,f})^{-1/2}$	
Average skin-friction coefficient $\overline{C}_{x,f}$	$\overline{C}_{x,f}=1.328\mathrm{Re}_{x,f}^{-1/2}$	

Table 6.3 (continued)

Equations related to heat transfer	
Wall dimensionless temperature gradient $\left[-\left(\frac{d\theta}{d\eta}\right)_{\eta=0}\right]_{f,Ec=0}$	For $-0.1 \leq \mathrm{Ec} \leq 0.1$
	$\left[\left(-\frac{d\theta(\eta)}{d\eta}\right)_{\eta=0}\right]_{f,Ec} = a \cdot \mathrm{Ec} + 0.4696\mathrm{Pr}_f{}^{1/3}$ $(-0.1 \leq \mathrm{E_c} \leq 0.1)$ and $(0.3 \leq \mathrm{Pr_f} \leq 3)$
	$\left[\left(-\frac{d\theta(\eta)}{d\eta}\right)_{\eta=0}\right]_{f,Ec} = b \cdot \mathrm{Ec}^2 + c \cdot \mathrm{Ec}$ $+0.4696\mathrm{Pr}_f{}^{1/3}(-0.1 \leq \mathrm{E_c} \leq 0.1)$ and $(3 \leq \mathrm{Pr_f} \leq 100)$
	$a = 0.0138\mathrm{Pr_f}^2 - 0.2305\mathrm{Pr_f} - 0.0164$ $b = (-0.0007\mathrm{Pr_f} + 0.002)\mathrm{Pr_f}^{1.75}$ $(3 \leq \mathrm{Pr_f} \leq 10)$ $b = 9\text{E-}05\mathrm{Pr_f}^2 - 0.08\mathrm{Pr_f} + 0.5278$ $(10 \leq \mathrm{Pr_f} \leq 100)$ $c = (-0.2027\mathrm{Pr_f} - 0.0929)/\mathrm{Pr_f}^{0.15}$ $(3 \leq \mathrm{Pr_f} \leq 10)$ $c = (-0.3525Pr_f + 0.2819)/\mathrm{Pr_f}^{1/3}$ $(10 \leq \mathrm{Pr_f} \leq 100)$
Local values $(Nu_{x,f})_{\mathrm{Ec}} = \frac{[\alpha_{x,f}]_{\mathrm{Ec}} \cdot x}{\lambda_f}$	$(q_{x,f})_{\mathrm{Ec}} = \lambda_f(t_w - t_\infty)\left(\frac{1}{2}\mathrm{Re}_{x,f}\right)^{1/4} x^{-1}\left[-\left(\frac{d\theta(\eta)}{d\eta}\right)_{\eta=0}\right]_{f,\mathrm{Ec}}$ $(Nu_{x,f})_{\mathrm{Ec}} = \left(\frac{1}{2}\mathrm{Re}_{x,f}\right)^{1/2}\left[-\left(\frac{\mathrm{d}\theta(\eta)}{\mathrm{d}\eta}\right)_{\eta=0}\right]_{f,\mathrm{Ec}}$
Average values $(\overline{Nu_{x,f}})_{\mathrm{Ec}} = \frac{(\overline{\alpha_{x,f}})_{\mathrm{Ec}} \cdot x}{\lambda_f}$	$(\overline{Q}_{x,f})_{\mathrm{Ec}} = \sqrt{2}x^{-1}\lambda_f(t_w - t_\infty)\mathrm{Re}_{x,f}{}^{1/2}\left[-\left(\frac{\mathrm{d}\theta(\eta)}{\mathrm{d}\eta}\right)_{\eta=0}\right]_{f,\mathrm{Ec}}$ $([\overline{Nu_{x,f}})_{\mathrm{Ec}} = \sqrt{2}\mathrm{Re}_{x,f}{}^{1/2}\left[-\left(\frac{\mathrm{d}\theta(\eta)}{\mathrm{d}\eta}\right)_{\eta=0}\right]_{f,\mathrm{Ec}}$
Total value	$(Q_{x,f})_{\mathrm{Ec}} = \sqrt{2}b\lambda_f(t_w - t_\infty)\mathrm{Re}_{x,f}{}^{1/2}\left[-\left(\frac{\mathrm{d}\theta(\eta)}{\mathrm{d}\eta}\right)_{\eta=0}\right]_{f,\mathrm{Ec}}$
Heat transfer prediction deviation caused by ignoring viscous thermal dissipation, $E_{(Q_{x,f})_{\mathrm{Ec}=0}}$	$E_{(Q_{x,f})_{\mathrm{Ec}=0}} = -\frac{a \cdot \mathrm{Ec}}{0.4696\mathrm{Pr}_f{}^{1/3}}$ $(-0.1 \leq Ec \leq 0.1)$ and $(0.3 \leq \mathrm{Pr_f} \leq 3)$ $E_{(Q_{x,f})_{\mathrm{Ec}=0}} = -\frac{b \cdot \mathrm{Ec}^2 + c \cdot \mathrm{Ec}}{0.4696\mathrm{Pr}_f{}^{1/3}}$ $(-0.1 \leq \mathrm{Ec} \leq 0.1)$ and $(3 \leq \mathrm{Pr_f} \leq 100)$ $a = 0.0138\mathrm{Pr_f}^2 - 0.2305\mathrm{Pr_f} - 0.0164$ $b = (-0.0007\mathrm{Pr_f} + 0.002)\mathrm{Pr_f}^{1.75}$ $(3 \leq \mathrm{Pr_f} \leq 10)$ $b = 9\text{E-}05\mathrm{Pr_f}^2 - 0.08\mathrm{Pr_f} + 0.5278$ $(10 \leq \mathrm{Pr_f} \leq 100)$ $c = (-0.2027\mathrm{Pr_f} - 0.0929)/\mathrm{Pr_f}^{0.15}$ $(3 \leq \mathrm{Pr_f} \leq 10)$ $c = (-0.3525Pr_f + 0.2819)/\mathrm{Pr_f}^{1/3}$ $(10 \leq \mathrm{Pr_f} \leq 100)$
Adiabatic Eckert number	
Adiabatic Eckert number $\mathrm{Ec_a}$	$\mathrm{Ec_a} = 2\mathrm{Pr}_f{}^{[0.0135 \cdot \ln(\mathrm{Pr}_f) - 0.4996]}$ $(0.3 \leq \mathrm{Pr}_f \leq 100)$

6.10 Remarks

The work in this chapter is an extensive study for effect of the viscous thermal dissipation on heat transfer of laminar forced convection. Based on the new similarity analysis method, a system of similarity governing mathematical model is provided for laminar forced convection with consideration of viscous thermal dissipation and constant physical properties for the extensive investigation.

A system of numerical solution is rigorously found out together with variation of Eckert number and average Prandtl number. With these numerical solutions, it is clear that: (i) Increasing the Eckert number causes decrease of the positive heat transfer (from the wall to flow), and causes increase of the negative heat transfer (from the flow to the wall); (ii) With increasing the Prandtl number, the effect of the viscous thermal dissipation on heat transfer will increase.

The system of key numerical solutions on wall dimensionless temperature gradient are formulated with the viscous thermal dissipation and average Prandtl number by using a curve-fitting approach, and the formulated equations are so reliable that they are very coincident to the related key numerical solutions. Then, combined with the theoretical equations on heat transfer, the simple and reliable prediction of heat transfer is realized by using these formulated equations on the wall dimensionless temperature gradient. Furthermore, deviation of heat transfer predicted by ignoring the viscous thermal dissipation is formulated, which is used for judgement of validity of ignoring the viscous thermal dissipation in practical calculation.

The adiabatic Eckert number $\mathrm{Ec_a}$ is provided for judgement of heat flux direction, that is positive (from wall to the flow bulk) with $\mathrm{Ec} < \mathrm{Ec_a}$ and negative (from flow bulk to wall) with $\mathrm{Ec} > \mathrm{Ec_a}$. Increasing the absolute value of the increment, $|\mathrm{Ec} - \mathrm{Ec_a}|$, will cause increase of heat transfer no matter heat flux direction. A system of numerical solutions on the adiabatic Eckert numbers is provided and then formulated numerically with variation of average Prandtl number. It is clear that the adiabatic Eckert number will increase at an accelerative pace with decrease of the average Prandtl number, especially in the region with lower Prandtl number.

6.11 Calculation Examples

Example 1 A flat plate with $b = 1\,\mathrm{m}$ in width and $x = 0.25\,\mathrm{m}$ in length is horizontally located in water flowing with $w_{x,\infty} = 1.3\,\mathrm{m/s}$. The water temperature is $t_\infty = 10°\mathrm{C}$, and surface temperature of the plate is $\mathrm{t_w} = 70°\mathrm{C}$. If we ignore variable physical properties but consider viscous thermal dissipation, please solve the following questions:

(1) judge if the forced convection is laminar;
(2) calculate the Eckert number;
(3) predict heat transfer prediction deviation caused by ignoring the viscous thermal dissipation.
(4) calculate the forced convection heat transfer on the plate;
(5) evaluate the adiabatic Eckert numbers $\mathrm{Ec_a}$.

Solution 1: The boundary temperature will be

$$t_f = \frac{t_w + t_\infty}{2} = \left(\frac{70 + 10}{2}\right)^\circ \mathrm{C} = 40^\circ\mathrm{C}$$

The related water properties at the average temperature t_f are listed in Table 6.4.

Table 6.4 Physical properties of water at the average temperature t_f

	t_w, °C	t_∞, °C	t_f, °C			
Temperature	70	10	40			
Physical properties			Pr_f	c_{pf}	λ_f, W/(m°C)	ν_f, m²/s
			4.32	4.179	0.63	0.6564×10^{-6}

The local Reynolds number with ignoring the variable physical properties will be

$$\mathrm{Re}_{x,f} = \frac{w_{x,\infty} x}{\nu_f} = \frac{1.3 \times 0.25}{0.6564 \times 10^{-6}} = 495124.9 < 5 \times 10^6$$

Then the flow can be regarded as laminar flow.

Solution 2: With (6.11), the Eckert number is

$$Ec = \frac{{w_{x,\infty}}^2}{c_{pf}(t_w - t_\infty)} = \frac{1.3^2}{4179 \times (70 - 10)} = 6.740 \times 10^{-6}$$

Solution 3: With $\mathrm{Pr}_f = 4.32$, (6.28) is taken for evaluation of heat transfer prediction deviation caused by ignoring the viscous thermal dissipation, i.e.

$$E_{(Q_{x,f})_{\mathrm{Ec}=0}} = -\frac{b \cdot \mathrm{Ec}^2 + c \cdot \mathrm{Ec}}{0.4696 {\mathrm{Pr}_f}^{1/3}} \qquad (-0.1 \le \mathrm{Ec} \le 0.1) \qquad (6.28)$$

where

$$b = (-0.0007\mathrm{Pr_f} + 0.002){\mathrm{Pr_f}}^{1.75} \qquad (3 \le \mathrm{Pr_f} \le 10)$$

$$c = (-0.2027\mathrm{Pr_f} - 0.0929)/{\mathrm{Pr_f}}^{0.15} \qquad (3 \le \mathrm{Pr_f} \le 10)$$

For $\mathrm{Pr}_f = 4.32$

$$b = (-0.0007 \times 4.32 + 0.002) \times 4.32^{1.75} = -0.01326$$

$$c = (-0.2027 \times 4.32 - 0.0929)/4,32^{0.15} = -0.77769$$

Then, heat transfer prediction deviation caused by ignoring the viscous thermal dissipation is

$$E_{(Q_{x,f})_{\mathrm{Ec}=0}} = -\frac{-0.01326 \times (6.740 \times 10^{-6})^2 - 0.77769 \times 6.740 \times 10^{-6}}{0.4696 \times 4.32^{1/3}}$$

$$= -6.85 \times 10^{-6}.$$

Solution 4: First, we can evaluate the wall dimensionless temperature gradient with ignoring variable physical properties but consideration of viscous thermal dissipation. With $\mathrm{Pr}_f = 4.32$, we have

$$\left[\left(-\frac{\mathrm{d}\theta(\eta)}{\mathrm{d}\eta}\right)_{\eta=0}\right]_{f,\mathrm{Ec}} = b\cdot\mathrm{Ec}^2 + c\cdot\mathrm{Ec} + \left[\left(-\frac{\mathrm{d}\theta(\eta)}{\mathrm{d}\eta}\right)_{\eta=0}\right]_{f,\mathrm{Ec}=0}, \quad (3 \le \mathrm{Pr_f} \le 100) \tag{6.20}$$

where

$$b = (-0.0007\mathrm{Pr_f} + 0.002)\mathrm{Pr_f}^{1.75}$$

$$c = (-0.2027\mathrm{Pr_f} - 0.0929)/\mathrm{Pr_f}^{0.15}$$

For $\mathrm{Pr}_f = 4.32$

$$b = (-0.0007 \times 4.32 + 0.002) \times 4.32^{1.75} = -0.01326$$

$$c = (-0.2027 \times 4.32 - 0.0929)/4,32^{0.15} = -0.77769$$

$$\left[\left(-\frac{\mathrm{d}\theta(\eta)}{\mathrm{d}\eta}\right)_{\eta=0}\right]_{f,\mathrm{Ec}}$$

$$= -0.01326 \times (6.740 \times 10^{-6})^2 - 0.77769 \times 6.740 \times 10^{-6} + 0.4696 \times 4.32^{1/3}$$

$$= 0.7648$$

Plate heat transfer rate with ignoring variable physical properties but consideration of viscous thermal dissipation can be calculated by (6.14), i.e.

$$(Q_{x,f})_{\mathrm{Ec}} = \sqrt{2}b\lambda_f(t_w - t_\infty)\mathrm{Re}_{x,f}{}^{1/2}\left[\left(-\frac{\mathrm{d}\theta(\eta)}{\mathrm{d}\eta}\right)_{\eta=0}\right]_{f,\mathrm{Ec}}. \tag{6.14}$$

Then

$$(Q_{x,f})_{\text{Ec}} = \sqrt{2} \times 1 \times 0.63 \times (70 - 10) \times 495124.9^{1/2} \times 0.764737 = 28765.8\ \text{W}$$

Solution 5: According to (6.29), adiabatic Eckert numbers Ec_a is expressed as

$$\text{Ec}_\text{a} = 2\text{Pr}_f{}^{[0.0135 \cdot \ln(\text{Pr}_f) - 0.4996]}$$

With $\text{Pr}_f = 4.32$, the adiabatic Eckert numbers Ec_a is

$$\text{Ec}_\text{a} = 2 \times \text{Pr}_f{}^{[0.0135 \cdot \ln(4.32) - 0.4996]} = 0.99105$$

Example 2 Only the water flow is replaced by air flow, all other conditions are same as those in example 1. If the ignoring variable physical properties but consideration of viscous thermal dissipation, please

(1) judge if the forced convection is laminar;
(2) calculate the Eckert number;
(3) predict heat transfer prediction deviation caused by ignoring the viscous thermal dissipation.
(4) calculate the forced convection heat transfer on the plate;
(5) evaluate the adiabatic Eckert numbers Ec_a.

Solution 1: The related air properties at the average temperature t_f are listed in Table 6.5.

Table 6.5 Physical properties of air at the average temperature t_f

	t_w,°C	t_∞,°C	t_f,°C			
Temperature	70	10	40			
Physical properties			Pr_f	c_{pf}	λ_f, W/(m°C)	ν_f, m²/s
			0.7	1.007	0.02726	17.20×10^{-6}

The local Reynolds number with ignoring the variable physical properties will be

$$\text{Re}_{x,f} = \frac{w_{x,\infty} x}{\nu_f} = \frac{1.3 \times 0.25}{17.20 \times 10^{-6}} = 18895.3 < 5 \times 10^5$$

Then, the flow can be regarded as laminar flow.

Solution 2: With (6.11), the Eckert number is

$$Ec = \frac{w_{x,\infty}{}^2}{c_{pf}(t_w - t_\infty)} = \frac{1.3^2}{1007 \times (70 - 10)} = 2.797 \times 10^{-5}$$

Solution 3: With $\mathrm{Pr}_f = 0.7$, (6.27) is taken for predict heat transfer prediction deviation caused by ignoring the viscous thermal dissipation, i.e.

$$E_{(Q_{x,f})\mathrm{Ec}=0} = -\frac{a \cdot \mathrm{Ec}}{0.4696 {\mathrm{Pr}_f}^{1/3}} \quad (-0.1 \le \mathrm{Ec} \le 0.1) \text{ and } (0.3 \le \mathrm{Pr_f} \le 3) \quad (6.27)$$

where

$$a = 0.0138 {\mathrm{Pr_f}}^2 - 0.2305 \mathrm{Pr_f} - 0.0164$$

Then

$$a = 0.0138 \times 0.7^2 - 0.2305 \times 0.7 - 0.0164 = -0.17099$$

$$E_{(Q_{x,f})\mathrm{Ec}=0} = -\frac{-0.17099 \times 2.797 \times 10^{-5}}{0.4696 \times 0.7^{1/3}} = -1.147 \times 10^{-5}$$

Solution 4: First, we can evaluate the wall dimensionless temperature gradient with ignoring variable physical properties but consideration of viscous thermal dissipation. With $\mathrm{Pr}_f = 0.7$, we have

$$\left[\left(-\frac{\mathrm{d}\theta(\eta)}{\mathrm{d}\eta}\right)_{\eta=0}\right]_{f,\mathrm{Ec}} = a \cdot \mathrm{Ec} + 0.4696 {\mathrm{Pr}_f}^{1/3} \quad (0.3 \le \mathrm{Pr_f} \le 3) \quad (6.19)$$

Then,

$$\left[\left(-\frac{\mathrm{d}\theta(\eta)}{\mathrm{d}\eta}\right)_{\eta=0}\right]_{f,\mathrm{Ec}} = -0.17099 \times 2.797 \times 10^{-5} + 0.4696 \times 0.7^{1/3} = 0.416955$$

With (6.14), the plate heat transfer rate with ignoring variable physical properties but consideration of viscous thermal dissipation will be

$$(Q_{x,f})_{\mathrm{Ec}} = \sqrt{2} b \lambda_f (t_w - t_\infty) {\mathrm{Re}_{x,f}}^{1/2} \left[\left(-\frac{\mathrm{d}\theta(\eta)}{\mathrm{d}\eta}\right)_{\eta=0}\right]_{f,\mathrm{Ec}} \quad (6.14)$$

Then,

$$(Q_{x,f})_{\mathrm{Ec}} = \sqrt{2} \times 1 \times 0.02726 \times (70 - 10) \times 18895.3^{1/2} \times 0.416955 = 132.57\,\mathrm{W}$$

Solution 5: According to (6.29), adiabatic Eckert numbers $\mathrm{Ec_a}$ is expressed as

$$\mathrm{Ec_a} = 2 {\mathrm{Pr}_f}^{[0.0135 \cdot \ln(\mathrm{Pr}_f) - 0.4996]}$$

With $\mathrm{Pr}_f = 0.7$, the adiabatic Eckert numbers $\mathrm{Ec_a}$ is

$$\mathrm{Ec_a} = 2 {\mathrm{Pr}_f}^{[0.0135 \cdot \ln(8.13) - 0.4996]} = 2 \times 0.7^{[0.0135 \cdot \ln(0.7) - 0.4996]} = 2.394225$$

Exercise

1. Follow example 1, change t_∞ to 30°C, but keep all other conditions.
   If we ignore variable physical properties but consider viscous thermal dissipation, please solve the following:
   (1) judge if the forced convection is laminar;
   (2) calculate the Eckert number;
   (3) predict heat transfer prediction deviation caused by ignoring the viscous thermal dissipation;
   (4) calculate the forced convection heat transfer on the plate;
   (5) evaluate the adiabatic Eckert numbers Eca.

References

1. H.H. Winter, Viscous dissipation in shear flows of molten polymers. Adv Heat Transfer **13**, 205–267 (1977).
2. H.H. Winter, Temperature induced pressure gradient in the clearance between screw flight and barrel of a single screw extruder. Polym Eng Sci **20**, 406–412 (1980)
3. V.D. Dang, Heat transfer of a power-law fluid at low Peclet number flow. J. Heat Transfer **105**, 542–549 (1983)
4. T. Basu, D.N. Roy, Laminar heat transfer in a tube with viscous dissipation. Int J Heat Mass Transfer **28**, 699–701 (1985)
5. A. Lawal, A.S. Mujumdar, The effects of viscous dissipation on heat transfer to power law fluids in arbitrary cross-sectional ducts, Warme Stoffübertragung **27**, 437–446 (1992)
6. Berardi PG, Cuccurullo G, Acierno D, Russo P. Viscous dissipation in duct flow of molten polymers. *Proceedings of Eurotherm Seminar 46* (Pisa, Italy, July 1995), pp. 39–43
7. E. Zanchini, Effect of viscous dissipation on the asymptotic behavior of laminar forced convection in circular tubes. Int J Heat Mass Transfer **40**, 169–178 (1997)
8. O. Aydin, Effects of viscous dissipation on the heat transfer in a forced pipe flow. Part 2: Thermally developing flow. Energy Convers Manage **46**(18–19), 3091–3102 (2005)
9. O. Aydin, Effects of viscous dissipation on the heat transfer in a forced pipe flow. Part 1. Both hydrodynamically and thermally fully developed flow, Energy Convers Manage **46**, 757–769 (2005)
10. E.R.G. Eckert, R.M. Drake, *Analysis of Heat and Mass Transfer* (McGraw-Hill, NY, 1972)
11. H. Schlichting, J. Kestin, *Boundary Layer Theory* (McGraw-Hill, NewYork, NY 1955)

Chapter 7
Heat Transfer of Gas Laminar Forced Convection with Consideration of Variable Physical Properties

Abstract The new similaritye analysis method is applied to develop a novel similarity analysis model of laminar forced convection with consideration of variable physical properties. It is a better alternative to the traditional Falkner-Skan type transformation for this issue. Meanwhile, an advanced approach on treatment of gas variable physical properties, the temperature parameter method, is first applied for convenient and reliable treatment of variable thermophysical properties in extensive study on gas laminar forced convection. Theoretical equations on skin-friction coefficient and heat transfer with consideration of variable physical properties are obtained, where only the dimensionless wall velocity and temperature gradients are no-given variables, respectively, both of which depend on Prandtl number, boundary temperature ratio, and gas temperature parameters. The numerical solutions on the dimensionless wall velocity and temperature gradients are found out rigorously on laminar forced convection on horizontal flat plate. A system of the wall dimensionless temperature gradient is formulated by a curve-fitting method, and then a difficult point was resolved for simple and reliable prediction of heat transfer with consideration of variable physical properties. It is found that not only Prandtl number but also gas temperature parameters together with the boundary temperature ratio have significant influences on velocity and temperature fields as well as heat transfer of gas laminar forced convection. Furthermore, the following regulations for the effect of the variable physical properties on heat transfer coefficient are clarified: (i) Effect of the gas variable physical properties on the heat transfer coefficient are dominated by the gas temperature parameters and boundary temperature ration; (ii) The effect of variable thermal conductivity on heat transfer coefficient is larger than that of the variable absolute viscosity, although the effect of the viscosity can never be ignored; and (iii) Increasing the boundary temperature ratio causes increase of the effect of the variable physical properties on heat transfer.

7.1 Introduction

Heat transfer of forced convection exists widely in every field of engineering practice and human life, so it is important to rigorously investigate this phenomenon. On the early time, the single Blasius ordinary equation transformed from the

D.Y. Shang, *Theory of Heat Transfer with Forced Convection Film Flows*,
Heat and Mass Transfer, DOI 10.1007/978-3-642-12581-2_7,

momentum partial differential equations was applied for prediction of its velocity field [1]. Pohlhausen [2] used Blasius transformation system to further calculate heat transfer with constant property assumption for laminar forced convection on a horizontal flat plate. After their studies, the investigations have been done for the effect of the variable thermophysical properties on compressible forced boundary layer (see Schlichting [3]). For incompressible forced boundary layer, there have been some studies with consideration of variable liquid viscosity, for instance, the studies of Ling and Dybbs [4, 5], who proposed an assumed equation for approximate simulation of the viscosity of the fluid, where the viscosity of the fluid is an inverse linear function of temperature. Recently, Postelnicu et al. [6] extended Ling and Dybbs' problem with an internal heat generation placed into the fluid-saturated porous medium. However, there still has been a requirement for an extensive investigation of coupled effect of the variable physical properties, such as density, thermal conductivity, and viscosity on heat transfer of forced convection. In my recent book [7], the coupled effect of these variable physical properties on heat transfer with free convection and free film flows were systematically reported. These studies demonstrated that the variable physical properties have obvious effect on free convection and frees film flows, and then can never be ignored for reliable prediction of the related heat transfer. In view of the importance of heat transfer of forced convection, since this Chapter, I will report successively the coupled effect of variable physical properties on heat transfer of laminar forced convection boundary layer and force film condensation.

Here, it is necessary to mention again the subscripts for expression of ignoring the viscous thermal dissipation. In Chaps. 5 and 6 for laminar forced convection for consideration of constant physical properties, I gave the subscripts $\mathrm{Ec} = 0$ and Ec in the related heat transfer variables for noting the cases with ignoring and considering viscous thermal dissipation respectively. According to these studies and the additional examples calculation results, it can be found that in the range of laminar forced convection the Eckert number is very small generally, and then ignoring the effect of the viscous thermal dissipation will not cause obvious influence on laminar forced convection and heat transfer. Then, from now on in this book, my work will focus on the coupled effects of variable physical properties on laminar forced convection and film condensation. Although the viscous thermal dissipation will not be considered in the successive studies, the subscript $\mathrm{Ec} = 0$ will never be noted anymore for saving space.

7.2 Governing Equations

7.2.1 Governing Partial Differential Equations

The physical model and coordinate system of boundary layer with two-dimensional laminar forced convection shown in Fig. 5.1 schematically. A flat plate is horizontally located in parallel fluid flow with its mainstream velocity $w_{x,\infty}$. The plate surface temperature is t_w and the fluid bulk temperature is t_∞. Then a velocity boundary layer is produced near the plate. If t_w is not equal to t_∞, a temperature boundary layer will occur near the plate. We assume that the velocity boundary layer is laminar. The only difference of the present issue from that in Chap. 5 is that the variable physical properties are considered here, but were never considered in Chap. 5. Then if we ignore the viscous thermal dissipation, from Chap. 2 the governing partial differential equations for fluid laminar forced convection boundary layer are expressed as below with consideration of these variable physical properties:

$$\frac{\partial}{\partial x}(\rho w_x) + \frac{\partial}{\partial y}(\rho w_y) = 0, \tag{7.1}$$

$$\rho\left(w_x \frac{\partial w_x}{\partial x} + w_y \frac{\partial w_x}{\partial y}\right) = \frac{\partial}{\partial y}\left(\mu \frac{\partial w_x}{\partial y}\right), \tag{7.2}$$

$$\rho\left[w_x \frac{\partial (c_p t)}{\partial x} + w_y \frac{\partial (c_p t)}{\partial y}\right] = \frac{\partial}{\partial y}\left(\lambda \frac{\partial t}{\partial y}\right), \tag{7.3}$$

with the boundary conditions

$$\mathrm{y} = 0\text{: } \mathrm{w_x} = 0,\ w_y = 0,\ t = t_w \tag{7.4}$$

$$\mathrm{y} \to \infty\text{:}\quad w_x = w_{x,\infty}\ (\text{constant}),\ t = t_\infty \tag{7.5}$$

Here, temperature-dependent physical properties density ρ, absolute viscosity μ, thermal conductivity λ, and specific heat c_p are taken into account.

7.2.2 Similarity Transformation Variables

The similarity transformation variables used in Chap. 5 for laminar forced convection without consideration of variable physical properties can be taken as those with consideration of variable physical properties in this chapter. Of course, the average value on physical properties should be changed to the related local value; and therefore, the local Reynolds number $\mathrm{Re}_{x,f} = \frac{w_{x,\infty} x}{\nu_f}$ with the average kinetic viscosity ν_f is changed to the local Reynolds number $\mathrm{Re}_{x,\infty} = \frac{w_{x,\infty} x}{\nu_\infty}$ with the local kinetic viscosity ν_∞. In this case, for the new similarity analysis method, a set of the related dimensional analysis variables can be shown as follows for consideration of variable physical properties:

$$\eta = \frac{y}{x}\left(\frac{1}{2}\mathrm{Re}_{x,\infty}\right)^{1/2}, \tag{7.6}$$

$$\mathrm{Re}_{x,\infty} = \frac{w_{x,\infty} x}{\nu_\infty}, \tag{7.7}$$

$$w_x = w_{x,\infty} W_x(\eta), \tag{7.8}$$

$$w_y = w_{x,\infty} \left(\frac{1}{2} \mathrm{Re}_{x,\infty} \right)^{-1/2} W_y(\eta), \tag{7.9}$$

$$\theta(\eta) = \frac{t - t_\infty}{t_w - t_w}, \tag{7.10}$$

where η is dimensionless coordinate variable, $\mathrm{Re}_{x,\infty}$ is local Reynolds number with consideration of variable physical properties, $W_x(\eta)$ and $W_y(\eta)$ are dimensionless velocity components in x and y coordinates, respectively, and $\theta(\eta)$ denotes dimensionless temperature.

7.2.3 Similarity Transformation of the Governing Partial Differential Equations

With (7.6), (7.7), (7.8), (7.9), and (7.10) for the dimensionless similarity variables η, $\mathrm{Re}_{x,\infty}$, $W_x(\eta)$, $W_y(\eta)$, and $\theta(\eta)$, the governing partial differential equations (7.1), (7.2), and (7.3) are transformed as below respectively.

7.2.3.1 Transformation of (7.1)

Equation (7.1) is changed to the following form

$$\rho \frac{\partial w_x}{\partial x} + \rho \frac{\partial w_y}{\partial y} + w_x \frac{\partial \rho}{\partial x} + w_y \frac{\partial \rho}{\partial y} = 0. \tag{7.1a}$$

With (7.8) we have

$$\frac{\partial w_x}{\partial x} = w_{x,\infty} \frac{\mathrm{d}W_x(\eta)}{\mathrm{d}\eta} \frac{\partial \eta}{\partial x}.$$

Where with (7.6) we have

$$\frac{\partial \eta}{\partial x} = -\frac{1}{2} \eta x^{-1}. \tag{a}$$

Then,

$$\frac{\partial w_x}{\partial x} = -\frac{1}{2} \eta x^{-1} w_{x,\infty} \frac{\mathrm{d}W_x(\eta)}{\mathrm{d}\eta}. \tag{b}$$

With (7.9) we have

$$\frac{\partial w_y}{\partial y} = w_{x,\infty} \left(\frac{1}{2} \mathrm{Re}_{x,\infty} \right)^{-1/2} \frac{\mathrm{d}W_y(\eta)}{\mathrm{d}\eta} \frac{\partial \eta}{\partial y},$$

where

$$\frac{\partial \eta}{\partial y} = x^{-1}\left(\frac{1}{2}\mathrm{Re}_{x,\infty}\right)^{1/2}. \tag{c}$$

Then

$$\frac{\partial w_y}{\partial y} = w_{x,\infty}\frac{\mathrm{d}W_y(\eta)}{\mathrm{d}\eta}x^{-1}. \tag{d}$$

In addition,

$$\frac{\partial \rho}{\partial x} = \frac{\mathrm{d}\rho}{\mathrm{d}\eta}\frac{\partial \eta}{\partial x} = -\frac{1}{2}\frac{\mathrm{d}\rho}{\mathrm{d}\eta}\eta x^{-1}, \tag{e}$$

$$\frac{\partial \rho}{\partial y} = \frac{\mathrm{d}\rho}{\mathrm{d}\eta}\frac{\partial \eta}{\partial y} = \frac{\mathrm{d}\rho}{\mathrm{d}\eta}x^{-1}\left(\frac{1}{2}\mathrm{Re}_{x,\infty}\right)^{1/2}. \tag{f}$$

With (b), (d), (e), and (f), (7.1a) is changed to

$$\rho\left[-\frac{1}{2}\eta x^{-1}w_{x,\infty}\frac{\mathrm{d}W_x(\eta)}{\mathrm{d}\eta} + w_{x,\infty}\frac{\mathrm{d}W_y(\eta)}{\mathrm{d}\eta}x^{-1}\right]$$

$$-\frac{1}{2}\frac{\mathrm{d}\rho}{\mathrm{d}\eta}\eta x^{-1}w_{x,\infty}W_x(\eta) + \frac{\mathrm{d}\rho}{\mathrm{d}\eta}x^{-1}\left(\frac{1}{2}\mathrm{Re}_{x,\infty}\right)^{1/2}w_{x,\infty}\left(\frac{1}{2}\mathrm{Re}_{x,\infty}\right)^{-1/2}W_y(\eta) = 0.$$

The above equation is divided by $\frac{1}{2}x^{-1}w_{x,\infty}$, and we have

$$\rho\left[-\eta\frac{\mathrm{d}W_x(\eta)}{\mathrm{d}\eta} + 2\frac{\mathrm{d}W_y(\eta)}{\mathrm{d}\eta}\right] - \frac{\mathrm{d}\rho}{\mathrm{d}\eta}\eta \cdot W_x(\eta) + 2\frac{\mathrm{d}\rho}{\mathrm{d}\eta}W_y(\eta) = 0.$$

The above equation is simplified to

$$-\eta\frac{\mathrm{d}W_x(\eta)}{\mathrm{d}\eta} + 2\frac{\mathrm{d}W_y(\eta)}{\mathrm{d}\eta} + \frac{1}{\rho}\frac{\mathrm{d}\rho}{\mathrm{d}\eta}[-\eta \cdot W_x(\eta) + 2W_y(\eta)] = 0. \tag{7.11}$$

7.2.3.2 Transformation of (7.2)

Equation (7.2) is changed to

$$\rho\left(w_x\frac{\partial w_x}{\partial x} + w_y\frac{\partial w_x}{\partial y}\right) = \mu\frac{\partial^2 w_x}{\partial y^2} + \frac{\partial \mu}{\partial y}\frac{\partial w_x}{\partial y}. \tag{7.2a}$$

With (7.8) we have

$$\frac{\partial w_x}{\partial y} = w_{x,\infty}\frac{\mathrm{d}W_x(\eta)}{\mathrm{d}\eta}\frac{\partial \eta}{\partial y}.$$

With (c), the above equation is changed to

$$\frac{\partial w_x}{\partial y} = w_{x,\infty} \frac{\mathrm{d}W_x(\eta)}{\mathrm{d}\eta} x^{-1} \left(\frac{1}{2}\mathrm{Re}_{x,\infty}\right)^{1/2} \tag{g}$$

and

$$\frac{\partial^2 w_x}{\partial y^2} = w_{x,\infty} \frac{\mathrm{d}^2 W_x(\eta)}{\mathrm{d}\eta^2} x^{-1} \left(\frac{1}{2}\mathrm{Re}_{x,\infty}\right)^{1/2} \frac{\partial \eta}{\partial y},$$

i.e.

$$\frac{\partial^2 w_x}{\partial y^2} = w_{x,\infty} \frac{\mathrm{d}^2 W_x(\eta)}{\mathrm{d}\eta^2} x^{-2} \left(\frac{1}{2}\mathrm{Re}_{x,\infty}\right). \tag{h}$$

Additionally,

$$\frac{\partial \mu}{\partial y} = \frac{\mathrm{d}\mu}{\mathrm{d}\eta} x^{-1} \left(\frac{1}{2}\mathrm{Re}_{x,\infty}\right)^{1/2}. \tag{i}$$

With (7.8), (7.9), (b), (g), (h), and (i), (7.2a) is changed to

$$\begin{aligned}
&\rho\left[-w_{x,\infty} W_x(\eta) \frac{1}{2}\eta x^{-1} w_{x,\infty} \frac{\mathrm{d}W_x(\eta)}{\mathrm{d}\eta} + w_{x,\infty}\left(\frac{1}{2}\mathrm{Re}_{x,\infty}\right)^{-1/2}\right.\\
&\left. W_y(\eta)\cdot w_{x,\infty} \frac{\mathrm{d}W_x(\eta)}{\mathrm{d}\eta} x^{-1}\left(\frac{1}{2}\mathrm{Re}_{x,\infty}\right)^{1/2}\right]\\
&= \mu w_{x,\infty} \frac{\mathrm{d}^2 W_x(\eta)}{\mathrm{d}\eta^2} x^{-2}\left(\frac{1}{2}\mathrm{Re}_{x,\infty}\right) + \frac{\mathrm{d}\mu}{\mathrm{d}\eta} x^{-1}\left(\frac{1}{2}\mathrm{Re}_{x,\infty}\right)^{1/2}\\
&w_{x,\infty} \frac{\mathrm{d}W_x(\eta)}{\mathrm{d}\eta} x^{-1}\left(\frac{1}{2}\mathrm{Re}_{x,\infty}\right)^{1/2}
\end{aligned}.$$

The above equation is divided by $\frac{1}{2}$, and simplified to

$$\begin{aligned}
&\rho\left[-w_{x,\infty} W_x(\eta)\cdot \eta x^{-1} w_{x,\infty} \frac{\mathrm{d}W_x(\eta)}{\mathrm{d}\eta} + 2w_{x,\infty} W_y(\eta)\cdot w_{x,\infty} \frac{\mathrm{d}W_x(\eta)}{\mathrm{d}\eta} x^{-1}\right]\\
&= \mu w_{x,\infty} \frac{\mathrm{d}^2 W_x(\eta)}{\mathrm{d}\eta^2} x^{-2}(\mathrm{Re}_{x,\infty}) + \frac{d\mu}{d\eta} x^{-1}(\mathrm{Re}_{x,\infty}) w_{x,\infty} \frac{\mathrm{d}W_x(\eta)}{\mathrm{d}\eta} x^{-1}
\end{aligned}.$$

With definition of the local Reynolds number, the above equation is changed to

$$\begin{aligned}
&\rho\left[-w_{x,\infty} W_x(\eta)\cdot \eta x^{-1} w_{x,\infty} \frac{\mathrm{d}W_x(\eta)}{\mathrm{d}\eta} + 2w_{x,\infty} W_y(\eta)\cdot w_{x,\infty} \frac{\mathrm{d}W_x(\eta)}{\mathrm{d}\eta} x^{-1}\right]\\
&= \mu w_{x,\infty} \frac{\mathrm{d}^2 W_x(\eta)}{\mathrm{d}\eta^2} x^{-2}\left(\frac{w_{x,\infty} x}{\nu_\infty}\right) + \frac{\mathrm{d}\mu}{\mathrm{d}\eta} x^{-1}\left(\frac{w_{x,\infty} x}{\nu_\infty}\right) w_{x,\infty} \frac{\mathrm{d}W_x(\eta)}{\mathrm{d}\eta} x^{-1}
\end{aligned}.$$

The above equation is divided by $x^{-1}w_{x,\infty}^2\frac{\mu}{\nu_\infty}$, and simplified to

$$\frac{\nu_\infty}{\nu}(-\eta W_x(\eta)+2W_y(\eta))\frac{\mathrm{d}W_x(\eta)}{\mathrm{d}\eta}=\frac{\mathrm{d}^2W_x(\eta)}{\mathrm{d}\eta^2}+\frac{1}{\mu}\frac{\mathrm{d}\mu}{\mathrm{d}\eta}\frac{\mathrm{d}W_x(\eta)}{\mathrm{d}\eta}. \tag{7.12}$$

7.2.3.3 Transformation of (7.3)

Equation (7.3) is changed to

$$\rho\left(w_xc_p\frac{\partial t}{\partial x}+w_xt\frac{\partial c_p}{\partial x}+w_yc_p\frac{\partial t}{\partial y}+w_yt\frac{\partial c_p}{\partial y}\right)=\lambda\frac{\partial^2 t}{\partial y^2}+\frac{\partial\lambda}{\partial y}\frac{\partial t}{\partial y}. \tag{7.3a}$$

With (7.6), (7.10), and (a), we have

$$\begin{aligned}\frac{\partial t}{\partial x}&=(t_w-t_\infty)\frac{\mathrm{d}\theta(\eta)}{\mathrm{d}\eta}\frac{\partial\eta}{\partial x},\\&=(t_w-t_\infty)\frac{\mathrm{d}\theta(\eta)}{\mathrm{d}\eta}\left(-\frac{1}{2}\eta x^{-1}\right),\\&=-\frac{1}{2}\eta x^{-1}(t_w-t_\infty)\frac{\mathrm{d}\theta(\eta)}{\mathrm{d}\eta}.\end{aligned} \tag{j}$$

With (7.6), (7.10), and (c), we have

$$\begin{aligned}\frac{\partial t}{\partial y}&=(t_w-t_\infty)\frac{\mathrm{d}\theta(\eta)}{\mathrm{d}\eta}\frac{\partial\eta}{\partial y},\\&=(t_w-t_\infty)\frac{\mathrm{d}\theta(\eta)}{\mathrm{d}\eta}x^{-1}\left(\frac{1}{2}\mathrm{Re}_{x,\infty}\right)^{1/2},\\&=x^{-1}\left(\frac{1}{2}\mathrm{Re}_{x,\infty}\right)^{1/2}(t_w-t_\infty)\frac{\mathrm{d}\theta(\eta)}{\mathrm{d}\eta},\end{aligned} \tag{k}$$

$$\frac{\partial^2 t}{\partial y^2}=x^{-2}\left(\frac{1}{2}\mathrm{Re}_{x,\infty}\right)(t_w-t_\infty)\frac{\mathrm{d}^2\theta(\eta)}{\mathrm{d}\eta^2}. \tag{l}$$

Similar to (e), we have

$$\frac{\partial c_p}{\partial x}=-\frac{1}{2}\frac{\mathrm{d}c_p}{\mathrm{d}\eta}\eta x^{-1}. \tag{m}$$

Similar to (f), we have

$$\frac{\partial\lambda}{\partial y}=\frac{\partial\lambda}{\partial\eta}x^{-1}\left(\frac{1}{2}\mathrm{Re}_{x,\infty}\right)^{1/2}, \tag{n}$$

$$\frac{\partial c_p}{\partial y} = \frac{\partial c_p}{\partial \eta} x^{-1} \left(\frac{1}{2} \mathrm{Re}_{x,\infty} \right)^{1/2}. \tag{o}$$

With (7.8), (7.9), (i) to (o), (7.3a) is changed to

$$\begin{aligned}
&\rho \Bigg[w_{x,\infty} W_x(\eta) c_p \left(-\tfrac{1}{2} \eta x^{-1} (t_w - t_\infty) \tfrac{\mathrm{d}\theta(\eta)}{\mathrm{d}\eta} \right) + w_{x,\infty} W_x(\eta) t \left(-\tfrac{1}{2} \eta x^{-1} \tfrac{\mathrm{d}c_p}{\mathrm{d}\eta} \right) \\
&+ w_{x,\infty} \left(\tfrac{1}{2} \mathrm{Re}_{x,\infty} \right)^{-1/2} W_y(\eta) c_p x^{-1} \left(\tfrac{1}{2} \mathrm{Re}_{x,\infty} \right)^{1/2} (t_w - t_\infty) \tfrac{\mathrm{d}\theta(\eta)}{\mathrm{d}\eta} \\
&+ w_{x,\infty} \left(\tfrac{1}{2} \mathrm{Re}_{x,\infty} \right)^{-1/2} W_y(\eta) t \tfrac{\mathrm{d}c_p}{\mathrm{d}\eta} x^{-1} \left(\tfrac{1}{2} \mathrm{Re}_{x,\infty} \right)^{1/2} \Bigg] \\
&= \lambda x^{-2} \left(\frac{1}{2} \mathrm{Re}_{x,\infty} \right) (t_w - t_\infty) \frac{\mathrm{d}^2\theta(\eta)}{\mathrm{d}\eta^2} + \frac{\partial \lambda}{\partial \eta} x^{-1} \left(\frac{1}{2} \mathrm{Re}_{x,\infty} \right)^{1/2} \\
&\quad x^{-1} \left(\frac{1}{2} \mathrm{Re}_{x,\infty} \right)^{1/2} (t_w - t_\infty) \frac{\mathrm{d}\theta(\eta)}{\mathrm{d}\eta}.
\end{aligned}$$

The above equation is simplified to

$$\begin{aligned}
&\rho \Big[w_{x,\infty} W_x(\eta) c_p \left(-\tfrac{1}{2} \eta x^{-1} (t_w - t_\infty) \tfrac{\mathrm{d}\theta(\eta)}{\mathrm{d}\eta} \right) + w_{x,\infty} W_x(\eta) t \left(-\tfrac{1}{2} \eta x^{-1} \tfrac{\mathrm{d}c_p}{\mathrm{d}\eta} \right) \\
&+ w_{x,\infty} W_y(\eta) c_p x^{-1} (t_w - t_\infty) \tfrac{\mathrm{d}\theta(\eta)}{\mathrm{d}\eta} + w_{x,\infty} W_y(\eta) t \tfrac{\mathrm{d}c_p}{\mathrm{d}\eta} x^{-1} \Big] \\
&= \lambda x^{-2} \left(\frac{1}{2} \mathrm{Re}_{x,\infty} \right) (t_w - t_\infty) \frac{\mathrm{d}^2\theta(\eta)}{\mathrm{d}\eta^2} + \frac{\partial \lambda}{\partial \eta} x^{-1} \left(\frac{1}{2} \mathrm{Re}_{x,\infty} \right) x^{-1} (t_w - t_\infty) \frac{\mathrm{d}\theta(\eta)}{\mathrm{d}\eta}
\end{aligned}.$$

With definition of $\mathrm{Re}_{x,\infty}$, the above equation is changed to

$$\begin{aligned}
&\rho \Big[w_{x,\infty} W_x(\eta) c_p \left(-\tfrac{1}{2} \eta x^{-1} (t_w - t_\infty) \tfrac{\mathrm{d}\theta(\eta)}{\mathrm{d}\eta} \right) + w_{x,\infty} W_x(\eta) t \left(-\tfrac{1}{2} \eta x^{-1} \tfrac{\mathrm{d}c_p}{\mathrm{d}\eta} \right) \\
&+ w_{x,\infty} W_y(\eta) c_p x^{-1} (t_w - t_\infty) \tfrac{\mathrm{d}\theta(\eta)}{\mathrm{d}\eta} + w_{x,\infty} W_y(\eta) t \tfrac{\mathrm{d}c_p}{\mathrm{d}\eta} x^{-1} \Big] \\
&= \lambda x^{-2} \left(\frac{1}{2} \frac{w_{x,\infty} x}{\nu_\infty} \right) (t_w - t_\infty) \frac{\mathrm{d}^2\theta(\eta)}{\mathrm{d}\eta^2} + \frac{\partial \lambda}{\partial \eta} x^{-1} \left(\frac{1}{2} \frac{w_{x,\infty} x}{\nu_\infty} \right) x^{-1} (t_w - t_\infty) \frac{\mathrm{d}\theta(\eta)}{\mathrm{d}\eta}
\end{aligned}.$$

The above equation is divided by $\frac{1}{2}(t_w - t_\infty) \frac{c_p w_{x,\infty} \lambda}{x \nu_\infty}$, and changed to

$$\begin{aligned}
&\frac{\rho \nu_\infty}{\lambda} \left[W_x(\eta) \left(-\eta \frac{\mathrm{d}\theta(\eta)}{\mathrm{d}\eta} \right) + W_x(\eta) \frac{t}{t_w - t_\infty} \left(-\eta \frac{1}{c_p} \frac{\mathrm{d}c_p}{\mathrm{d}\eta} \right) \right. \\
&\quad \left. + 2W_y(\eta) \frac{\mathrm{d}\theta(\eta)}{\mathrm{d}\eta} + 2W_y(\eta) \frac{t}{t_w - t_\infty} \frac{1}{c_p} \frac{\mathrm{d}c_p}{\mathrm{d}\eta} \right] \\
&= \frac{1}{c_p} \frac{\mathrm{d}^2\theta(\eta)}{\mathrm{d}\eta^2} + \frac{1}{c_p} \frac{1}{\lambda} \frac{\partial \lambda}{\partial \eta} \frac{\mathrm{d}\theta(\eta)}{\mathrm{d}\eta},
\end{aligned} \tag{7.3b}$$

where

$$\frac{t}{t_w - t_\infty} = \frac{(t_w - t_\infty)\cdot\theta(\eta) + t_\infty}{t_w - t_\infty} = \theta(\eta) + \frac{t_\infty}{t_w - t_\infty}. \qquad \text{(p)}$$

With (p), (7.3b) is changed to

$$\frac{\rho\nu_\infty}{\lambda}\left[W_x(\eta)\left(-\eta\frac{\mathrm{d}\theta(\eta)}{\mathrm{d}\eta}\right) + W_x(\eta)\left(\theta + \frac{t_\infty}{t_w - t_\infty}\right)\left(-\eta\frac{1}{c_p}\frac{\mathrm{d}c_p}{\mathrm{d}\eta}\right) + 2W_y(\eta)\frac{\mathrm{d}\theta(\eta)}{\mathrm{d}\eta}\right.$$
$$\left. + 2W_y(\eta)\left(\theta(\eta) + \frac{t_\infty}{t_w - t_\infty}\right)\frac{1}{c_p}\frac{\mathrm{d}c_p}{\mathrm{d}\eta}\right] = \frac{1}{c_p}\frac{\mathrm{d}^2\theta(\eta)}{\mathrm{d}\eta^2} + \frac{1}{c_p}\frac{1}{\lambda}\frac{\partial\lambda}{\partial\eta}\frac{\mathrm{d}\theta(\eta)}{\mathrm{d}\eta},$$

or

$$\mathrm{Pr}\,\frac{\lambda}{\mu c_p}c_p\frac{\rho\nu_\infty}{\lambda}\left[-\eta W_x(\eta)\frac{\mathrm{d}\theta(\eta)}{\mathrm{d}\eta} - \eta W_x(\eta)\left(\theta(\eta) + \frac{t_\infty}{t_w - t_\infty}\right)\frac{1}{c_p}\frac{\mathrm{d}c_p}{\mathrm{d}\eta}\right.$$
$$\left. + 2W_y(\eta)\frac{\mathrm{d}\theta(\eta)}{\mathrm{d}\eta} \times 2W_y(\eta)\left(\theta(\eta) + \frac{t_\infty}{t_w - t_\infty}\right)\frac{1}{c_p}\frac{\mathrm{d}c_p}{\mathrm{d}\eta}\right] = \frac{\mathrm{d}^2\theta(\eta)}{\mathrm{d}\eta^2} + \frac{1}{\lambda}\frac{\partial\lambda}{\partial\eta}\frac{\mathrm{d}\theta(\eta)}{\mathrm{d}\eta}.$$

The above equation is further changed to

$$\mathrm{Pr}\,\frac{\nu_\infty}{\nu}\left[-\eta W_x(\eta)\frac{\mathrm{d}\theta(\eta)}{\mathrm{d}\eta} + 2W_y(\eta)\frac{\mathrm{d}\theta(\eta)}{\mathrm{d}\eta} - \eta W_x(\eta)\left(\theta(\eta) + \frac{t_\infty}{t_w - t_\infty}\right)\frac{1}{c_p}\frac{\mathrm{d}c_p}{\mathrm{d}\eta}\right.$$
$$\left. + 2W_y(\eta)\times\left(\theta(\eta) + \frac{t_\infty}{t_w - t_\infty}\right)\frac{1}{c_p}\frac{\mathrm{d}c_p}{\mathrm{d}\eta}\right] = \frac{\mathrm{d}^2\theta(\eta)}{\mathrm{d}\eta^2} + \frac{1}{\lambda}\frac{\partial\lambda}{\partial\eta}\frac{\mathrm{d}\theta(\eta)}{\mathrm{d}\eta},$$

i.e.

$$\mathrm{Pr}\,\frac{\nu_\infty}{\nu}\left[(-\eta W_x(\eta) + 2W_y(\eta))\frac{\mathrm{d}\theta(\eta)}{\mathrm{d}\eta} + (-\eta W_x(\eta) + 2W_y(\eta))\right.$$
$$\left.\times\left(\theta(\eta) + \frac{t_\infty}{t_w - t_\infty}\right)\frac{1}{c_p}\frac{\mathrm{d}c_p}{\mathrm{d}\eta}\right] = \frac{\mathrm{d}^2\theta(\eta)}{\mathrm{d}\eta^2} + \frac{1}{\lambda}\frac{\partial\lambda}{\partial\eta}\frac{\mathrm{d}\theta(\eta)}{\mathrm{d}\eta}. \qquad (7.13)$$

For summary, governing partial differential equations (7.1), (7.2), and (7.3) have been transformed to the following governing ordinary differential equations:

$$-\eta\frac{\mathrm{d}W_x(\eta)}{\mathrm{d}\eta} + 2\frac{\mathrm{d}W_y(\eta)}{\mathrm{d}\eta} + \frac{1}{\rho}\frac{\mathrm{d}\rho}{\mathrm{d}\eta}[-\eta\cdot W_x(\eta) + 2W_y(\eta)] = 0, \qquad (7.11)$$

$$\frac{\nu_\infty}{\nu}(-\eta W_x(\eta) + 2W_y(\eta))\frac{\partial W_x(\eta)}{\partial\eta} = \frac{\partial^2 W_x(\eta)}{\partial\eta^2} + \frac{1}{\mu}\frac{\partial\mu}{\partial\eta}\frac{\partial W_x(\eta)}{\partial\eta}, \qquad (7.12)$$

$$\Pr \frac{\nu_\infty}{\nu}\left[(-\eta W_x(\eta)+2W_y(\eta))\frac{\mathrm{d}\theta(\eta)}{\mathrm{d}\eta}+(-\eta W_x(\eta)+2W_y(\eta))\left(\theta(\eta)+\frac{t_\infty}{t_w-t_\infty}\right)\frac{1}{c_p}\frac{\mathrm{d}c_p}{\mathrm{d}\eta}=\frac{\mathrm{d}^2\theta(\eta)}{\mathrm{d}\eta^2}+\frac{1}{\lambda}\frac{\partial\lambda}{\partial\eta}\frac{\mathrm{d}\theta(\eta)}{\mathrm{d}\eta}\right], \tag{7.13}$$

with the boundary condition equations

$$\eta=0: \quad W_x(\eta)=0,\ W_y(\eta)=0,\quad \theta(\eta)=1 \tag{7.14}$$

$$\eta\to\infty: \quad W_x(\eta)=1,\ \theta(\eta)=0. \tag{7.15}$$

Equations (7.11), (7.12), and (7.13) are complete ordinary differential equations of laminar forced convection. These transformed governing ordinary differential equations are completely dimensionless because of the followings: (i) They involve dimensionless velocity components $W_x(\eta)$ and $W_y(\eta)$, dimensionless temperature $\theta(\eta)$, as well as their dimensionless derivatives, which constitute whole unknown variables of the governing equations; (ii) All physical properties exist in form of the dimensionless physical property factors, such as $\frac{1}{\rho}\frac{\mathrm{d}\rho}{\mathrm{d}\eta}$, $\frac{1}{\mu}\frac{\mathrm{d}\mu}{\mathrm{d}\eta}$, $\frac{1}{\lambda}\frac{\mathrm{d}\lambda}{\mathrm{d}\eta}$, $\frac{1}{c_p}\frac{\mathrm{d}c_p}{\mathrm{d}\eta}$, and $\frac{\nu_\infty}{\nu}$ as well as the temperature ratio $\frac{t_\infty}{t_w-t_\infty}$. The coupled effect of the variable physical properties on laminar forced convection is dominated by these dimensionless physical property factors.

7.3 Treatment of Gas Variable Physical Properties

As illustrated in our research [8, 9], gas variable physical properties have significant impacts on heat transfer of free convection and free film flows. As a result, they need to be appropriately treated to ensure accurate and reliable solution of the governing equations. In this work, I will apply this method to treat the gas variable physical properties for laminar forced convection. With this method, the viscosity, thermal conductivity, and specific heat are expressed as simple power law, i.e. $\mu \approx T^{n_\mu}$, $\lambda \approx T^{n_\lambda}$, and $c_p \approx c_p^{n_\lambda}$, respectively. The absolute temperature of the gas far away from the wall, T_∞, can be taken as the reference temperature for forced convection analysis. So we assume as follows at atmospheric pressure:

$$\frac{\mu}{\mu_\infty}=\left(\frac{T}{T_\infty}\right)^{n_\mu}, \tag{7.16}$$

$$\frac{\lambda}{\lambda_\infty}=\left(\frac{T}{T_\infty}\right)^{n_\lambda}, \tag{7.17}$$

$$\frac{c_p}{c_{p\infty}}=\left(\frac{T}{T_\infty}\right)^{n_{c_p}}. \tag{7.18}$$

While the change of gas density with absolute temperature at atmospheric pressure is expressed as

$$\frac{\rho}{\rho_\infty} = \left(\frac{T}{T_\infty}\right)^{-1}. \tag{7.19}$$

Combining (16) with (19) we have

$$\frac{\nu}{\nu_\infty} = \left(\frac{T}{T_\infty}\right)^{(n_\mu+1)}, \tag{7.20}$$

where the exponents n_μ, n_λ, and n_{c_p} are viscosity parameter, thermal conductivity parameter, and specific heat parameter of gas, respectively. According to the study in [8, 9], the values of n_μ, n_λ, and $n_{c_p}$are listed in Tables 7.1 and 7.2, respectively, for different gases together with their Prandtl number Pr. These values of gas temperature parameters are also included in Appendix A of this book.

According to the study in [8, 9], for monatomic and diatomic gases, air, and water vapour the value of specific heat parameter n_{c_p} is so small that it can be ignored. Even for general polyatomic gases such as those listed in Table 7.2, the value of n_{c_p} can be also ignored without any obvious deviation on heat transfer result prediction.

Furthermore, with (7.10), (7.16), (7.17), (7.19), and (7.20), the related gas physical property factors in (7.11), (7.12), and (7.13) become the following ones, respectively with the dimensionless temperature variable $\theta(\eta)$ and its derivative $\mathrm{d}\theta(\eta)/\mathrm{d}\eta$, boundary temperature ratio, and the related temperature parameters n_μ and n_λ:

$$\frac{1}{\rho}\frac{\mathrm{d}\rho}{\mathrm{d}\eta} = -\frac{(T_w/T_\infty - 1)\mathrm{d}\theta(\eta)/\mathrm{d}\eta}{(T_w/T_\infty - 1)\theta(\eta) + 1}, \tag{7.21}$$

$$\frac{1}{\mu}\frac{\mathrm{d}\mu}{\mathrm{d}\eta} = \frac{n_\mu(T_w/T_\infty - 1)\mathrm{d}\theta(\eta)/\mathrm{d}\eta}{(T_w/T_\infty - 1)\theta(\eta) + 1}, \tag{7.22}$$

$$\frac{1}{\lambda}\frac{\mathrm{d}\lambda}{\mathrm{d}\eta} = \frac{n_\lambda(T_w/T_\infty - 1)\mathrm{d}\theta(\eta)/\mathrm{d}\eta}{(T_w/T_\infty - 1)\theta(\eta) + 1}, \tag{7.23}$$

Table 7.1 The value of parameters n_μ and n_λ with their deviations for monatomic and diatomic gases, air, and water vapour obtained according to the typical experimental data for μ and λ, cited from Shang and Wang [8]

Gas	T_0 K	n_μ	Temperature range K	Deviation %	n_λ	Temperature range K	Deviation %	Pr
Ar	273	0.72	220–1500	±3	0.73	210–1500	±4	0.622
He	273	0.66	273–873	±1	0.725	273–873	±0.3	0.675
H_2	273	0.68	80–1000	±2	0.8	220–700	±1	0.68
Air	273	0.68	220–1400	±3	0.81	220–1000	±2.5	0.7
CO	273	0.71	230–1500	±2	0.83	220–600	±2	0.72
N_2	273	0.67	220–1500	±3	0.76	220–1200	±3	0.71
O_2	273	0.694	230–2000	±2	0.86	220–600	±2	0.733
Water Vapour	380	1.04	380–1500	±4	1.185	380–800	±0.3	1

Note: The value of specific heat parameter n_{c_p} is so small for monatomic and diatomic gases, air, and water vapour that it is omitted

Table 7.2 The values of temperature parameters n_λ, n_μ, and n_{c_p} with their deviations for polyatomic gases obtained according to the experimental data, cited from Shang and Wang [9]

Gas	T_0 K	n_μ	Temperature Range K	Deviation %	n_λ	Temperature range K	Deviation %	n_{c_p}	Temperature range K	Deviation %	Pr
Gas mixture	273	0.75	273–1173	±3	1.02	273–1473	±1	0.134	273–1473	±2.5	0.63
CO_2	273	0.88	220–700	±2	1.3	220–720	±2	0.34	230–810	±1	0.73
CH_4	273	0.78	220–1000	±3	1.29	230–1000	±3.6	0.534	273–1033	±4.2	0.74
CCl_4	273	0.912	270–500	±1	1.29	260–400	±1	0.28	270–700	±2	0.8
SO_2	273	0.91	200–1250	±4	1.323	250–900	±4.5	0.257	273–1200	±3	0.81
H_2S	273	1	270–500	±1	1.29	260–400	±1	0.18	230–700	±2	0.85
NH_3	273	1.04	220–1000	±1	1.375	250–900	±2	0.34	230–1033	±0.36	0.87

Note: (1) n_{c_p} is specific heat parameter, and according to the study in [9], for general polyatomic gases it can be ignored without any obvious deviation on heat transfer results. (2) Note: component of the gas mixture: $CO_2 = 0.13$, $N_2 = 0.76$, and water vapour $= 0.11$

$$\frac{1}{c_p}\frac{\mathrm{d}c_p}{\mathrm{d}\eta} = \frac{n_{c_p}(T_w/T_\infty - 1)\mathrm{d}\theta(\eta)/\mathrm{d}\eta}{(T_w/T_\infty - 1)\theta(\eta) + 1}, \tag{7.24}$$

$$\frac{\nu_\infty}{\nu} = [(T_w/T_\infty - 1)\theta(\eta) + 1]^{-(n_\mu+1)}, \tag{7.25}$$

$$\mathrm{Pr} = \mathrm{Pr}_\infty[(T_w/T_\infty - 1)\theta(\eta) + 1]^{n_\mu - n_\lambda + n_{c_p}}. \tag{7.26}$$

The Prandtl number is defined as $\mathrm{Pr} = \mu c_p/\lambda$. Strictly speaking, Pr should also depend on temperature, and described as (7.26). However, generally the variation of Prandtl number of gas is slight that it can be taken as the constant shown as in Tables 7.1 and 7.2.

With the above dimensionless expressions for the dimensionless physical factors $\frac{1}{\rho}\frac{\mathrm{d}\rho}{\mathrm{d}\eta}$, $\frac{1}{\mu}\frac{\mathrm{d}\mu}{\mathrm{d}\eta}$, $\frac{1}{\lambda}\frac{\mathrm{d}\lambda}{\mathrm{d}\eta}$, $\frac{1}{c_p}\frac{\mathrm{d}c_p}{\mathrm{d}\eta}$, and $\frac{\nu_\infty}{\nu}$, the ordinary governing equations (7.11), (7.12), and (7.13) can be solved numerically with the related temperature parameters n_μ and n_λ, as well as the boundary temperature ratio T_w/T_∞.

7.4 Velocity and Temperature Fields

It is mentioned according to the study in [8] and [9] that for monatomic and diatomic gases, air, and water vapour, the specific heat parameter n_{c_p} can be regarded as zero due to its very small value, while for general polyatomic gases n_{c_p} can also be omitted, which will not cause any obvious deviation on heat transfer result prediction.

Since the physical properties vary with temperature, the governing equations (7.11), (7.11), and (7.13) are coupled each other. They are solved by a shooting method with fifth-order Runge-Kutta integration with the boundary equations (7.14) and (7.15). For solving the nonlinear problem, a variable mesh approach is applied to the numerical procedure. In this present work, forced convection for some gases is taken as example for numerical calculation. A system of accurate numerical results of the velocity and temperature fields is obtained as follows.

Since the governing equations of laminar forced convection are coupled each other for consideration of gas variable physical properties, both the velocity and the temperature fields depend on Prandtl number. In addition, they depend on the temperature boundary conditions and gas temperature parameters. The numerical results for the dimensionless velocity component $W_x(\eta)$ and dimensionless temperature $\theta(\eta)$ of air, a gas mixture, water vapour, and NH_3 laminar forced convection on a horizontal flat plate are obtained, and plotted in Figs. 7.1, 7.2, 7.3, and 7.4, respectively. It is really seen that not only Prandtl number but also temperature parameters n_μ and n_λ and the boundary temperature ratio $\frac{T_w}{T_\infty}$ influence the velocity and temperature fields of laminar forced convection. Especially, the effect of the boundary temperature ratio $\frac{T_w}{T_\infty}$ on the velocity and temperature fields of gas laminar forced convection is even obvious. While increasing the boundary temperature ratio $\frac{T_w}{T_\infty}$ will cause increase of the momentum and temperature boundary layer thicknesses. Furthermore, increasing the overall temperature parameter ($n_{\mu\lambda} = 0.45n_\mu + 0.55n_\lambda$)

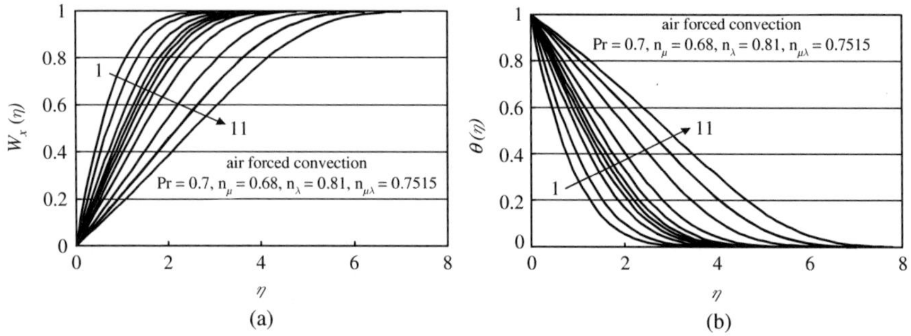

Fig. 7.1 Velocity and temperature profiles of air laminar forced convection on a horizontal plate with variation of the boundary temperature ratio $\frac{T_w}{T_\infty}$
Note: 1. Lines 1 to 11 for $\frac{T_w}{T_\infty} = 0.3, 0.5, 0.75, 0.9, \to 1, 1.1, 1.25, 1.5, 2, 2.5$ and 3; 2. (a) denotes velocity profile, and (b) denotes temperature profile

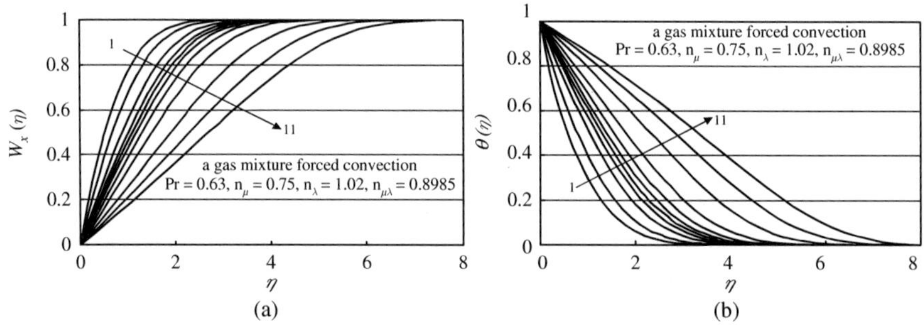

Fig. 7.2 Velocity and temperature profiles of a gas mixture laminar forced convection on a horizontal plate with variation of the boundary temperature ratio $\frac{T_w}{T_\infty}$
Note: 1. Lines 1 to 11: for $\frac{T_w}{T_\infty} = 0.3, 0.5, 0.75, 0.9, \to 1, 1.1, 1.25, 1.5, 2, 2.5$, and 3; 2. Components of the gas mixture are $CO_2 = 0.13$, water vapour $= 0.11$ and $N_2 = 0.76$ with $\Pr = 0.63$, $n_\mu = 0.75$ and $n_\lambda = 1.02$; 3. (a) denotes velocity profile, and (b) denotes temperature profile

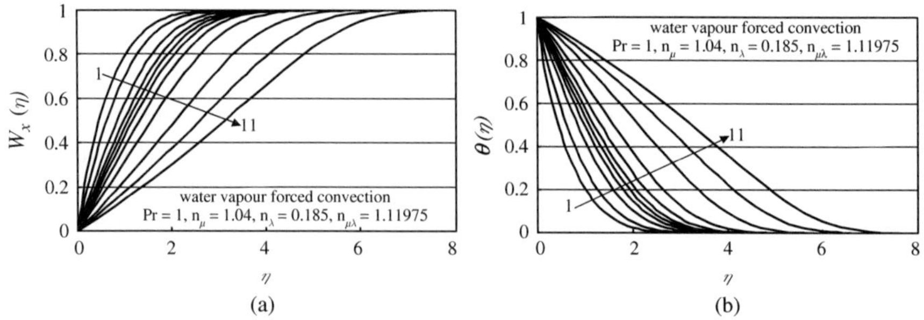

Fig. 7.3 Velocity and temperature profiles of water vapour laminar forced convection on a horizontal plate with variation of the boundary temperature ratio $\frac{T_w}{T_\infty}$
Note: 1. Lines 1 to 11: for $\frac{T_w}{T_\infty} = 0.3, 0.5, 0.75, 0.9, \to 1, 1.1, 1.25, 1.5, 2, 2.5$, and 3; 2. (a) denotes velocity profile, and (b) denotes temperature profile

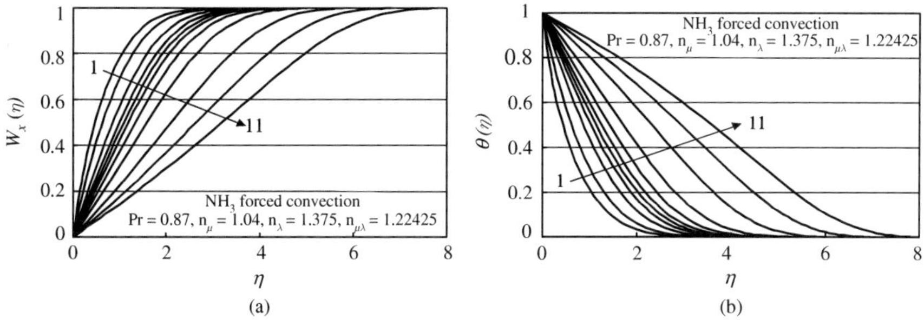

Fig. 7.4 Velocity profile of NH_3 laminar forced convection on a horizontal plate with variation of the boundary temperature ratio $\frac{T_w}{T_\infty}$
Note: 1. Lines 1 to 11: for $\frac{T_w}{T_\infty} = 0.3, 0.5, 0.75, 0.9, \rightarrow 1, 1.1, 1.25, 1.5, 2, 2.5$, and 3; 2. (a) denotes velocity profile, and (b) denotes temperature profile

causes increasing effect of boundary layer temperature ratio on the momentum and temperature fields. Actually, the effect of the boundary temperature ratio $\frac{T_w}{T_\infty}$ on gas laminar forced convection demonstrates the effect of the gas temperature parameter n_μ and n_λ, i.e. the effect of the variable physical properties of gases. Such variable velocity and temperature fields dominated not only by Prandtl number but also by the boundary temperature ratio and gas temperature parameters are quite different from the Blasius and Pohlhausen solutions [1, 2]. While Pohlhausen solutions only depend on fluid Prandtl number, and Blasius solutions even does not depend on Prandtl number.

7.5 Skin-Friction Coefficient with Consideration of Variable Physical Properties

In Chap. 5, I presented the skin-friction coefficient on laminar forced flow with ignoring variable physical properties. In this chapter I will further investigate the skin-friction coefficient with consideration of variable physical properties. The velocity gradient at the wall is important characteristic of the solution, and the local skin-friction coefficient $C_{w,x}$ with consideration of variable physical properties is a dimensionless measure of the shear stress at the wall, i.e.

$$C_{x,w} = \frac{\tau_{x,w}}{\frac{1}{2}\rho_w w_{x,\infty}{}^2} = \frac{\mu_w \left(\frac{\partial w_x}{\partial y}\right)_{y=0}}{\frac{1}{2}\rho_w w_{x,\infty}{}^2}. \tag{7.27}$$

With (4.23), we have

$$\left(\frac{\partial w_x}{\partial y}\right)_{y=0} = w_{x,\infty}\left(\frac{dW_x(\eta)}{d\eta}\right)_{\eta=0}\left(\frac{\partial \eta}{\partial y}\right)_{y=0}.$$

With (4.12), we have

$$\left(\frac{\partial \eta}{\partial y}\right)_{y=0} = x^{-1}\left(\frac{1}{2}\mathrm{Re}_{x,\infty}\right)^{1/2}.$$

Then

$$\left(\frac{\partial w_x}{\partial y}\right)_{y=0} = x^{-1}\left(\frac{1}{2}\mathrm{Re}_{x,\infty}\right)^{1/2} w_{x,\infty}\left(\frac{\mathrm{d}W_x(\eta)}{\mathrm{d}\eta}\right)_{\eta=0}.$$

Therefore, the local skin-friction coefficient $C_{m,x}$ is expressed as

$$\begin{aligned}
C_{x,w} &= \frac{\mu_w x^{-1}\left(\frac{1}{2}\mathrm{Re}_{x,\infty}\right)^{1/2} w_{x,\infty}\left(\frac{\mathrm{d}W_x(\eta)}{\mathrm{d}\eta}\right)_{\eta=0}}{\frac{1}{2}\rho_w w_{x,\infty}{}^2} \\
&= \sqrt{2}\frac{\nu_\infty}{\nu_\infty}\frac{\nu_w}{x w_{x,\infty}}(\mathrm{Re}_{x,\infty})^{1/2}\left(\frac{\mathrm{d}W_x(\eta)}{\mathrm{d}\eta}\right)_{\eta=0} \\
&= \sqrt{2}\frac{\nu_w}{\nu_\infty}(\mathrm{Re}_{x,\infty})^{-1/2}\left(\frac{\mathrm{d}W_x(\eta)}{\mathrm{d}\eta}\right)_{\eta=0}
\end{aligned} \quad (7.28)$$

It should be indicated that the numerical solution $\left(\frac{\mathrm{d}W_x(\eta)}{\mathrm{d}\eta}\right)_{\eta=0}$ here for consideration of variable physical properties is different from that in Chap. 5 for ignoring the variable physical properties. The former numerical solution $\left(\frac{\mathrm{d}W_x(\eta)}{\mathrm{d}\eta}\right)_{\eta=0}$ is variable with the boundary temperature ratio and gas temperature parameters.

The average skin friction coefficient with consideration of variable physical properties from $x = 0$ to x is described as

$$\begin{aligned}
\overline{C}_{x,w} &= \frac{1}{x}\int_0^x C_{x,w}\mathrm{d}x \\
&= \sqrt{2}\frac{1}{x}\int_0^x \frac{\nu_w}{\nu_\infty}(\mathrm{Re}_{x,\infty})^{-1/2}\left(\frac{\mathrm{d}W_x(\eta)}{\mathrm{d}\eta}\right)_{\eta=0}\mathrm{d}x \\
&= 2\times\sqrt{2}\frac{\nu_w}{\nu_\infty}(\mathrm{Re}_{x,\infty})^{-1/2}\left(\frac{\mathrm{d}W_x(\eta)}{\mathrm{d}\eta}\right)_{\eta=0}
\end{aligned} \quad (7.29)$$

For evaluation of the Skin-friction coefficient with consideration of variable physical properties, the numerical solutions on the wall dimensionless velocity gradient $\left(\frac{\mathrm{d}W_x(\eta)}{\mathrm{d}\eta}\right)_{\eta=0}$ are plotted in Fig. 7.5, and listed in Table 7.3 for different gas laminar forced convection on a horizontal flat plate. It is seen that $\left(\frac{\mathrm{d}W_x(\eta)}{\mathrm{d}\eta}\right)_{\eta=0}$ depends on Prandtl number Pr, gas temperature parameters n_μ and n_λ, as well as the boundary temperature ratio T_w/T_∞. Especially, with increasing the boundary temperature

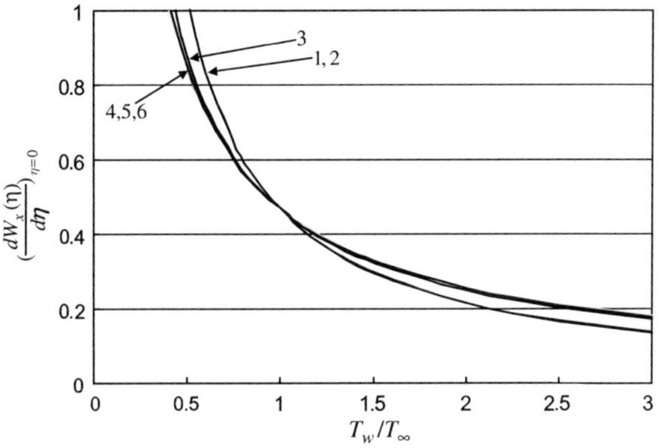

Fig. 7.5 Wall dimensionless velocity gradient $\left(\frac{\mathrm{d}W_x(\eta)}{\mathrm{d}\eta}\right)_{\eta=0}$ on gas laminar forced convection on a horizontal flat plate with consideration of variable physical properties. Lines $1 \rightarrow 6$ denote water vapour, NH_3, a gas mixture, O_2, H_2, and air laminar forced convection

ratio T_w/T_∞, the wall dimensionless velocity gradient $\left(\frac{\mathrm{d}W_x(\eta)}{\mathrm{d}\eta}\right)_{\eta=0}$ will decrease obviously. When $T_w/T_\infty \rightarrow 1$, the value of $\left(\frac{\mathrm{d}W_x(\eta)}{\mathrm{d}\eta}\right)_{\eta=0}$ returned to 0.4696 corresponding to that with ignoring variable physical properties.

7.6 Heat Transfer

In view of that there has been a lack of theoretical investigation on rigorous prediction of gas laminar forced convection and its heat transfer with consideration of

Table 7.3 Wall dimensionless velocity gradient $\left(\frac{\mathrm{d}W_x(\eta)}{\mathrm{d}\eta}\right)_{\eta=0}$ on gas laminar forced convection with consideration of variable physical properties

Term	Gas					
	Gas mixture	H_2	Air	O_2	NH_3	Water vapour
Pr	0.63	0.68	0.7	0.733	0.87	1
n_μ	0.75	0.68	0.68	0.694	1.04	1.04
n_λ	1.02	0.8	0.81	0.86	1.375	1.185
$n_{\mu\lambda}$	0.8985	0.746	0.7515	0.7853	1.22425	1.11975
T_w/T_∞	$\left(\frac{\mathrm{d}W_x(\eta)}{\mathrm{d}\eta}\right)_{\eta=0}$					
0.3	1.382528	1.300493	1.299159	1.313086	1.817683	1.824162
0.5	0.880549	0.849632	0.849283	0.854787	1.027351	1.028683
0.75	0.610734	0.601708	0.601706	0.603379	0.650841	0.651107
0.9	0.517166	0.514357	0.514361	0.514891	0.529107	0.529338
$\rightarrow 1$	0.469605	0.469601	0.469605	0.469605	0.469645	0.469645
1.1	0.430271	0.432386	0.432405	0.431999	0.421661	0.421603
1.25	0.382555	0.386951	0.386970	0.386114	0.364832	0.364872
1.5	0.323407	0.330136	0.330154	0.32882	0.296973	0.296738
2	0.248056	0.256783	0.256789	0.255013	0.215251	0.214757
2.5	0.201973	0.211257	0.211245	0.209326	0.167744	0.167107
3	0.170843	0.180128	0.180113	0.178156	0.137025	0.136291

variable thermophysical properties due to its complexity, such difficult point will be addressed here in this work.

7.6.1 Heat Transfer Analysis

The heat transfer analysis on laminar forced convection boundary layer with consideration of variable physical properties is done as follows.

The local heat transfer rate $q_{x,w}$ at position x per unit area from the surface of the plate to the fluid with consideration of variable physical properties and ignoring viscous thermal dissipation is calculated by Fourier's law as $q_{x,w} = -\lambda_w(\frac{\partial t}{\partial y})_{y=0}$.

While, with (k) we have

$$-\left(\frac{\partial t}{\partial y}\right)_{y=0} = x^{-1}\left(\frac{1}{2}\mathrm{Re}_{x,\infty}\right)^{1/2}(t_w - t_\infty)\left(-\frac{\mathrm{d}\theta(\eta)}{\mathrm{d}\eta}\right)_{\eta=0}.$$

Then,

$$q_{x,w} = \lambda_w(t_w - t_\infty)\left(\frac{1}{2}\mathrm{Re}_{x,\infty}\right)^{1/2} x^{-1}\left(-\frac{\mathrm{d}\theta(\eta)}{\mathrm{d}\eta}\right)_{\eta=0}, \tag{7.30}$$

where $\left(-\frac{\mathrm{d}\theta(\eta)}{\mathrm{d}\eta}\right)_{\eta=0}$ is wall dimensionless temperature gradient with consideration of variable physical properties.

The local heat transfer coefficient $\alpha_{x,w}$, defined as $q_{x,w} = \alpha_{x,w}(t_w - t_\infty)$, will be given by

$$\alpha_{x,w} = \lambda_w\left(\frac{1}{2}\mathrm{Re}_{x,\infty}\right)^{1/2} x^{-1}\left(-\frac{\mathrm{d}\theta(\eta)}{\mathrm{d}\eta}\right)_{\eta=0}. \tag{7.31}$$

The local Nusselt number defined by $Nu_{x,w} = \frac{\alpha_{x,w}\cdot x}{\lambda_w}$ will be

$$Nu_{x,w} = \left(\frac{1}{2}\mathrm{Re}_{x,\infty}\right)^{1/2}\left(-\frac{\mathrm{d}\theta(\eta)}{\mathrm{d}\eta}\right)_{\eta=0}. \tag{7.32}$$

Total heat transfer rate for position $x = 0$ to x with width of b on the plate is a integration $Q_{x,w} = \iint\limits_A q_{x,w}\mathrm{d}A = \int_0^x q_{x,w} b\, \mathrm{d}x$, and hence

$$Q_{x,w} = 2b\lambda_w(t_w - t_\infty)\left(\frac{1}{2}\mathrm{Re}_{x,\infty}\right)^{1/2}\left(-\frac{\mathrm{d}\theta(\eta)}{\mathrm{d}\eta}\right)_{\eta=0}. \tag{7.33}$$

The average heat transfer rate defined as $\overline{Q}_{x,w} = \frac{Q_{x,w}}{b \times x}$ is

$$\overline{Q}_{x,w} = 2x^{-1}\lambda_w(t_w - t_\infty)\left(\frac{1}{2}\mathrm{Re}_{x,\infty}\right)^{1/2}\left(-\frac{d\theta(\eta)}{d\eta}\right)_{\eta=0}. \tag{7.34}$$

The average heat transfer coefficient $\overline{\alpha_{x,w}}$ defined as $\overline{Q}_{x,w} = \overline{\alpha_{x,w}}(t_w - t_\infty)$ is expressed as

$$\overline{\alpha_{x,w}} = 2\lambda_w\left(\frac{1}{2}\mathrm{Re}_{x,\infty}\right)^{1/2} x^{-1}\left(-\frac{d\theta(\eta)}{d\eta}\right)_{\eta=0}. \tag{7.35}$$

The average Nusselt number is defined as $\overline{Nu}_{x,w} = \frac{\overline{\alpha_{x,w}}\cdot x}{\lambda_w}$, and hence

$$\overline{Nu}_{x,w} = \sqrt{2}(\mathrm{Re}_{x,\infty})^{1/2}\left(-\frac{\mathrm{d}\theta(\eta)}{\mathrm{d}\eta}\right)_{\eta=0}. \tag{7.36}$$

It should be indicated that in this chapter for consideration of variable physical properties, $q_{x,w}$, $\alpha_{x,w}$, $Nu_{x,w}$, $Q_{x,w}$, $\overline{Q}_{x,w}$, $\overline{\alpha}_{x,w}$, and $\overline{Nu}_{x,w}$ not only express the case with consideration of variable physical properties but also denote the case with ignoring viscous thermal dissipation, although the subscript Ec $= 0$ is omitted. According to the studies as well as the example calculation results in Chaps. 5 and 6 of this book, it can be found that in the range of laminar forced convection the Eckert number is very small generally, and then ignoring the effect of the viscous thermal dissipation will not cause obvious influence on laminar forced convection and heat transfer.

From (7.30), (7.31), (7.32), (7.33), (7.34), (7.35), and (7.36), it is seen that the dimensionless wall temperature gradient $\left(-\frac{\mathrm{d}\theta(\eta)}{\mathrm{d}\eta}\right)_{\eta=0}$ with consideration of variable physical properties is the only one no-given variable for prediction of heat transfer. In the next Section, I will focus on the investigation of the simple and reliable prediction of $\left(-\frac{\mathrm{d}\theta(\eta)}{\mathrm{d}\eta}\right)_{\eta=0}$, so that the above heat transfer equations can be available for heat transfer prediction.

7.6.2 Dimensionless Wall Temperature Gradient

Since dimensionless wall temperature gradient is the key solution dominating the evaluation of heat transfer, it should be investigated more carefully. For this purpose, the numerical results of a system of the dimensionless wall temperature gradients of a gas mixture, hydrogen, air, oxygen, NH_3, and water vapour laminar forced convection on a horizontal flat plate are plotted in Fig. 7.6, with variation of the boundary temperature ratio $\frac{T_w}{T_\infty}$ from 0.3 to 2, and some of the typical values are listed in Table 7.4. The numerical results demonstrate that the wall temperature gradient $\left(-\frac{\mathrm{d}\theta(\eta)}{\mathrm{d}\eta}\right)_{\eta=0}$ not only depends on Prandtl number Pr but also on viscosity parameter n_μ and thermal conductivity parameter n_λ together with the boundary

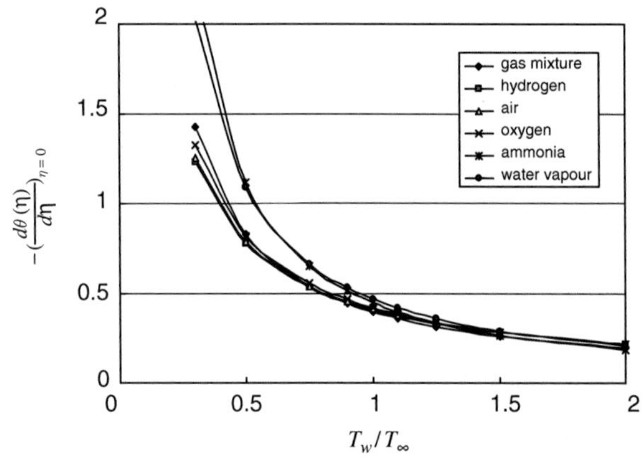

Fig. 7.6 Numerical results on wall temperature gradient $\left(-\frac{\mathrm{d}\theta(\eta)}{\mathrm{d}\eta}\right)_{\eta=0}$ of laminar forced convection on the horizontal plate with consideration of variable physical properties

temperature ratio T_w/T_∞. Decreasing the boundary temperature ratio $\frac{T_w}{T_\infty}$ will cause increase of the wall temperature gradient $\left(-\frac{\mathrm{d}\theta(\eta)}{\mathrm{d}\eta}\right)_{\eta=0}$ in accelerative pace. In addition, different gas convection with different Prandtl numbers and overall temperature parameters has different curve of the wall temperature gradient. All these demonstrate the effect of the variable physical properties.

It is also found that only in the case $\frac{T_w}{T_\infty} \to 1$ the calculated results on the wall temperature gradient $\left(-\frac{\mathrm{d}\theta(\eta)}{\mathrm{d}\eta}\right)_{\mu=0}$ here is identical to that without consideration of variable physical properties shown in Chap. 5. It is because only in the case $\frac{T_w}{T_\infty} \to 1$, the physical properties of the boundary layer can be regarded as constant.

By using a curve-fitting approach, the following equations of $\left(-\frac{\mathrm{d}\theta(\eta)}{\mathrm{d}\eta}\right)_{\eta=0}$ are obtained from its system of numerical results.

$$\left(-\frac{\mathrm{d}\theta(\eta)}{\mathrm{d}\eta}\right) = \left[\left(-\frac{\mathrm{d}\theta(\eta)}{\mathrm{d}\eta}\right)_{\mu=0}\right]_{T_w/T_\infty\to 1} \cdot \left(\frac{T_w}{T_\infty}\right)^{-m}, \tag{7.37}$$

where $\left[\left(-\frac{\mathrm{d}\theta(\eta)}{\mathrm{d}\eta}\right)_{\eta=0}\right]_{T_w/T_\infty\to 1}$ is the reference wall dimensionless temperature gradient with $T_w/T_\infty \to 1$ equivalent to the Pohlhausen equation [2], i.e.

$$\left[\left(-\frac{\mathrm{d}\theta(\eta)}{\mathrm{d}\eta}\right)_{\eta=0}\right]_{T_w/T_\infty\to 1} = 0.4696\,{\Pr_f}^{1/3}. \qquad (0.3 \le \Pr \le 1) \tag{7.38}$$

Table 7.4 Wall temperature gradient $\left(-\frac{d\theta(\eta)}{d\eta}\right)_{\eta=0}$ of laminar forced convection on the horizontal plate with consideration of variable physical properties. A: numerical results; B: predicted results by (7.39); C: predicted deviation

Gas		Gas mixture	H_2	Air	O_2	NH_3	Water vapour
Pr		0.63	0.68	0.7	0.733	0.87	1
n_μ		0.75	0.68	0.68	0.694	1.04	1.04
n_λ		1.02	0.8	0.81	0.86	1.375	1.185
T_w/T_∞		$\left(-\frac{d\theta(\eta)}{d\eta}\right)_{\eta=0}$					
0.3	A	1.426838	1.232269	1.253236	1.322974	2.217698	2.027272
	B	1.418567	1.246733	1.265876	1.330267	2.197733	2.070775
	C	−0.0058	0.0117	0.0101	0.0055	−0.00903	0.0215
0.5	A	0.829365	0.774664	0.785671	0.815515	1.116327	1.088393
	B	0.831311	0.780136	0.790243	0.818456	1.119558	1.103374
	C	0.0023	0.0071	0.0058	0.0036	0.0029	0.0138
0.75	A	0.539962	0.534171	0.540901	0.554359	0.652639	0.665542
	B	0.543934	0.537726	0.543671	0.556622	0.655445	0.669429
	C	0.0074	0.0067	0.0051	0.0041	0.0043	0.0058
0.9	A	0.445409	0.451519	0.456614	0.465518	0.513033	0.533474
	B	0.449480	0.454875	0.459516	0.468027	0.515212	0.534713
	C	0.0091	0.0074	0.0064	0.0054	0.0042	0.0023
→ 1	A	0.398737	0.409666	0.413982	0.420815	0.447221	0.469645
	B	0.402570	0.412950	0.416960	0.423412	0.448299	0.469600
	C	0.0096	0.0080	0.0072	0.0062	0.0024	−0.0001
1.1	A	0.360705	0.375176	0.378990	0.384211	0.395269	0.418749
	B	0.364764	0.378133	0.381659	0.386660	0.397157	0.419043
	C	0.0113	0.0079	0.0070	0.0064	0.0048	0.0007
1.25	A	0.315595	0.333396	0.336543	0.340007	0.335059	0.359249
	B	0.319571	0.335999	0.338960	0.342328	0.337603	0.359674
	C	0.0126	0.0078	0.0072	0.0068	0.0076	0.0012
1.5	A	0.261011	0.281755	0.284165	0.285725	0.265436	0.288754
	B	0.264629	0.283898	0.286192	0.287747	0.267779	0.289253
	C	0.0139	0.0076	0.0071	0.0071	0.0088	0.0017
2	A	0.194046	0.216157	0.217738	0.217385	0.185277	0.205496
	B	0.19650	0.217621	0.219127	0.218771	0.185778	0.205098
	C	0.0126	0.0068	0.0064	0.0064	0.0027	−0.0019
2.5	A	0.1545	0.176127	0.177196	0.17607	0.140552	0.15805
	B	0.155987	0.177069	0.178136	0.176876	0.139905	0.157088
	C	0.0096	0.0053	0.0053	0.0046	−0.0046	−0.0061
3	A	0.128883	0.149113	0.149969	0.148376	0.112553	0.127796
	B	0.129169	0.149612	0.150404	0.148674	0.110969	0.126331
	C	0.0022	0.0033	0.0029	0.0020	−0.0141	−0.0115

Then

$$\left(-\frac{\mathrm{d}\theta(\eta)}{\mathrm{d}\eta}\right)_{\eta=0} = 0.4696\ \mathrm{Pr}_f{}^{1/3} \cdot (\frac{T_w}{T_\infty})^{-m}, \tag{7.39}$$

with

$$m = 0.8419 n_{\mu\lambda} + 0.2897 \qquad (0.3 \le T_w/T_\infty \le 1)$$
$$m = 0.725 n_{\mu\lambda} + 0.3833 \quad (1 \le T_w/T_\infty \le 3)$$

Here, the overall temperature parameter $n_{\mu\lambda}$ is expressed by a weighted sum of n_μ and n_λ, as given by

$$n_{\mu\lambda} = 0.45 n_\mu + 0.55 n_\lambda.$$

From (7.39) and the equations with the overall temperature parameter, the following effect of the overall temperature parameter on the wall dimensionless temperature gradient is found: (1) For $0.3 \le T_w/T_\infty \le 1$, the wall dimensionless temperature gradient will increase with increasing overall temperature parameter; (2) for $1 \le T_w/T_\infty \le 3$, the wall dimensionless temperature gradient will decrease with increasing overall temperature parameter. A system of the wall dimensionless temperature gradients $\left(-\frac{\mathrm{d}\theta(\eta)}{\mathrm{d}\eta}\right)_{\eta=0}$ of laminar forced convection of a gas mixture, hydrogen, air, oxygen, NH_3, and water vapour on a horizontal flat plate are evaluated by using (7.38) for different boundary temperature ratio $\frac{T_w}{T_\infty}$, and listed in Table 7.4 with their evaluated deviation. It is seen that the evaluated values of $-\left(\frac{\mathrm{d}\theta(\eta)}{\mathrm{d}\eta}\right)_{\eta=0}$ are very well coincident to the related numerical solutions.

7.6.3 Prediction Equations on Heat Transfer

With (7.39), (7.30) to (7.36) are available for prediction of heat transfer. For example, (7.33) and (7.36) will become the following corresponding equations for simple and reliable prediction of heat transfer rate and average Nusselt number for laminar gas forced convection on horizontal flat plate for consideration of coupled effects of variable physical properties:

$$Q_{x,w} = 0.664 b \lambda_w (t_w - t_\infty) \mathrm{Re}_{x,\infty}{}^{1/2} \mathrm{Pr}_f{}^{1/3} \left(\frac{T_w}{T_\infty}\right)^{-m}, \quad (0.3 \le \mathrm{Pr} \le 1) \tag{7.40}$$

$$\overline{Nu}_{x,w} = 0.664 \mathrm{Re}_{x,\infty}{}^{1/2} \mathrm{Pr}_f{}^{1/3} \left(\frac{T_w}{T_\infty}\right)^{-m}, \quad (0.3 \le \mathrm{Pr} \le 1) \tag{7.41}$$

with

$$
\begin{aligned}
m &= 0.8419n_{\mu\lambda} + 0.2897 \ (0.3 \le T_w/T_\infty \le 1) \\
m &= 0.725n_{\mu\lambda} + 0.3833 \ (1 \le T_w/T_\infty \le 3) \\
n_{\mu\lambda} &= 0.45n_\mu + 0.55n_\lambda .
\end{aligned}
$$

Equations (7.40) and (7.41) are used for prediction of heat transfer result of gas laminar forced convection on a horizontal flat plate with consideration of the coupled effect of variable thermophysical properties. Obviously, the accuracy of (7.40) and (7.41), dependent on accuracy of reliable equation (7.39), is reliable for practical prediction.

7.7 Remarks

A set of dimensionless similarity mathematical models of gas laminar forced convection with consideration of variable physical properties is developed by the new similarity analysis method. With this new similarity analysis method, the two-dimensional velocity components are directly transformed to the related dimensionless velocity components, which leads to a convenience for dealing with the variable physical properties.

The advanced method, viz. the temperature parameter method for treatment of gas variable physical properties, is first applied in extensive study on gas laminar forced convection. With this method a simple power law is applied to treat the variations of density, viscosity, specific heat, and thermal conductivity of gases with temperature parameters, conveniently and reliably. However, for general gas laminar convection, the variation of specific heat with temperature can be ignored without any obvious deviation caused on heat transfer results, then the variable physical properties, which have significant effect on heat transfer, are density, thermal conductivity, and viscosity.

Based on the present new similarity analysis method, the equation on skin-friction coefficient for laminar forced convection is derived for consideration of variable physical properties. In the equation on the skin-friction coefficient, the wall dimensionless velocity component gradient as the dominating variable is the only no-given variable which is dependent on variable physical properties. However, the wall dimensionless velocity component gradient is a constant with ignoring the variable physical properties.

Based on the present new similarity analysis method, the theoretical equations on heat transfer are obtained through heat transfer analysis. In the theoretical equations on heat transfer, the wall dimensionless temperature gradient exists as only one no-given variable. The wall dimensionless temperature gradient depends on Prandtl number Pr, boundary temperature ratio, and the temperature parameters n_μ and n_λ for gas laminar forced convection for consideration of variable physical properties. A system of numerical solutions of wall dimensionless temperature gradient for gas laminar forced convection on a horizontal flat plate is formulated by a curve-fitting approach. With the formulated equation of wall dimensionless temperature gradient, the theoretical equations on heat transfer become available for simple and reliable prediction of heat transfer.

It is found that not only Prandtl number but also the temperature parameters together with the boundary temperature ratio have significant influences on velocity and temperature fields, as well as heat transfer of gas forced convection. These influences are actually caused by the variable physical properties. Increasing the gas temperature parameters will cause decrease of heat transfer coefficient when the boundary temperature ratio is larger than unity, and cause increase of heat transfer coefficient when the boundary temperature ratio is less than the unity. The effect of the variable thermal conductivity on the heat transfer coefficient is slightly larger than that of the variable viscosity. With decreasing the boundary layer temperature ratio, the heat transfer coefficient will increase at an accelerative pace. Only in the special case when the boundary temperature ratio tends to unity, the result of wall dimensionless temperature gradient is identical to that for ignoring the variable physical properties. The traditional Blasius solutions on velocity field and Pohlhausen equation on heat transfer are only the special cases for consideration the variable physical properties.

7.8 Calculation Examples

Example 1 A flat plate with $b = 1$ m in width and $x = 0.5$ m in length is horizontally located in air flowing with the velocity $w_{x,\infty} = 7.5$ m/s. The surface temperature of the plate is $t_w = 427°C$, and air temperature is $t_\infty = 27°C$. Please solve the followings questions:

(1) evaluate the related local Reynolds numbers for determination of the flow types;
(2) calculate the related Eckert numbers;
(3) determine the related heat transfer prediction deviation caused by ignoring the viscous thermal dissipation;
(4) calculate the related heat transfer results with consideration of coupled effect of the variable physical properties;
(5) evaluate the prediction deviation of heat transfer caused by ignoring the variable physical properties.

Solution 1: $t_f = \frac{t_w+t_\infty}{2} = \frac{427+27}{2} = 227°C$, and $t_w - t_\infty = 427 - 27 = 400°C$. Then, the related physical properties for air are obtained and listed in Table 7.5.

For ignoring variable physical properties, related local Reynolds number is described with

$$Re_{x,f} = \frac{w_{x,\infty}x}{\nu_f} = \frac{7.5 \times 0.5}{38.785 \times 10^{-6}} = 96686.86 < 5 \times 10^5.$$ Then the flow is regarded as laminar.

Solution 2: The Eckert number is evaluated as

Table 7.5 The related physical properties for air

Air,	$n_\mu = 0.68, n_\mu = 0.81$						
	t'_wK		t_∞, K	t_f, K			
Temperature	427°C		27°C	227°C			
Physical properties	λ_w W/(m°C)	ν_w m^2/s	ν_∞m^2/s	$\Pr_f$	c_{pf} kJ/(kg°C)	λ_f W/(m°C)	ν_f m^2/s
	0.0524	68.1×10^{-6}	15.895×10^{-6}	0.7	1.030×10^3	0.0407	38.785×10^{-6}

$$Ec = \frac{w_{x,\infty}{}^2}{c_{p_f}(t_w - t_\infty)} = \frac{7.5^2}{1030 \times 400} = 0.000137.$$

With the so small Eckert number, the effect of viscous thermal dissipation can be ignored.

Solution 3: It is well known that air Prandtl number is 0.7, then here we take the air average Prandtl number is $\Pr_f = 0.7$. In this case, (6.27) is used for evaluation of heat transfer prediction deviations caused by ignoring the viscous thermal dissipation, i.e.

$$E_{(Q_{x,f})_{\mathrm{Ec}=0}} = E_{(\overline{Nu}_{x,f})_{\mathrm{Ec}=0}} = -\frac{a \cdot \mathrm{Ec}}{0.4696\,\Pr_f{}^{1/3}}$$
$$(-0.1 \le \mathrm{Ec} \le 0.1) \text{ and } (0.3 \le \Pr_\mathrm{f} \le 3) \qquad (6.27)$$

where

$$a = 0.0138\,\Pr_\mathrm{f}{}^2 - 0.2305\,\Pr_\mathrm{f} - 0.0164 = 0.0138 \times 0.7^2 - 0.2305 \times 0.7 - 0.0164$$
$$= -0.17099.$$

Then

$$E_{(Q_{x,f})_{\mathrm{Ec}=0}} = -\frac{-0.17099 \times 0.000137}{0.4696 \times 0.7^{1/3}} = 5.618 \times 10^{-5}.$$

Obviously, the relative deviation on heat transfer prediction caused by ignoring the viscous thermal dissipation is very slight, so the effect of the viscous thermal dissipation can be neglected.

Solution 4: The local Reynolds number with consideration of variable physical properties will be

$$\mathrm{Re}_{x,\infty} = \frac{w_{x,\infty}x}{\nu_\infty} = \frac{7.5 \times 0.5}{15.895 \times 10^{-6}} = 235923.2.$$

With $\frac{T_w}{T_\infty} = \frac{700}{300} = 2.333333$, the heat transfer results on the plate with consideration of coupled effect of the variable physical properties is evaluated by (7.40), i.e.

$$Q_{x,w} = 0.664 b \lambda_w (t_w - t_\infty) \mathrm{Re}_{x,\infty}{}^{1/2} \mathrm{Pr}_f{}^{1/3} (\frac{T_w}{T_\infty})^{-m} \quad (0.3 \le \mathrm{Pr} \le 1), \quad (7.40)$$

with

$$m = 0.725 n_{\mu\lambda} + 0.3833 \ (1 \le T_w/T_\infty \le 3),$$
$$n_{\mu\lambda} = 0.45 n_\mu + 0.55 n_\lambda.$$

Then

$$n_{\mu\lambda} = 0.45 n_\mu + 0.55 n_\lambda = 0.45 \times 0.68 + 0.55 \times 0.81 = 0.7515,$$
$$m = 0.725 n_{\mu\lambda} + 0.3833 = 0.725 \times 0.7515 + 0.3833 = 0.928138,$$
$$Q_{x,w} = 0.664 \times 1 \times 0.05240 \times 400 \times 235923.2^{1/2} \times 0.7^{1/3} \times 2.333333^{-0.928138}$$

Thus

$$Q_{x,w} = 2733.9 \, \mathrm{W}.$$

Solution 5: For ignoring the variable physical properties and viscous thermal dissipation, (5.13) can be used for evaluation of the plate heat transfer result for laminar forced convection on a horizontal flat plate, i.e.

$$[Q_{x,f}]_{\mathrm{Ec}=0} = 0.664 b \lambda_f (t_w - t_\infty) \mathrm{Re}_{x,f}{}^{1/2} \mathrm{Pr}_f{}^{1/3} \ (0.3 \le \mathrm{Pr}_f \le 100). \quad (5.13)$$

Where from solution (1), we have for ignoring variable physical properties, related local Reynolds number is $\mathrm{Re}_{x,f} == 96686.86$.

Then

$$[Q_{x,f}]_{\mathrm{Ec}=0} = 0.664 \times 1 \times 0.0407 \times 400 \times 96686.86^{1/2} \times 0.7^{1/3} = 2984.5 \, \mathrm{W}.$$

The prediction deviation for heat transfer caused by ignoring the variable physical properties is calculated as

$$\frac{[Q_{x,f}]_{\mathrm{Ec}=0} - Q_{x,w}}{Q_{x,w}} = \frac{2984.5 - 2733.9}{2733.9} = 9.166\%$$

Here, we can find that the big deviation is caused on heat transfer prediction result of laminar forced convection, if the couple affect of variable physical properties is ignored.

Example 2 Continued to example 1, and solve the following questions:

(1) calculate the average skin-friction coefficient with consideration of variable physical properties;
(2) calculate the average skin-friction coefficient with ignoring variable physical properties.

Solution 1: With (7.29) the average skin-friction coefficient with consideration of variable physical properties is as follows:

$$\overline{C}_{x,w} = 2 \times \sqrt{2}\frac{\nu_w}{\nu_\infty}(\mathrm{Re}_{x,\infty})^{-1/2}\left(\frac{\mathrm{d}W_x(\eta)}{\mathrm{d}\eta}\right)_{\eta=0}.$$

From example 1, we have $\mathrm{Re}_{x,\infty} = 235923.2$ and $\frac{T_w}{T_\infty} = 2.333333$.

For $t_w = 427°C$ and $t_\infty = 27\,°\mathrm{C}$, from Fig. 7.6 we have

$$\nu_w = 68.1 \times 10^{-6}\mathrm{m}^2/\mathrm{s},\ \nu_\infty = 15.895 \times 10^{-6}\mathrm{m}^2/\mathrm{s}.$$

With $\frac{T_w}{T_\infty} = 2.333333$, by using interpolation approach, we have the following value of $\left(\frac{\mathrm{d}W_x(\eta)}{\mathrm{d}\eta}\right)_{\eta=0}$ from Table 7.3:

$$\left(\frac{\mathrm{d}W_x(\eta)}{\mathrm{d}\eta}\right)_{\eta=0} = 0.226426.$$

Then, for consideration of variable physical properties, the average skin friction is

$$\begin{aligned}\overline{C}_{x,w} &= 2 \times \sqrt{2} \times \frac{68.1}{15.895} \times 235923.2^{-1/2} \times 0.226426\\ &= 0.005649.\end{aligned}$$

Solution 2: With (5.2), the average skin friction with ignoring variable physical properties is

$$\overline{C}_{x,f} = 1.328\mathrm{Re}_{x,f}{}^{-1/2}.$$

Then

$$C_{x,f} = 1.328 \times 96686.86^{-1/2} = 0.00427.$$

Then the relative deviation of the average skin-friction with ignoring variable physical properties to that with consideration of variable physical properties is

$$\frac{\overline{C}_{x,f} - \overline{C}_{x,w}}{\overline{C}_{x,w}} = \frac{0.00427 - 0.005649}{0.005649} = -0.244 = -24.4\%.$$

These examples demonstrate that ignoring the variable physical properties will cause very large deviation on prediction of heat transfer and momentum issue.

Exercises

1. According to this chapter, try to explain the importance and necessity for consideration of variable physical properties in theoretical investigation on heat transfer of gas laminar forced convection.
2. Please try to use Falkner–Skan-type transformation reviewed in Chap. 3, also transform similarly the governing partial differential equations (7.1), (7.2), and (7.3) as well as their boundary condition equations (7.4) and (7.5) to the corresponding governing ordinary differential equations.
3. On the basis of solution of Example 1, please compare in detail the similarity governing ordinary equations with consideration of variable physical properties based on the present new similarity analysis method and Falkner-Skan type transformation. What conclusions would you have from such comparison?
4. Follow Example 1, the air is replaced by a typical waste gas of the furnace whose Prandtl number, viscous parameter, and thermal conductivity parameter are 0.63, 0.75, and 1.02, respectively, and keep all other conditions. Please solve the following:

    (1) evaluate the related local Reynolds numbers for determination of the flow types;
    (2) calculate the related Eckert numbers;
    (3) determine the related heat transfer prediction deviation caused by ignoring the viscous thermal dissipation;
    (4) calculate the related heat transfer results with consideration of coupled effect of the variable physical properties;
    (5) evaluate the prediction deviation on heat transfer caused by ignoring the variable physical properties.
    (6) calculate the average skin-friction coefficient with consideration of variable physical properties;
    (7) calculate the average skin-friction coefficient with ignoring variable physical properties.

References

1. H. Blasius, Grenzschichten in Flussigkeiten mit kleiner Reibung, Z. Math. Phys. **56**, 1–37 (1908)
2. E. Pohlhausen, Der Wärmeaustausch zwischen festen Körpern und Flüssigkeiten mit kleiner Reibung und kleiner Wärmeleitung. Z. Angew. Math. Mech. **1**, 115–121 (1921)
3. H. Schlichting, J. Kestin, *Boundary Layer Theory* (McGraw-Hill, New York, NY, 1955)
4. J.X. Ling, A. Dybbs, Forced convection over a flat plate submersed in a porous medium: variable viscosity case. Paper 87-WA/HT-23, New York, ASME, (1987)

5. J.X. Ling, A. Dybbs, The effect of variable viscosity on forced convection. J. Heat Transfer **114**, 1063 (1992)
6. A. Postelnicu, T. Grosan, I. Pop, The effect of variable viscosity on forced convection flow past a horizontal flat plate in a porous medium with internal heat generation. Mech. Res. Commun. **28**(3), 331–337 (2001)
7. D.Y. Shang, *Free Convection Film Flows and Heat Transfer* (Springer, Berlin, Heidelberg and New York, NY, 2006)
8. D.Y. Shang, B.X. Wang, Effect of variable thermophysical properties on laminar free convection of gas. Int. J. Heat Mass Transfer **33**(7), 1387–1395 (1990)
9. D.Y. Shang, B. X. Wang, Effect of variable thermophysical properties on laminar free convection of polyatomic gas. Int. J. Heat Mass Transfer **34**(3), 749–755 (1991)

Chapter 8
Heat Transfer of Liquid Laminar Forced Convection with Consideration of Variable Physical Properties

Abstract The new similarity analysis method is applied to develop a novel system of analysis and transformation models of liquid laminar forced convection with consideration of variable physical properties. As an example, the formulated temperature-dependent equations on density, thermal conductivity, absolute viscosity of water, and specific heat are applied, and the reliable numerical solutions on water laminar forced convection are obtained for investigation of the coupled effect of these variable physical properties. Theoretical equations on heat transfer are obtained based on the new similarity model, where the wall dimensionless temperature gradient is the no-given variable. The system of key numerical solutions on the wall dimensionless temperature gradient is formulated by a curve-fitting approach. Combined with the formulated equations of the wall dimensionless temperature gradient, the theoretical equations on heat transfer become available for simple and reliable prediction of heat transfer of water laminar forced convection with consideration of coupled effect of the variable physical properties.

8.1 Introduction

In Chap. 7, we investigated heat transfer of gas laminar forced convection with consideration of the coupled effect of variable thermophysical properties. In that investigation, the temperature-dependent physical properties at atmospheric pressure, such as density, thermal conductivity, specific heat, and absolute viscosity, are treated by a simple power law with the related temperature parameters. While, for general gases, the specific heat parameter is so small compared with other temperature parameters, their specific heat can be regarded as constant with temperature. Based on the new similarity analysis method, in the transformed governing ordinary differential equations the variable physical properties exist in form of the dimensionless physical property factors, which are late treated to the function of the dimensionless temperature variable, related temperature parameters, and boundary temperature ratio. On this basis, the developed dimensionless governing differential equations are solved

D.Y. Shang, *Theory of Heat Transfer with Forced Convection Film Flows*, Heat and Mass Transfer, DOI 10.1007/978-3-642-12581-2_8,

for different gases with different boundary temperature ratios. Furthermore, with the reliable formulated equation of the wall dimensionless temperature gradient based on its system of key numerical, the heat transfer theoretical equations obtained under the new similarity model were available for simple and reliable prediction on heat transfer of laminar forced convection of different gases on the horizontal flat plate.

However, the temperature-dependent physical properties of liquids are quite different from those of gas; therefore, heat transfer of liquid laminar forced convection are separately investigated in this chapter with consideration of variable physical properties.

8.2 Governing Equations

8.2.1 Governing Partial Differential Equations

The physical model and coordinate system of boundary layer with two-dimensional laminar forced convection has been shown in Fig. 5.1 schematically. A flat plate is horizontally located in parallel fluid flow with its main stream velocity $w_{x,\infty}$. The plate surface temperature is t_w and the liquid bulk temperature is t_∞. Then a velocity boundary layer is produced near the plate. If t_w is not equal to t_∞, a temperature boundary layer will occur near the plate also. We assume that the velocity boundary layer is laminar. Then, based on Chap. 7, the governing partial differential equations for laminar boundary layer with forced convection are expressed as below with consideration of these variable physical properties:

$$\frac{\partial}{\partial x}(\rho w_x) + \frac{\partial}{\partial y}(\rho w_y) = 0, \tag{7.1}$$

$$\rho\left(w_x \frac{\partial w_x}{\partial x} + w_y \frac{\partial w_x}{\partial y}\right) = \frac{\partial}{\partial y}\left(\mu \frac{\partial w_x}{\partial y}\right), \tag{7.2}$$

$$\rho\left[w_x \frac{\partial (c_p t)}{\partial x} + w_y \frac{\partial (c_p t)}{\partial y}\right] = \frac{\partial}{\partial y}\left(\lambda \frac{\partial t}{\partial y}\right), \tag{7.3}$$

with the boundary conditions

$$y = 0: \qquad w_x = 0,\ w_y = 0, \qquad t = t_w \tag{7.4}$$

$$y \to \infty: \qquad w_x = w_{x,\infty}\ (\text{constant}),\ t = t_\infty, \tag{7.5}$$

where temperature-dependent physical properties density ρ, absolute viscosity μ, specific heat, and thermal conductivity λ are taken into account.

8.2.2 Similarity Transformation Variables

According to Chap. 7 for the new similarity analysis method, the set of the related dimensional analysis variables are shown as follows:

$$\eta = \frac{y}{x}\left(\frac{1}{2}\mathrm{Re}_{x,\infty}\right)^{1/2}, \tag{7.6}$$

$$\mathrm{Re}_{x,\infty} = \frac{w_{x,\infty}x}{\nu_\infty}, \tag{7.7}$$

$$w_x = w_{x,\infty}W_x(\eta), \tag{7.8}$$

$$w_y = w_{x,\infty}\left(\frac{1}{2}\mathrm{Re}_{x,\infty}\right)^{-1/2} W_y(\eta), \tag{7.9}$$

$$\theta(\eta) = \frac{t - t_\infty}{t_w - t_\infty}, \tag{7.10}$$

where η is dimensionless coordinate variable, $\mathrm{Re}_{x,\infty}$ is local Reynolds number, $W_x(\eta)$ and $W_y(\eta)$ are dimensionless velocity components in x and y coordinates respectively, and $\theta(\eta)$ denotes dimensionless temperature.

8.2.3 Governing Ordinary Differential Equations

Consulting the derivation in Chap. 7, the transformed governing ordinary differential equations for laminar forced convection are

$$-\eta\frac{\mathrm{d}W_x(\eta)}{\mathrm{d}\eta} + 2\frac{\mathrm{d}W_y(\eta)}{\mathrm{d}\eta} + \frac{1}{\rho}\frac{\mathrm{d}\rho}{\mathrm{d}\eta}[-\eta \cdot W_x(\eta) + 2W_y(\eta)] = 0, \tag{7.11}$$

$$\frac{\nu_\infty}{\nu}(-\eta W_x(\eta) + 2W_y(\eta))\frac{\mathrm{d}W_x(\eta)}{\mathrm{d}\eta} = \frac{\mathrm{d}^2 W_x(\eta)}{\mathrm{d}\eta^2} + \frac{1}{\mu}\frac{\mathrm{d}\mu}{\mathrm{d}\eta}\frac{\mathrm{d}W_x(\eta)}{\mathrm{d}\eta}, \tag{7.12}$$

$$\begin{aligned}\mathrm{Pr}\,\frac{\nu_\infty}{\nu}&\left[(-\eta W_x(\eta) + 2W_y(\eta))\frac{\mathrm{d}\theta(\eta)}{\mathrm{d}\eta} + \big(-\eta W_x(\eta) + 2W_y(\eta)\big)\right.\\ &\left.\times\left(\theta(\eta) + \frac{t_\infty}{t_w - t_\infty}\right)\frac{1}{c_p}\frac{\mathrm{d}c_p}{\mathrm{d}\eta}\right] = \frac{\mathrm{d}^2\theta(\eta)}{\mathrm{d}\eta^2} + \frac{1}{\lambda}\frac{\partial\lambda}{\partial\eta}\frac{\mathrm{d}\theta(\eta)}{\mathrm{d}\eta}\end{aligned} \tag{7.13}$$

with the boundary condition equations

$$\eta = 0: \quad W_x(\eta) = 0,\ W_y(\eta) = 0, \quad \theta(\eta) = 1 \tag{7.14}$$

$$\eta \to \infty: \quad W_x(\eta) = 1,\ \theta(\eta) = 0 \tag{7.15}$$

8.3 Treatment of Liquid Variable Physical Properties

We take water as an example to introduce the treatment of variable physical properties of liquid. In fact, the variation regulation of the physical properties of water can represent those of most of liquids, and water forced convection exists universally in various industrial departments and human life.

The dimensionless physical property factors $\Pr = \frac{\mu \cdot c_p}{\lambda}, \frac{\nu_\infty}{\nu}, \frac{1}{\rho}\frac{\mathrm{d}\rho}{\mathrm{d}\eta}, \frac{1}{\mu}\frac{\mathrm{d}\mu}{\mathrm{d}\eta}, \frac{1}{c_p}\frac{\mathrm{d}c_p}{\mathrm{d}\eta}$, and $\frac{1}{\lambda}\frac{\mathrm{d}\lambda}{\mathrm{d}\eta}$ with the variable physical properties in the governing dimensionless equations (7.11), (7.12), and (7.13) need to be dealt with for solution of the governing equations.

From [1, 2], the temperature-dependent expressions of density, thermal conductivity, and absolute viscosity of water at atmospheric pressure with the temperature range between t = 0 to 100°C are respectively expressed by polynomials as

$$\rho = -4.48 \times 10^{-3} t^2 + 999.9, \tag{8.1}$$

$$\lambda = -8.01 \times 10^{-6} t^2 + 1.94 \times 10^{-3} t + 0.563, \tag{8.2}$$

$$\mu = \exp\left(-1.6 - \frac{1150}{T} + \left(\frac{690}{T}\right)^2\right) \times 10^{-3}. \tag{8.3}$$

With (7.10), here, the absolute temperature in (8.1), (8.2), and (8.3) are expressed as

$$t = (t_w - t_\infty) \cdot \theta(\eta) + t_\infty.$$

In addition, the water-specific heat is generally a constant with the temperature range between t = 0 to 100°C, and then we can regarded as

$$\frac{1}{c_p}\frac{dc_p}{d\eta} = 0. \tag{8.4}$$

With (8.1), (8.2), and (8.3), the physical property factors Pr, $\frac{\nu_\infty}{\nu}$, $\frac{1}{\rho}\frac{\mathrm{d}\rho}{\mathrm{d}\eta}$, $\frac{1}{\mu}\frac{\mathrm{d}\mu}{\mathrm{d}\eta}$, and $\frac{1}{\lambda}\frac{\mathrm{d}\lambda}{\mathrm{d}\eta}$ can be expressed as follows, respectively.

For Prandtl number Pr

$$\Pr = \frac{\mu \cdot c_p}{\lambda},$$
$$= \frac{\left[\exp\left(-1.6 - \frac{1150}{T} + \left(\frac{690}{T}\right)^2\right) \times 10^{-3}\right] \times 4200}{-8.01 \times 10^{-6} t^2 + 1.94 \times 10^{-3} t + 0.563}. \tag{8.5}$$

While the physical property factors $\frac{1}{\rho}\frac{\mathrm{d}\rho}{\mathrm{d}\eta}$, $\frac{1}{\mu}\frac{\mathrm{d}\mu}{\mathrm{d}\eta}$ and $\frac{1}{\lambda}\frac{\mathrm{d}\lambda}{\mathrm{d}\eta}$ are treated as below:

For physical property factor $\frac{1}{\rho}\frac{\mathrm{d}\rho}{\mathrm{d}\eta}$

The physical property factor $\frac{1}{\rho_l}\frac{d\rho}{d\eta}$ is expressed as

$$\frac{1}{\rho}\frac{\mathrm{d}\rho}{\mathrm{d}\eta} = \frac{1}{\rho}\frac{\mathrm{d}\rho}{\mathrm{d}t}\frac{\mathrm{d}t}{\mathrm{d}\eta}. \tag{a}$$

With (8.1), the following equation is obtained

$$\frac{\mathrm{d}\rho}{\mathrm{d}t} = -2 \times 4.48 \times 10^{-3} t. \tag{b}$$

With (7.10) we obtain

$$\frac{\mathrm{d}t}{\mathrm{d}\eta} = (t_w - t_\infty)\frac{\mathrm{d}\theta(\eta)}{\mathrm{d}\eta}. \tag{c}$$

Then

$$\frac{1}{\rho}\frac{\mathrm{d}\rho}{\mathrm{d}\eta} = \frac{(-2 \times 4.48 \times 10^{-3} t)(t_w - t_\infty)\frac{\mathrm{d}\theta(\eta)}{\mathrm{d}\eta}}{-4.48 \times 10^{-3} t^2 + 999.9}. \tag{8.6}$$

For physical property factor $\frac{1}{\lambda}\frac{\mathrm{d}\lambda}{\mathrm{d}\eta}$

The physical property factor $\frac{1}{\lambda}\frac{\mathrm{d}\lambda}{\mathrm{d}\eta}$ is expressed as

$$\frac{1}{\lambda}\frac{\mathrm{d}\lambda}{\mathrm{d}\eta} = \frac{1}{\lambda}\frac{\mathrm{d}\lambda}{\mathrm{d}t}\frac{\mathrm{d}t}{\mathrm{d}\eta}. \tag{d}$$

With (8.2), the following equation is obtained

$$\frac{\mathrm{d}\lambda}{\mathrm{d}t} = -2 \times 8.01 \times 10^{-6} t + 1.94 \times 10^{-3}. \tag{e}$$

Then (d) becomes

$$\frac{1}{\lambda}\frac{\mathrm{d}\lambda}{\mathrm{d}\eta} = \frac{(-2 \times 8.01 \times 10^{-6} t + 1.94 \times 10^{-3})(t_w - t_\infty)\frac{\mathrm{d}\theta(\eta)}{\mathrm{d}\eta}}{-8.01 \times 10^{-6} t^2 + 1.94 \times 10^{-3} t + 0.563}. \tag{8.7}$$

For physical property facto $\frac{1}{\mu}\frac{\mathrm{d}\mu}{\mathrm{d}\eta}$

The physical property factor $\frac{1}{\mu}\frac{\mathrm{d}\mu}{\mathrm{d}\eta}$ is expressed as

$$\frac{1}{\mu}\frac{\mathrm{d}\mu}{\mathrm{d}\eta} = \frac{1}{\mu}\frac{\mathrm{d}\mu}{\mathrm{d}T}\frac{\mathrm{d}T}{\mathrm{d}\eta}. \tag{f}$$

With (8.3), we have

$$\frac{d\mu}{dT} = \exp\left[-1.6 - \frac{1150}{T} + \left(\frac{690}{T}\right)^2\right] \times 10^{-3} \cdot \left(\frac{1150}{T^2} - 2 \times \frac{690^2}{T^3}\right). \quad \text{(g)}$$

Then, (f) becomes

$$\frac{1}{\mu}\frac{d\mu}{d\eta} = \left(\frac{1150}{T^2} - 2 \times \frac{690^2}{T^3}\right)(T_w - T_\infty)\frac{d\theta(\eta)}{d\eta}. \quad (8.8)$$

For physical property facto $\frac{\nu_\infty}{\nu}$

$\frac{\nu_\infty}{\nu}$ is expressed as

$$\frac{\nu_\infty}{\nu} = \left(\frac{\mu_\infty}{\rho_\infty}\right) \Big/ \left(\frac{\mu}{\rho}\right) = \left(\frac{\mu_\infty}{\mu}\right)\left(\frac{\rho}{\rho_\infty}\right),$$

$$= \frac{\exp\left[-1.6 - \frac{1150}{T_\infty} + \left(\frac{690}{T_\infty}\right)^2\right]}{\exp\left[-1.6 - \frac{1150}{T} + \left(\frac{690}{T}\right)^2\right]} \times \frac{-4.48 \times 10^{-3}t^2 + 999.9}{-4.48 \times 10^{-3}t_\infty^2 + 999.9}, \quad (8.9)$$

where

$$t = (t_w - t_\infty) \cdot \theta(\eta) + t_\infty \qquad T = t + 273.$$

Thus, with (8.5), (8.6), (8.7), (8.8), and (8.9), all physical property factors in governing ordinary differential equations (7.11) and (7.13) become function of the dimensionless temperature $\theta(\eta)$.

8.4 Velocity and Temperature Fields

Due to the temperature-dependent physical properties, the governing equations (7.11) to (7.13) are coupled with each other, and both velocity and temperature field depend on Prandtl number and the temperature boundary conditions. The above coupled equations (7.11), (7.12), and (7.13) with the boundary equations (14) and (15) are solved by a shooting method program with fifth-order Runge-Kutta integration. For solving the nonlinear problem, a special variable mesh approach is applied to the numerical procedure. In this present work, water laminar forced convection on a horizontal flat plate is taken as example for numerical calculation, and a system of rigorous numerical results of the velocity and temperature fields are obtained and plotted in Figs. 8.1, 8.2, 8.3, and 8.4, respectively, for the bulk temperatures

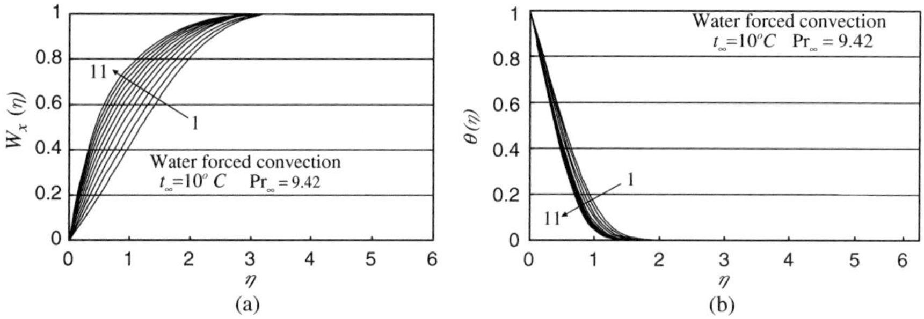

Fig. 8.1 (**a**) Velocity and (**b**) temperature profiles of water laminar forced convection on a horizontal plate with the bulk temperature $t_\infty = 10°C$ and variation of the wall temperature t_w Note: Lines 1 to 11: for $t_w = 0, \rightarrow 10, 20, 30, 40, 50, 60, 70, 80, 90$ and 100°C (i.e. $t_w - t_\infty = -10$, $\rightarrow$ 0, 10, 20, 30, 40, 50, 60, 70, 80, and 90 °C)

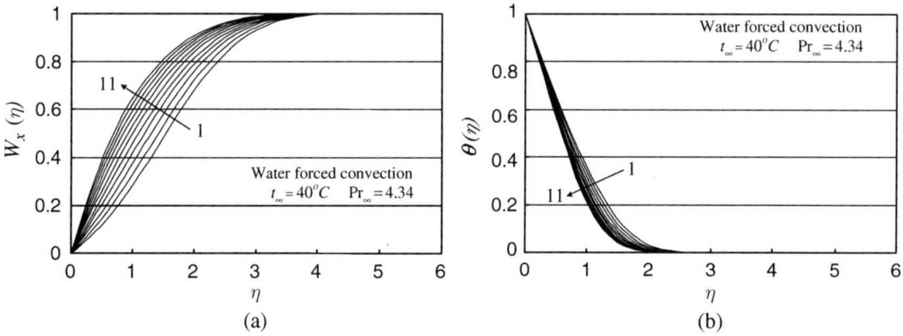

Fig. 8.2 (**a**) Velocity and (**b**) temperature profiles of water laminar forced convection on a horizontal plate with the bulk temperature $t_\infty = 40°C$ and variation of the wall temperature t_w Note: Lines 1 to 11: for $t_w = 0, 10, 20, 30, \rightarrow 40, 50, 60, 70, 80, 90$ and 100 °C (i.e. $t_w - t_\infty = -40$, -30, -20, -10, $\rightarrow$ 0, 10, 20, 30, 40, 50, and 60 °C)

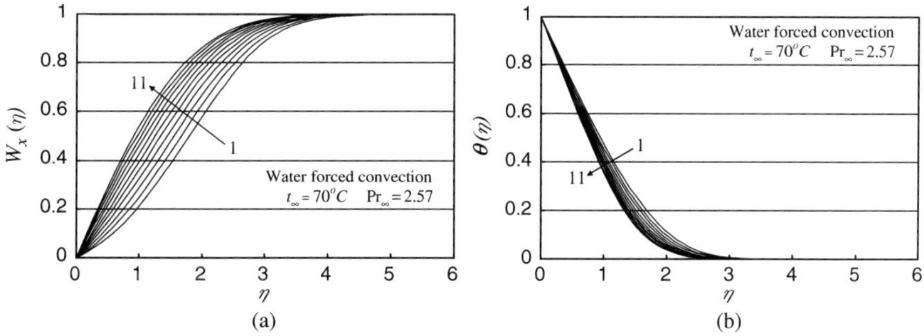

Fig. 8.3 (**a**) Velocity and (**b**) temperature profiles of water laminar forced convection on a horizontal plate with the bulk temperature $t_\infty = 70$°C and variation of the wall temperature t_w Note: Lines 1 to 11: for $t_w = 0, 10, 20, 30, 40, 50, 60, \rightarrow 70, 80, 90$ and 100°C (i.e. $t_w - t_\infty = -70$, -60, -50, -40, -30, -20, -10, $\rightarrow$ 0, 10, 20, and 30°C)

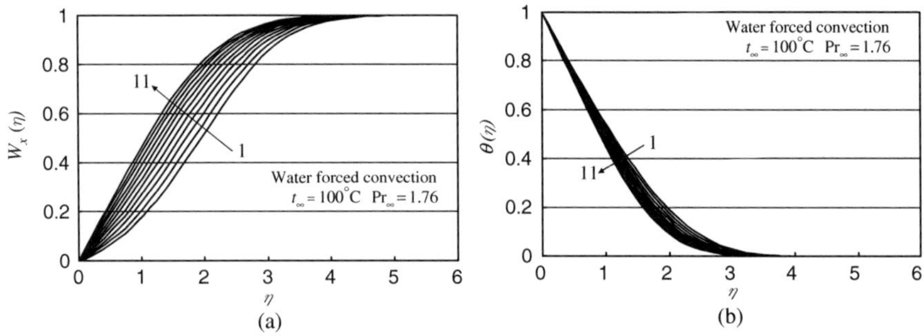

Fig. 8.4 (**a**) Velocity and (**b**) temperature profiles of water laminar forced convection on a horizontal plate with the bulk temperature $t_\infty = 100^\circ$C and variation of the wall temperature t_w Note: Lines 1 to 11: for $t_w = 0$, 10, 20, 30, 40, 50, 60, 70, 80, 90 and $\rightarrow$ 100°C (i.e. $t_w - t_\infty = -100, -90, -80, -70, -60, -50, -40, -30, -20, -10$, and $\rightarrow$ 0°C)

$t_\infty = 10^\circ$C, 40°C, 70°C), and 100°C with variations of the wall temperature respectively, where the following phenomena are found.

It is seen that, for special bulk temperature t_∞ (or special Pr$_\infty$), increasing the boundary temperature difference $t_w - t_\infty$ (i.e. increasing t_w or decreasing Pr$_w$) will cause increase of dimensionless velocity field level and wall dimensionless temperature gradient.

It is obvious that the velocity boundary layer thickness is larger than that of temperature boundary thickness for liquid laminar forced convection. This phenomenon is different from that for gas laminar forced convection, since the Prandtl number of liquid is much larger than that of gas. For the same reason, the difference between the velocity and temperature boundary layer thicknesses is even obvious with decreasing t_∞ (i.e. increasing Pr$_\infty$) for most of laminar forced convection of liquid, whose Prandtl number increases with decreasing temperature.

All above phenomena demonstrate the effect of variable physical properties on liquid forced convection. These phenomena are different from those expressed by Blasius and Pohlhausen solutions [3, 4] with ignoring the variable physical properties. Obviously, only in the case for $t_w - t_\infty \rightarrow 0$, Blasius and Pohlhausen solutions are valid.

8.5 Skin-Friction Coefficient with Consideration of Variable Physical Properties

Following Chap. 7, the equations of local and average skin-friction coefficient with consideration of variable physical properties for liquid laminar forced convection are respectively

$$C_{x,w} = \sqrt{2}\frac{\nu_w}{\nu_\infty}(\mathrm{Re}_{x,\infty})^{-1/2}\left(\frac{\mathrm{d}W_x(\eta)}{\mathrm{d}\eta}\right)_{\eta=0} \quad (7.28)$$

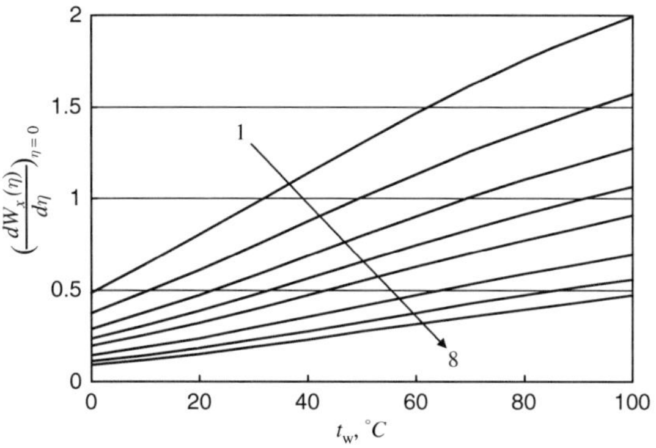

Fig. 8.5 Wall velocity gradient $\left(\frac{\mathrm{d}W_x(\eta)}{\mathrm{d}\eta}\right)_{\eta=0}$ of water laminar forced convection on the horizontal plate. Lines 1 → 8 denotes $t_\infty = 0$, 10 20, 30, 40, 60, 80, and 100°C respectively

and

$$\overline{C}_{x,w} = 2 \times \sqrt{2}\frac{\nu_w}{\nu_\infty}(\mathrm{Re}_{x,\infty})^{-1/2}\left(\frac{\mathrm{d}W_x(\eta)}{\mathrm{d}\eta}\right)_{\eta=0}. \tag{7.29}$$

It is seen from (7.28) and (7.29) that the wall dimensionless velocity gradient $\left(\frac{\mathrm{d}W_x(\eta)}{\mathrm{d}\eta}\right)_{\eta=0}$ is the only one no-given variable for evaluation of the skin-friction coefficient with consideration of variable physical properties. The numerical solutions on the wall dimensionless velocity gradient $\left(\frac{\mathrm{d}W_x(\eta)}{\mathrm{d}\eta}\right)_{\eta=0}$ of water laminar forced convection on a flat plate are plotted in Fig. 8.5, and some of the typical valued are listed in Table 7.3 for temperature boundary conditions. It is seen that $\left(\frac{\mathrm{d}W_x(\eta)}{\mathrm{d}\eta}\right)_{\eta=0}$ depends on bulk Prandtl number Pr_∞, and boundary temperatures t_w and t_∞. Increasing the temperature t_w will cause increasing the wall dimensionless velocity gradient $\left(\frac{\mathrm{d}W_x(\eta)}{\mathrm{d}\eta}\right)_{\eta=0}$ proportionally. Decreasing the temperature t_∞ will cause increasing the wall dimensionless velocity gradient $\left(\frac{\mathrm{d}W_x(\eta)}{\mathrm{d}\eta}\right)_{\eta=0}$ at accelerative pace.

8.6 Heat Transfer Analysis

So far, there has been a lack of theoretical investigation on heat transfer for liquid laminar forced convection with consideration of coupled effect of variable physical properties. This issue will be addressed here in this chapter.

First, it is necessary to review the heat transfer theoretical equations as below for laminar forced convection developed in Chap. 7 based on the present similarity analysis method.

The local heat transfer rate $q_{x,w}$ at position x per unit area from the surface of the plate to the fluid with consideration of variable physical properties is calculated by Fourier's law as $q_{x,w} = -\lambda_w \left(\frac{\partial t}{\partial y}\right)_{y=0}$.

While, with (k) we have

$$-\left(\frac{\partial t}{\partial y}\right)_{y=0} = x^{-1}\left(\frac{1}{2}\text{Re}_{x,\infty}\right)^{1/2}(t_w - t_\infty)\left(-\frac{\text{d}\theta(\eta)}{\text{d}\eta}\right)_{\eta=0}.$$

Then

$$\text{q}_{\text{x,w}} = \lambda_w(t_w - t_\infty)\left(\frac{1}{2}\text{Re}_{x,\infty}\right)^{1/2} x^{-1}\left(-\frac{\text{d}\theta(\eta)}{\text{d}\eta}\right)_{\eta=0}, \tag{7.30}$$

where $\left(-\frac{\text{d}\theta(\eta)}{\text{d}\eta}\right)_{\eta=0}$ is wall dimensionless temperature gradient with consideration of variable physical properties.

The local heat transfer coefficient $\alpha_{x,w}$, defined as $q_{x,w} = \alpha_{x,w}(t_w - t_\infty)$, will be given by

$$\alpha_{x,w} = \lambda_w\left(\frac{1}{2}\text{Re}_{x,\infty}\right)^{1/2} x^{-1}\left(-\frac{\text{d}\theta(\eta)}{\text{d}\eta}\right)_{\eta=0}. \tag{7.31}$$

The local Nusselt number defined by $Nu_{x,w} = \dfrac{\alpha_{x,w} \cdot x}{\lambda_w}$ will be

$$Nu_{x,w} = \left(\frac{1}{2}\text{Re}_{x,\infty}\right)^{1/2}\left(-\frac{\text{d}\theta(\eta)}{\text{d}\eta}\right)_{\eta=0} \tag{7.32}$$

Total heat transfer rate for position $x = 0$ to x with width of b on the plate is a integration $Q_{x,w} = \iint\limits_A q_{x,w}\text{d}A = \int_0^x q_{x,w}b\text{d}x$, and hence

$$Q_{x,w} = 2b\lambda_w(t_w - t_\infty)\left(\frac{1}{2}\text{Re}_{x,\infty}\right)^{1/2}\left(-\frac{\text{d}\theta(\eta)}{\text{d}\eta}\right)_{\eta=0}. \tag{7.33}$$

The average heat transfer rate defined as $\overline{Q}_{x,w} = \dfrac{Q_{x,w}}{b \times x}$ is

$$\overline{Q}_{x,w} = 2x^{-1}\lambda_w(t_w - t_\infty)\left(\frac{1}{2}\text{Re}_{x,\infty}\right)^{1/2}\left(-\frac{\text{d}\theta(\eta)}{\text{d}\eta}\right)_{\eta=0},$$

i.e.

$$\overline{Q}_{x,w} = \sqrt{2}x^{-1}\lambda_w(t_w - t_\infty)\text{Re}_{x,\infty}{}^{1/2}\left(-\frac{\text{d}\theta(\eta)}{\text{d}\eta}\right)_{\eta=0} \tag{7.34}$$

The average heat transfer coefficient $\overline{\alpha_{x,w}}$ defined as $\overline{Q}_{x,w} = \overline{\alpha_{x,w}}(t_w - t_\infty)$ is expressed as

$$\overline{\alpha_{x,w}} = 2\lambda_w \left(\frac{1}{2}\mathrm{Re}_{x,\infty}\right)^{1/2} x^{-1} \left(-\frac{d\theta(\eta)}{d\eta}\right)_{\eta=0},$$

i.e.

$$\overline{\alpha_{x,w}} = \sqrt{2}\lambda_w \mathrm{Re}_{x,\infty}{}^{1/2} x^{-1} \left(-\frac{d\theta(\eta)}{d\eta}\right)_{\eta=0}. \tag{7.35}$$

The average Nusselt number is defined as $\overline{Nu}_{x,w} = \dfrac{\overline{\alpha_{x,w}} \cdot x}{\lambda_w}$, and hence

$$\overline{Nu}_{x,w} = -\sqrt{2}(\mathrm{Re}_{x,\infty})^{1/2} \left(-\frac{d\theta(\eta)}{d\eta}\right)_{\eta=0}. \tag{7.36}$$

It should be indicated that in this chapter for consideration of variable physical properties, $q_{x,w}$, $\alpha_{x,w}$, $Nu_{x,w}$, $Q_{x,w}$, $\overline{Q}_{x,w}$, $\overline{\alpha}_{x,w}$, and $\overline{Nu}_{x,w}$ not only express the case with consideration of variable physical properties but also denote the case with ignoring viscous thermal dissipation, although the subscript Ec $= 0$ is omitted. According to the studies as well as the example calculation results in Chaps. 5 and 6 of this book, it can be found that in the range of laminar forced convection the Eckert number is very small generally, and then ignoring the effect of the viscous thermal dissipation will not cause obvious influence on laminar forced convection and heat transfer.

From (7.30), (7.31), (7.32), (7.33), (7.34), (7.35), and (7.36), it is seen that the dimensionless wall temperature gradient $\left(-\frac{d\theta(\eta)}{d\eta}\right)_{\eta=0}$ with consideration of variable physical properties is the only one no-given variable for prediction of heat transfer. In the next section, I will focus on the investigation of the simple and reliable prediction of $\left(-\frac{d\theta(\eta)}{d\eta}\right)_{\eta=0}$ for liquid laminar forced convection with consideration of variable physical properties, so that the above heat transfer equations can be available for heat transfer prediction.

8.7 Dimensionless Wall Temperature Gradient

Since dimensionless wall temperature gradient is necessary for evaluation of heat transfer, it should be investigated more carefully. For this purpose, the numerical results of a system of key solutions on the dimensionless wall temperature gradient of water forced convection with consideration of coupled effect of variable physical properties on a horizontal flat plate are obtained, plotted in Fig. 8.5, and some of the typical values are listed in Table 8.1 with different bulk temperatures and variation

Table 8.1 Wall velocity gradient $(\mathrm{d}W_x(\eta)/\mathrm{d}\eta)_{\eta=0}$ of water laminar forced convection on the horizontal plate

$t_\infty = 0°\mathrm{C}$, $\mathrm{Pr}_\infty = 13.6$						
$t_w, °\mathrm{C}$	$\to 0$	20	40	60	80	100
$t_w - t_\infty, °\mathrm{C}$	$\to 0$	20	40	60	80	100
$\left(\frac{\mathrm{d}W_x(\eta)}{\mathrm{d}\eta}\right)_{\eta=0}$	0 .482926	0.801547	1.139358	1.460847	1.756183	1.994678
$t_\infty = 10°\mathrm{C}$, $\mathrm{Pr}_\infty = 9.42$						
$t_w, °\mathrm{C}$	0	20	40	60	80	100
$t_w - t_\infty$, °C	−10	10	30	50	70	90
$\left(\frac{\mathrm{d}W_x(\eta)}{\mathrm{d}\eta}\right)_{\eta=0}$	0.371807	0.607875	0.872556	1.131842	1.368404	1.570470
$t_\infty = 20°\mathrm{C}$, $\mathrm{Pr}_\infty = 6.99$						
$t_w, °\mathrm{C}$	0	$\to 20$	40	60	80	100
$t_w - t_\infty$, °C	−20	$\to 0$	20	40	60	80
$\left(\frac{\mathrm{d}W_x(\eta)}{\mathrm{d}\eta}\right)_{\eta=0}$	0.289907	0.472510	0.689262	0.903547	1.104420	1.277460
$t_\infty = 30°\mathrm{C}$, $\mathrm{Pr}_\infty = 5.42$						
$t_w, °\mathrm{C}$	0	20	40	60	80	100
$t_w - t_\infty$, °C	−30	−10	10	30	50	70
$\left(\frac{\mathrm{d}W_x(\eta)}{\mathrm{d}\eta}\right)_{\eta=0}$	0.233311	0.386157	0.562114	0.743621	0.914734	1.065005
$t_\infty = 40°\mathrm{C}$, $\mathrm{Pr}_\infty = 4.34$						
$t_w, °\mathrm{C}$	0	20	$\to 40$	60	80	100
$t_w - t_\infty$, °C	−40	−20	$\to 0$	20	0	60
$\left(\frac{\mathrm{d}W_x(\eta)}{\mathrm{d}\eta}\right)_{\eta=0}$	0.192887	.321361	0.469976	0.626241	0.774164	0.906800
$t_\infty = 60°\mathrm{C}$, $\mathrm{Pr}_\infty = 3$						
$t_w, °\mathrm{C}$	0	20	40	$\to 60$	80	100
$t_w - t_\infty$, °C ($\mathrm{Pr}_\infty = 3$)	−60	−40	−20	$\to 0$	20	40
$\left(\frac{\mathrm{d}W_x(\eta)}{\mathrm{d}\eta}\right)_{\eta=0}$	0.140852	0.236804	0.350084	0.469705	0.586177	0.692322
$t_\infty = 80°\mathrm{C}$, $\mathrm{Pr}_\infty = 2.23$						
$t_w, °\mathrm{C}$	0	20	40	60	$\to 80$	100
$t_w - t_\infty$, °C	−80	−60	−40	−20	$\to 0$	20
$\left(\frac{\mathrm{d}W_x(\eta)}{\mathrm{d}\eta}\right)_{\eta=0}$	0.109858	0.185854	0.276632	0.373818	0.469652	0.558367
$t_\infty = 100°\mathrm{C}$, $\mathrm{Pr}_\infty = 1.76$						
$t_w, °\mathrm{C}$	0	20	40	60	80	$\to 100$
$t_w - t_\infty$, °C	−100	−80	−60	−40	−20	$\to 0$
$\left(\frac{\mathrm{d}W_x(\eta)}{\mathrm{d}\eta}\right)_{\eta=0}$	0.090085	0.153058	0.228934	0.311050	0.392915	.469564

of the wall temperatures. It is seen that for special bulk temperature t_∞, increasing the boundary layer temperature difference $t_w - t_\infty$ (i.e. increasing t_w or decreasing $\Pr_w$) causes increase of the wall temperature gradient. For special wall temperature, decreasing the bulk temperature t_∞ will cause increase of the wall dimensionless temperature gradient at an accelerative pace. These phenomena for liquid laminar forced convection are different from that for gas laminar forced convection shown in Chap. 7, where increasing the boundary layer temperature ratio T_w/T_∞ causes decrease of the wall temperature gradient. Such different variation regulations of the wall dimensionless temperature gradient with the temperature conditions must be caused by the different variation of the absolute viscosity of gas and liquid with temperature. For instance, generally, the absolute viscosity of gases increases with increasing the temperature, but the absolute viscosity of liquid decreases with increasing the temperature.

It should be indicated that the case $t_w - t_\infty \to 1$ shown in Fig. 8.6 is identical to that without consideration of variable physical properties, and is regarded as the special case for consideration of variable physical properties. Then, it follows that consideration of variable physical properties is very necessary for investigation on heat transfer of forced convection.

The numerical results shown in Fig. 8.6 and Table 8.2 demonstrate that the wall temperature gradient $\left(-\frac{d\theta(\eta)}{d\eta}\right)_{\eta=0}$ with consideration of coupled effect of variable physical properties not only depend on Prandtl number $\Pr_\infty$ but also depend on the boundary temperature t_w and t_∞. By using a curve-fitting approach, the following equations of $\left(-\frac{d\theta(\eta)}{d\eta}\right)_{\eta=0}$ are obtained from the numerical results for water laminar forced convection on a horizontal flat plate:

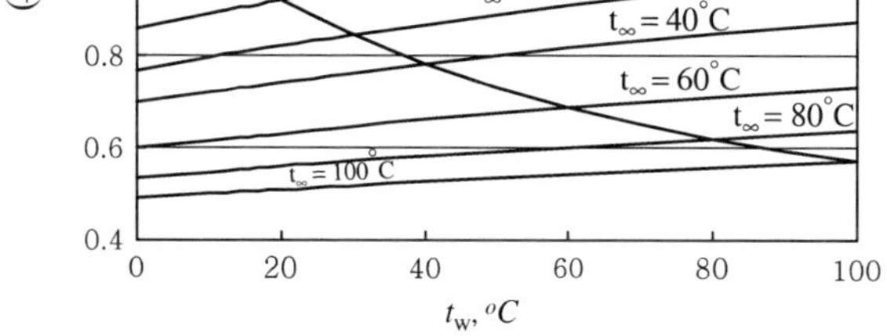

Fig. 8.6 Numerical results on wall temperature gradient $-\left(\frac{d\theta(\eta)}{d\eta}\right)_{\eta=0}$ of water laminar forced convection on horizontal flat plate (Lines with $t_w - t_\infty \to 0^\circ C$ is corresponding to the case without consideration of variable thermophysical properties)

Table 8.2 Wall temperature gradient $-(\mathrm{d}\theta(\eta)/\mathrm{d}\eta)_{\eta=0}$ of water laminar forced convection on the horizontal plate. a, numerical solution; b, predicted values by using (8.11) and (8.12); and c, predicted deviation by using (8.11) and (8.12)

$t_\infty = 0°\mathrm{C}$, $\mathrm{Pr}_\infty = 13.6$							
$t_w, °\mathrm{C}$		$\to 0$	20	40	60	80	100
$t_w - t_\infty, °\mathrm{C}$		$\to 0$	20	40	60	80	100
$-\left(\frac{\mathrm{d}\theta(\eta)}{\mathrm{d}\eta}\right)_{\eta=0}$	a	1.132134	1.242132	1.319886	1.384475	1.440437	1.485845
	b	1.120919	1.190233	1.263833	1.341984	1.424968	1.513083
	c	−0.0099	−0.0418	−0.04247	−0.0307	−0.0107	0.0183
$t_\infty = 10°\mathrm{C}$, $\mathrm{Pr}_\infty = 9.42$							
$t_w, °\mathrm{C}$		0	20	40	60	80	100
$t_w - t_\infty, °\mathrm{C}$		−10	10	30	50	70	90
$-\left(\frac{\mathrm{d}\theta(\eta)}{\mathrm{d}\eta}\right)_{\eta=0}$	a	0.974992	1.052693	1.119837	1.174426	1.220327	1.259959
	b	0.963558	1.020811	1.081466	1.145725	1.213802	1.285924
	c	−0.0117	−0.0303	−0.0343	−0.0244	−0.0053	0.0206
$t_\infty = 20°\mathrm{C}$, $\mathrm{Pr}_\infty = 6.99$							
$t_w, °\mathrm{C}$		0	$\to 20$	40	60	80	100
$t_w - t_\infty, °\mathrm{C}$		−20	$\to 0$	20	40	60	80
$-\left(\frac{\mathrm{d}\theta(\eta)}{\mathrm{d}\eta}\right)_{\eta=0}$	a	0.860088	0.920662	0.98086	1.025678	1.065989	1.100424
	b	0.849461	0.897885	0.949069	1.003171	1.060358	1.120804
	c	−0.0124	−0.0247	−0.0324	−0.0219	−0.0053	0.0185
$t_\infty = 30°\mathrm{C}$, $\mathrm{Pr}_\infty = 5.42$							
$t_w, °\mathrm{C}$		0	20	40	60	80	100
$t_w - t_\infty, °\mathrm{C}$		−30	−10	10	30	50	70
$-\left(\frac{\mathrm{d}\theta(\eta)}{\mathrm{d}\eta}\right)_{\eta=0}$	a	0.769577	0.822077	0.866916	0.909517	0.944213	0.974296
	b	0.761665	0.803250	0.847107	0.893357	0.942133	0.993572
	c	−0.0103	−0.0229	−0.0229	−0.0178	−0.0022	0.0198
$t_\infty = 40°\mathrm{C}$, $\mathrm{Pr}_\infty = 4.34$							
$t_w, °\mathrm{C}$		0	20	$\to 40$	60	80	100
$t_w - t_\infty, °\mathrm{C}$		−40	−20	$\to 0$	20	0	60
$-\left(\frac{\mathrm{d}\theta(\eta)}{\mathrm{d}\eta}\right)_{\eta=0}$	a	0.698560	0.742925	0.782495	0.817854	0.847823	0.874393
	b	0.691847	0.727976	0.765993	0.805994	0.848085	0.892373
	c	−0.0096	−0.0201	−0.0211	−0.0145	0.00030	0.02056
$t_\infty = 60°\mathrm{C}$, $\mathrm{Pr}_\infty = 3$							
$t_w, °\mathrm{C}$		0	20	40	$\to 60$	80	100
$t_w - t_\infty, °\mathrm{C}$		−60	−40	−20	$\to 0$	20	40
$-\left(\frac{\mathrm{d}\theta(\eta)}{\mathrm{d}\eta}\right)_{\eta=0}$	a	0.599352	0.631513	0.661232	0.687495	0.710692	0.731406
	b	0.589412	0.617356	0.646624	0.67728	0.70939	0.743022
	c	−0.0166	−0.0224	−0.0221	−0.0149	−0.0018	0.0159
$t_\infty = 80°\mathrm{C}$, $\mathrm{Pr}_\infty = 2.23$							
$t_w, °\mathrm{C}$		0	20	40	60	$\to 80$	100
$t_w - t_\infty, °\mathrm{C}$		−80	−60	−40	−20	$\to 0$	20
$-\left(\frac{\mathrm{d}\theta(\eta)}{\mathrm{d}\eta}\right)_{\eta=0}$	a	0.535462	0.559149	0.581670	0.602019	0.620184	0.636701
	b	0.519141	0.541279	0.564362	0.588428	0.613521	0.639685
	c	−0.0305	−0.0320	−0.0298	−0.0226	−0.0107	0.0047

Table 8.2 (continued)

$t_\infty = 100°C$, $Pr_\infty = 1.76$							
t_w, °C		0	20	40	60	80	→ 100
$t_w - t_\infty$, °C		−100	−80	−60	−40	−20	→ 0
$-\left(\frac{d\theta(\eta)}{d\eta}\right)_{\eta=0}$	a	0.492895	0.510384	0.527594	0.543529	0.558004	0.571331
	b	0.470746	0.488588	0.507106	0.526325	0.546273	0.566977
	c	−0.0449	−0.0427	−0.0388	−0.0317	−0.0210	−0.0076

$$-\left(\frac{d\theta(\eta)}{d\eta}\right)_{\eta=0} = \left[\left(-\frac{d\theta(\eta)}{d\eta}\right)_{\eta=0}\right]_{(t_w - t_\infty)\to 0} \exp[(-0.114 \times t_\infty + 30) \times 10^{-4} \times (t_w - t_\infty)], \tag{8.10}$$

where $\left[\left(-\frac{d\theta(\eta)}{d\eta}\right)_{\eta=0}\right]_{(t_w - t_\infty)\to 0}$ as the reference dimensionless temperature gradient for ignoring variable physical properties is equivalent to Pohlhausen equation [4] with the reference Prandtl number Pr_∞, i.e.

$$\left[\left(-\frac{d\theta(\eta)}{d\eta}\right)_{\eta=0}\right]_{(t_w - t_\infty)\to 0} = 0.4696\, {Pr_\infty}^{1/3}, \tag{8.11}$$

where Pr_∞ is reference Prandtl number at bulk temperature as reference temperature. Here, it is seen that (8.10) for consideration of variable thermophysical properties is the spread of the Poulhausen equation. Therefore, consideration of variable thermophysical properties is very necessary and important for reliable and practical investigation of heat transfer on liquid forced convection.

With (8.11), (8.10) becomes

$$\left(-\frac{d\theta(\eta)}{d\eta}\right)_{\eta=0} = 0.4696\, {Pr_\infty}^{1/3} \times \exp[(-0.114 \times t_\infty + 30) \times 10^{-4} \times (t_w - t_\infty)]. \tag{8.12}$$

A system of the wall dimensionless temperature gradients $\left(-\frac{d\theta(\eta)}{d\eta}\right)_{\eta=0}$ of water laminar forced convection on a horizontal flat plate is evaluated by using (8.12) with variation of the bulk Prandtl number, as well as the boundary temperatures t_w and t_∞, and listed in Table 8.2 also. Meanwhile, for the reference Prandtl number Pr_∞, the following data are recommended with the reference temperature t_∞ for water shown as in Table 8.3.

It is seen that the evaluated values of $\left(-\frac{d\theta(\eta)}{d\eta}\right)_{\eta=0}$ are pretty well coincident to the related numerical solutions. Then, it follows that (8.12) is valid for simple and reliable prediction of the wall dimensionless temperature gradients $\left(-\frac{d\theta(\eta)}{d\eta}\right)_{\eta=0}$

Table 8.3 water reference Prandtl number Pr_∞ with the reference temperature t_∞

t_∞, °C	Pr_∞	t_∞, °C	Pr_∞
0	13.44	60	3
10	9.42	70	2.57
20	6.99	80	2.23
30	5.42	90	1.97
40	4.34	100	1.76
50	3.57		

of water laminar forced convection on a horizontal flat plate with consideration of coupled effect of variable physical properties.

8.8 Prediction Equations on Heat Transfer

Combined with (8.12), heat transfer (7.30), (7.31), (7.32), (7.33), (7.34), (7.35), and (7.36) will be the available equations for simple and reliable prediction of heat transfer of water laminar forced convection on horizontal flat plate. For example, (7.32) and (7.35) will become

$$Q_{x,w} = 0.664b\lambda_w(t_w - t_\infty)\mathrm{Re}_{x,\infty}^{1/2}\mathrm{Pr}_\infty^{1/3}\exp[(-0.114 \times t_\infty + 30) \times 10^{-4} \times (t_w - t_\infty)]. \tag{8.13}$$

$$\overline{Nu_{x,w}} = 0.664\mathrm{Re}_{x,\infty}^{1/2}\mathrm{Pr}_\infty^{1/3}\exp[(-0.114 \times t_\infty + 30) \times 10^{-4} \times (t_w - t_\infty)]. \tag{8.14}$$

It is easy to know that the accuracy of (8.13) and (8.14) for prediction of heat transfer of water laminar forced convection on a horizontal flat plate with consideration of the coupled effect of variable physical properties depends on reliable (8.12), are reliable.

From (8.13) and (8.14) it is clear that heat transfer on liquid laminar forced convection is related to the local Reynolds number, bulk Prandtl number, as well as the boundary temperature conditions. However, the effect of variable physical properties on heat transfer seems mainly to be dominated by the boundary temperature conditions, because the variable physical properties are dependent on temperature.

8.9 Summary

Now, it is time to summarize in Table 8.4 the equations on dimensional similarity variables, governing partial differential equations, governing similarity ordinary differential equations, and related heat transfer equations of gas and liquid laminar forced convection on a horizontal flat plate, which are based on the present similarity analysis method and consideration of variable physical properties.

Table 8.4 Equations on dimensional similarity variables, governing partial differential equations, governing similarity ordinary differential equations, and related heat transfer equations of gas and liquid laminar forced convection on a horizontal flat plate, which are based on the present similarity analysis method and consideration of variable physical properties

Governing equations of laminar forced convection on a horizontal flat plate with consideration of variable physical properties	
Governing partial differential equations	
Mass equation	$\frac{\partial}{\partial x}(\rho w_x) + \frac{\partial}{\partial y}(\rho w_y) = 0$
Momentum equation	$\rho\left(w_x \frac{\partial w_x}{\partial x} + w_y \frac{\partial w_x}{\partial y}\right) = \frac{\partial}{\partial y}\left(\mu \frac{\partial w_x}{\partial y}\right)$
Energy equation	$\rho\left[w_x \frac{\partial (c_p t)}{\partial x} + w_y \frac{\partial (c_p t)}{\partial y}\right] = \frac{\partial}{\partial y}\left(\lambda \frac{\partial t}{\partial y}\right)$
Boundary conditions	$y = 0: \quad w_x = 0, \quad w_y = 0, \quad t = t_w$ $y \to \infty \quad w_x = w_{x,\infty}$ (constant), $t = t_\infty$
Equations on dimensionless similarity variables	
Local Reynolds number $\mathrm{Re}_{x,\infty}$	$\mathrm{Re}_{x,\infty} = \frac{w_{x,\infty} x}{\nu_\infty}$
Coordinate variable η	$\eta = \frac{y}{x}\left(\frac{1}{2}\mathrm{Re}_{x,\infty}\right)^{1/2}$
Temperature θ	$\theta = \frac{t - t_\infty}{t_w - t_\infty}$
Velocity component w_x	$w_x = w_{x,\infty} \cdot W_x(\eta)$
Velocity component w_y	$w_y = w_{x,\infty}\left(\frac{1}{2}\mathrm{Re}_{x,\infty}\right)^{-1/2} W_y(\eta)$
Governing ordinary differential equations	
Mass equation	$-\eta \frac{\mathrm{d}W_x(\eta)}{\mathrm{d}\eta} + 2\frac{\mathrm{d}W_y(\eta)}{\mathrm{d}\eta} + \frac{1}{\rho}\frac{\mathrm{d}\rho}{\mathrm{d}\eta}[-\eta \cdot W_x(\eta) + 2W_y(\eta)] = 0$
Momentum equation	$\frac{\nu_\infty}{\nu}(-\eta W_x(\eta) + 2W_y(\eta))\frac{\partial W_x(\eta)}{\partial \eta} = \frac{\partial^2 W_x(\eta)}{\partial \eta^2} + \frac{1}{\mu}\frac{\partial \mu}{\partial \eta}\frac{\partial W_x(\eta)}{\partial \eta}$
Energy equation	$\mathrm{Pr}\frac{\nu_\infty}{\nu}\left[(-\eta W_x(\eta) + 2W_y(\eta))\frac{\mathrm{d}\theta(\eta)}{\mathrm{d}\eta} + (-\eta W_x(\eta) + 2W_y(\eta))\left(\theta(\eta) + \frac{t_\infty}{t_w - t_\infty}\right)\frac{1}{c_p}\frac{\mathrm{d}c_p}{\mathrm{d}\eta}\right] = \frac{\mathrm{d}^2\theta(\eta)}{\mathrm{d}\eta^2} + \frac{1}{\lambda}\frac{\partial \lambda}{\partial \eta}\frac{\mathrm{d}\theta(\eta)}{\mathrm{d}\eta}$
Boundary conditions	$\eta = 0: \quad W_x(\eta) = 0,\ W_y(\eta) = 0,\ \theta(\eta) = 1$ $\eta \to \infty: \quad W_x(\eta) = 1, \quad \theta(\eta) = 0$
Physical property factors of the similarity governing mathematical model	
Related to gas laminar forced convection	$\frac{1}{\rho}\frac{\mathrm{d}\rho}{\mathrm{d}\eta} = -\frac{(T_w/T_\infty - 1)\mathrm{d}\theta(\eta)/\mathrm{d}\eta}{(T_w/T_\infty - 1)\theta(\eta) + 1}$ $\frac{1}{\mu}\frac{\mathrm{d}\mu}{\mathrm{d}\eta} = \frac{n_\mu(T_w/T_\infty - 1)\mathrm{d}\theta(\eta)/\mathrm{d}\eta}{(T_w/T_\infty - 1)\theta(\eta) + 1}$ $\frac{1}{\lambda}\frac{\mathrm{d}\lambda}{\mathrm{d}\eta} = \frac{n_\lambda(T_w/T_\infty - 1)\mathrm{d}\theta(\eta)/\mathrm{d}\eta}{(T_w/T_\infty - 1)\theta(\eta) + 1}$ $\frac{1}{c_p}\frac{\mathrm{d}c_p}{\mathrm{d}\eta} = \frac{n_{c_p}(T_w/T_\infty - 1)\mathrm{d}\theta(\eta)/\mathrm{d}\eta}{(T_w/T_\infty - 1)\theta(\eta) + 1}$ $\left(\frac{1}{c_p}\frac{\mathrm{d}c_p}{\mathrm{d}\eta}\right.$ can be regarded as zero for general gases$\left.\right)$ $\frac{\nu_\infty}{\nu} = [(T_w/T_\infty - 1)\theta(\eta) + 1]^{-(n_\mu + 1)}$ $\mathrm{Pr} = \mathrm{Pr}_\infty[(T_w/T_\infty - 1)\theta(\eta) + 1]^{n_\mu - n_\lambda + n_{c_p}}$

Table 8.4 (continued)

Related to liquid laminar forced convection	For water laminar forced convection $\Pr = \frac{\mu \cdot c_p}{\lambda} = \frac{\left[\exp\left(-1.6 - \frac{1150}{T} + \left(\frac{690}{T}\right)^2\right) \times 10^{-3}\right] \times 4200}{-8.01\times10^{-6}t^2+1.94\times10^{-3}t+0.563}$ $\frac{1}{\rho}\frac{d\rho}{d\eta} = \frac{(-2\times4.48\times10^{-3}t)(t_w - t_\infty)\frac{d\theta(\eta)}{d\eta}}{-4.48\times10^{-3}t^2+999.9}$ $\frac{1}{\lambda}\frac{d\lambda}{d\eta} = \frac{(-2\times8.01\times10^{-6}t+1.94\times10^{-3})(t_w - t_\infty)\frac{d\theta(\eta)}{d\eta}}{-8.01\times10^{-6}t^2+1.94\times10^{-3}t+0.563}$ $\frac{1}{\mu}\frac{d\mu}{d\eta} = \left(\frac{1150}{T^2} - 2\times\frac{690^2}{T^3}\right)(T_w - T_\infty)\frac{d\theta(\eta)}{d\eta}$ $\frac{\nu_\infty}{\nu} = \frac{\exp\left[-1.6 - \frac{1150}{T_\infty} + \left(\frac{690}{T_\infty}\right)^2\right]}{\exp\left[-1.6 - \frac{1150}{T} + \left(\frac{690}{T}\right)^2\right]} \times \frac{-4.48\times10^{-3}t^2+999.9}{-4.48\times10^{-3}t_\infty^2+999.9}$ Where $t = (t_w - t_\infty)\cdot\theta(\eta) + t_\infty \quad T = t + 273$
Skin-friction coefficient	
Local skin-friction coefficient $C_{f,x}$	$C_{x,w} = \sqrt{2}\frac{\nu_w}{\nu_\infty}(\mathrm{Re}_{x,\infty})^{-1/2}\left(\frac{dW_x(\eta)}{d\eta}\right)_{\eta=0}$
Average skin-friction coefficient $\overline{C}_{x,f}$	$\overline{C}_{x,w} = 2\times\sqrt{2}\frac{\nu_w}{\nu_\infty}(\mathrm{Re}_{x,\infty})^{-1/2}\left(\frac{dW_x(\eta)}{d\eta}\right)_{\eta=0}$
Equations on heat transfer	
Wall dimensionless temperature gradient $\left(-\frac{d\theta(\eta)}{d\eta}\right)_{\eta=0}$	For gas laminar forced convection on a horizontal flat plate $\left(-\frac{d\theta(\eta)}{d\eta}\right)_{\eta=0} = 0.4696\Pr_f^{1/3}\cdot\left(\frac{T_w}{T_\infty}\right)^{-m}$ $m = 0.8419n_{\mu\lambda} + 0.2897 \quad (0.3 \le T_w/T_\infty \le 1)$ $m = 0.725n_{\mu\lambda} + 0.3833 \quad (1 \le T_w/T_\infty \le 3)$ $n_{\mu\lambda} = 0.45n_\mu + 0.55n_\lambda$ For water laminar forced convection on a horizontal flat plate $\left(-\frac{d\theta(\eta)}{d\eta}\right)_{\eta=0} = 0.4696\Pr_\infty^{1/3}\times\exp[(-0.114\times t_\infty + 30)\times10^{-4}\times(t_w - t_\infty)]$
Local values on heat transfer $Nu_{x,w} = \frac{\alpha_{x,w}\cdot x}{\lambda_w}$	$q_{x,w} = \lambda_w(t_w - t_\infty)\left(\frac{1}{2}\mathrm{Re}_{x,\infty}\right)^{1/2}x^{-1}\left(-\frac{d\theta(\eta)}{d\eta}\right)_{\eta=0}$ $Nu_{x,w} = \left(\frac{1}{2}\mathrm{Re}_{x,\infty}\right)^{1/2}\left(-\frac{d\theta(\eta)}{d\eta}\right)_{\eta=0}$
Average values on heat transfer $\overline{Nu}_{x,f} = \frac{\overline{\alpha}_{x,f}\cdot x}{\lambda_f}$	$\overline{Q}_{x,w} = \sqrt{2}x^{-1}\lambda_w(t_w - t_\infty)\mathrm{Re}_{x,\infty}^{1/2}\left(-\frac{d\theta(\eta)}{d\eta}\right)_{\eta=0}$ $\overline{Nu}_{x,w} = -\sqrt{2}(\mathrm{Re}_{x,\infty})^{1/2}\left(\frac{d\theta(\eta)}{d\eta}\right)_{\eta=0}$
Total value on heat transfer	$Q_{x,w} = \sqrt{2}b\lambda_w(t_w - t_\infty)\mathrm{Re}_{x,\infty}^{1/2}\left(-\frac{d\theta(\eta)}{d\eta}\right)_{\eta=0}$

8.10 Remarks

Following the study in Chap. 7 on heat transfer of gas laminar forced convection for consideration of variable physical properties, in this chapter, heat transfer on liquid laminar forced convection is further investigated with consideration of variable physical properties. Meanwhile, water laminar forced convection on a horizontal flat plate was taken as an example for the extensive study. By means of the formulated equations of temperature dependent density, thermal conductivity, and absolute viscosity, the set of the physical property factors are transformed to the dimensionless temperature based on the new similarity analysis method. Then, the developed completely dimensionless system with the governing ordinary differential equations are conveniently coupled with the set of the transformed dimensionless physical property factors, and solved with the temperature boundary conditions.

The coupled effect of the variable physical properties, such as density, thermal conductivity, and viscosity on laminar liquid forced convection is reflected by the influence of the boundary temperatures, because these physical properties are temperature-dependent. For special bulk temperature t_∞ (i.e. special Pr_∞), increasing the boundary temperature difference $t_w - t_\infty$ (i.e. increasing t_w or decreasing Pr_w), causes increasing the velocity field level and the wall temperature gradient. In addition, for special wall temperaturet_w (i.e. special Pr_w), increasing the boundary temperature difference $t_w - t_\infty$ (i.e. decreasing t_∞ or increasing Pr_∞) causes increasing the velocity field level and the wall temperature gradient. All these phenomena demonstrate the coupled effect of variable physical properties on liquid laminar forced convection.

Generally, increasing the boundary temperature difference $t_w - t_\infty$ for liquid laminar forced convection causes increase of the wall temperature gradient, which is different from that for gas laminar forced convection, where increasing the boundary temperature ratio T_w/T_∞ causes decrease of the wall temperature gradient. Such reason is based on the different variation regulations between the liquid and gas viscosities with temperature.

Only the solution on case $t_w - t_\infty \to 0$ for liquid laminar forced convection with consideration of variable physical properties is identical to Blasius solution on the velocity field, and Pohlhausen equation on heat transfer. It demonstrates the limitations for consideration of constant physical properties for investigation of convection heat transfer.

The temperature boundary layer is much narrower than the related velocity boundary layer for liquid laminar forced convection, and increasing the bulk Prandtl number Pr_∞ causes decreasing the temperature boundary layer thickness.

The system of key solutions on the wall dimensionless temperature gradient for water laminar free convection on a horizontal flat plate was formulated by using a curve-fitting approach, and the formulated equation is a function of the bulk Prandtl number, as well as the boundary temperatures t_w and t_∞, which demonstrates the coupled effect of the variable thermophysical properties. The formulated equation on the wall dimensionless temperature gradient is so reliable that its predicted results are pretty well coincident with the related rigorous numerical solutions.

The theoretical equations for heat transfer rate, heat transfer coefficient, and Nessult number are reported based on the heat transfer analysis with the new similarity system method. In these theoretical equations, the only unknown variable is the wall dimensionless temperature gradient. Combined with the reliable formulated equations on the wall dimensionless temperature gradient, these analysis equations are available for simple and reliable prediction on heat transfer of water laminar forced convection. These prediction equations on heat transfer show that heat transfer for liquid laminar forced convection with consideration of variable thermophysical properties depend on local Reynolds number and local Prandtl number, as well as the boundary temperature t_w and t_∞, while heat transfer without consideration of thermophysical properties is only its special case with $t_w - t_\infty \to 0$.

Obviously, the analysis presented here extends the previous studies for liquid laminar forced convection.

8.11 Calculation Examples

Example 1 A flat plate with $b = 1\,\mathrm{m}$ in width and $x = 0.25\,\mathrm{m}$ in length is horizontally located in water flowing with the velocity $w_{x,\infty} = 1.3\,\mathrm{m/s}$. The surface temperature of the plate is $t_w = 70°\mathrm{C}$, and water temperature is $t_\infty = 10°\mathrm{C}$. Please solve the following questions:

(1) evaluate the related local Reynolds number for determination the flow types (laminar or turbulent flows);
(2) calculate the related Eckert number;
(3) determine the related heat transfer prediction deviation caused by ignoring the viscous thermal dissipation;
(4) calculate the related heat transfer results with consideration of coupled effect of the variable thermophysical properties;
(5) evaluate the heat transfer prediction deviation caused by ignoring the variable thermophysical properties.

Solution 1: The boundary layer average temperature $t_f = \frac{t_w + t_\infty}{2} = \frac{70+10}{2} = 40°\mathrm{C}$. The related physical properties are listed in Table 8.4.

Figure 8.4 Water physical properties

Temperature	t_w, °C		t_∞, °C.		t_f, °C			
	70		10		40			
Physical Properties	λ_w W/(m °C)	$\nu_w \times 10^{-6}$ m^2/s	$\nu_\infty \times 10^{-6}$ m^2/s	Pr_∞	Pr_f	c_p kJ/(kg °C)	$\nu_f \times 10^{-6}$ m^2/s	λ_f W/(m °C)
	0.662	0.4014	1.2996	9.42	4.32	4179	0.6564	0.63

Local Reynolds number with ignoring variable physical properties for water is evaluated as

$$\mathrm{Re}_{x,f} = \frac{w_{x,\infty}x}{\nu_f} = \frac{1.3 \times 0.25}{0.6564 \times 10^{-6}} = 495124.9 < 5\times 10^6.$$ Then, it can be regarded as laminar flow.

Solution 2: The Eckert number is evaluated as

$$Ec = \frac{{w_{x,\infty}}^2}{c_{p_f}(t_w - t_\infty)} = \frac{1.3^2}{4179 \times 60} = 6.74 \times 10^{-6}$$

With such small Eckert number, the effect of viscous thermal dissipation on heat transfer will be very small, and can be ignored.

Solution 3: With $\mathrm{Ec} = 6.74 \times 10^{-6}$ and $\mathrm{Pr}_f = 4.32$, the heat transfer prediction deviation caused by ignoring the viscous thermal dissipation can be evaluated as

$$E_{(Q_{x,f})Ec=0} = E_{(\overline{Nu}_{x,f})Ec=0} = -\frac{b \cdot Ec^2 + c \cdot Ec}{0.4696\,{\mathrm{Pr}_f}^{1/3}},$$

with

$$\mathrm{b} = (-0.0007\mathrm{Pr_f} + 0.002){\mathrm{Pr_f}}^{1.75} \text{ and } \mathrm{c} = (-0.2027\,\mathrm{Pr_f} - 0.0929)/{\mathrm{Pr_f}}^{0.15}.$$

Then

$$\mathrm{b} = (-0.0007 \times 4.32 + 0.002) \times 4.32^{1.75} = -0.01326,$$
$$\mathrm{c} = (-0.2027 \times 4.32 - 0.0929)/4.32^{0.15} = -0.77769,$$
$$E_{(Q_{x,f})Ec=0} = -\frac{-0.01326 \times (6.74 \times 10^{-6})^2 - 0.77769 \times 6.74 \times 10^{-6}}{0.4696 \times 4.32^{1/3}}$$
$$= 6.85 \times 10^{-6}.$$

It is seen that the effect of viscous thermal dissipation on heat transfer is very small in this case.

Solution 4: The local Reynolds number with consideration of variable physical properties will be

$$\mathrm{Re}_{x,\infty} = \frac{w_{x,\infty}x}{\nu_\infty} = \frac{1.3 \times 0.25}{1.2996 \times 10^{-6}} = 250076.9.$$

With (8.13), the heat transfer result of water laminar forced convection on the horizontal plate with consideration of coupled effect of the variable thermophysical properties is evaluated by the following equation:

$$Q_{x,w} = 0.664b\lambda_w(t_w - t_\infty)\mathrm{Re}_{x,\infty}^{1/2}\mathrm{Pr}_\infty^{1/3}\exp[(-0.114 \times t_\infty + 30) \times 10^{-4} \times (t_w - t_\infty)],$$

i.e.

$$Q_{x,w} = 0.664 \times 1 \times 0.662 \times 60 \times 250076.9^{1/2} \times 9.42^{1/3}\exp[(-0.114 \times 10 + 30) \times 10^{-4} \times (70 - 10)] = 33120.75\,\mathrm{W}.$$

Solution 5: With (5.13), heat transfer result of water laminar forced convection on a horizontal flat plate with ignoring the variable physical properties and viscous thermal dissipation is evaluated as

$$[Q_{x,f}]_{Ec=0} = 0.664b\lambda_f(t_w - t_\infty)\mathrm{Re}_{x,f}^{1/2}\mathrm{Pr}_f^{1/3} \quad (0.3 \le \mathrm{Pr}_f \le 100). \quad (5.13)$$

Then

$$[Q_{x,f}]_{Ec=0} = 0.664 \times 1 \times 0.63 \times 60 \times 495124.9^{1/2} \times 4.32^{1/3},$$

i.e.

$$[Q_{x,f}]_{Ec=0} = 28763.7\,\mathrm{W}.$$

Heat transfer prediction deviation caused by ignoring the variable thermophysical properties is evaluated as

$$\frac{[Q_{x,f}]_{Ec=0} - Q_{x,w}}{Q_{x,w}} = \frac{28763.7 - 33120.75}{28763.7} = -0.1515 = -15.15\%.$$

Such evaluated result for heat transfer prediction deviation caused by ignoring the variable physical properties reveals that the coupled effect of variable physical properties on forced convection heat transfer is so obvious that it can never be ignored.

Example 2 Continued to example 1, and solve the following questions:

(1) calculate the average skin-friction coefficient with consideration of variable physical properties;
(2) calculate the average skin-friction coefficient with ignoring variable physical properties.

Solution 1: With (7.29) the average skin-friction coefficient with consideration of variable physical properties is

$$\overline{C}_{x,w} = 2 \times \sqrt{2}\frac{\nu_w}{\nu_\infty}(\mathrm{Re}_{x,\infty})^{-1/2}\left(\frac{\mathrm{dW}_x(\eta)}{\mathrm{d}\eta}\right)_{\eta=0}$$

From example 1, we have $\mathrm{Re}_{x,\infty} = 250076.9$

For $t_w = 70°\mathrm{C}$ and $t_\infty = 10°\mathrm{C}$, from Fig. 7.6 we have

$\nu_w = 0.4014 \times 10^{-6}$, $\nu_\infty = 1.2996 \times 10^{-6}$.

With $t_w = 70°\mathrm{C}$ and $t_\infty = 10°\mathrm{C}$, by using interpolation approach we have the following value of $\left(\frac{\mathrm{dW}_x(\eta)}{\mathrm{d}\eta}\right)_{\eta=0}$ from Table 8.1:

$$\left(\frac{\mathrm{dW}_x(\eta)}{\mathrm{d}\eta}\right)_{\eta=0} = 1.25.$$

Then, for consideration of variable physical properties, the average skin friction is

$$\begin{aligned}\overline{C}_{x,w} &= 2 \times \sqrt{2} \times \frac{0.4014}{1.2996} \times 250076.9^{-1/2} \times 1.25 \\ &= 0.002184.\end{aligned}$$

Solution 2: From (5.4), we have the equation of the average skin friction with ignoring variable physical properties for laminar forced convection on an horizontal flat plate, i.e.

$$\overline{C}_{x,f} = 1.328\,\mathrm{Re}_{x,f}^{-1/2}. \tag{5.4}$$

Then

$$\overline{C}_{x,f} = 1.328 \times 495124.9^{-1/2} = 0.001887.$$

Then the relative deviation of the average skin-friction with ignoring variable physical properties to that with consideration of variable physical properties is

$$\frac{\overline{C}_{x,f} - \overline{C}_{x,w}}{\overline{C}_{x,f}} = \frac{0.001887 - 0.002184}{0.001887} = -0.1574 = -15.74\%$$

Exercises

1. Can you give an appropriate explanation on the necessity of consideration of effect of variable physical properties on liquid laminar forced convection?
2. Based on the general similarity governing mathematical model provided in this chapter, if you are interested in liquid laminar forced convection on a horizontal flat plate with consideration of coupled effect of variable physical properties, you can select any other liquid to investigate reliable heat transfer equations like (8.13) and (8.14) for their reliable application.

3. Based on the analysis approach provided in Chaps. 7 and 8 for derivation of the similarity analysis governing mathematical model of laminar forced convection on a horizontal flat plate, try to investigate the identical similarity analysis governing mathematical model of laminar forced convection with consideration of variable physical properties on an inclined flat plate with any inclined angle α.
4. Follow Example 1, the bulk velocity is replaced by 2.6 m/s, the wall temperature is replaced by 80°C, but all other conditions are kept. Please solve the following:

   (1) evaluate the related local Reynolds number for determination the flow types (laminar or turbulent flows);
   (2) calculate the related Eckert number;
   (3) determine the related heat transfer prediction deviation caused by ignoring the viscous thermal dissipation;
   (4) calculate the related heat transfer results with consideration of coupled effect of the variable physical properties;
   (5) evaluate the heat transfer prediction deviation caused by ignoring the variable physical properties.
   (6) calculate the average skin-friction coefficient with consideration of variable physical properties;
   (7) calculate the average skin-friction coefficient by ignoring variable physical properties.

References

1. D.Y. Shang, B.X. Wang, Y. Wang ,Y. Quan, Study on liquid laminar free convection with consideration of variable thermophysical properties. Int. J. Heat Mass Transfer **36**(14), 3411–3419 (1993)
2. D.Y. Shang, *Free Convection Film Flows and Heat Transfer* (Springer, Berlin, Heidelberg and New York, NY, 2006)
3. H. Blasius, Grenzschichten in Flussigkeiten mit kleiner Reibung. Z. Math. Phys. **56**, 1–37 (1908)
4. E. Pohlhausen, Der Wärmeaustausch zwischen festen Körpern und Flüssigkeiten mit kleiner Reibung und kleiner Wärmeleitung. Z. Angew. Math. Mech. **1**, 115–121 (1921)

Part III
Laminar Forced Film Condensation

Chapter 9
Complete Similarity Mathematical Models on Laminar Forced Film Condensation of Pure Vapour

Abstract The new similarity analysis method is successfully applied in this work, and a complete similarity mathematical model is developed on the laminar forced film condensation of superheated and saturated vapours with consideration of variable physical properties. The laminar forced convection film condensation (for short, the laminar forced film condensation, the same below) with two-phase film flow consists of the condensate liquid film and vapour film, and in both film flows, the variable physical properties are considered. For model of the condensate liquid film, the determination of the similarity variables and consideration of the variable physical properties are based on the approach in Chap. 8 for liquid laminar forced convection, while for the model of the vapour film the proposal of the similarity variables and consideration of the variable physical properties are based on Chap. 7 for gas laminar forced convection. The two models are combined with the five interfacial matching balance equations, respectively describing the velocity component continuity, mass flow rate continuity, the balance of shear force, the energy balance, and the temperature continuity. Based on the new similarity analysis method, the developed system of complete similarity dimensionless model provides convenience for treatment of variable physical properties and solution.

9.1 Introduction

Since the pioneer scientist Nusselt published his research [1] on film condensation, numerous studies have followed for laminar free or forced film condensation. For the laminar forced film condensation, only some of them, such as in [2–15] are listed here due to the space limitation. Meanwhile, Cess [2] treated the case where no body force was present and the free stream velocity is very large, and concluded that the resulting interfacial shear stress dominates the heat transfer. Koh [3] developed a complete numerical solution on the basis of the two-phase boundary-layer equations, using similarity methods. Shekriladze and Gomelauri [4] proposed unique approximate solutions for various cases of condensation, with an assumption that the local shearing stress at the interface of vapour and liquid is equal to the momentum given up by condensing vapour. Using an approximate integral method, Jakobs [5] obtained good agreement with his experimental results for Freon 113 at atmospheric

D.Y. Shang, *Theory of Heat Transfer with Forced Convection Film Flows*, Heat and Mass Transfer, DOI 10.1007/978-3-642-12581-2_9,

pressure. By using an assuming that vapour outside its boundary is potential flow, Denny and Mills [6] adopted the two-phase boundary layer equations. Beckett and Poots [7] presented the detailed attempt to study laminar film condensation in forced flows in the context of planar as well as curvilinear geometries. Fujii and Uehara [8] solved two-phase boundary layer equations of laminar film condensation with an approximate method due to Jacobs. Fujii et al. [9] studied the laminar film condensation of flowing vapour on a horizontal cylinder. Rose [10] developed theoretical equations related to the mass flux of vapour to the condensing surface. Rose [11] found that when $\frac{\rho_v w_{x,\infty}{}^2}{\rho_l g d} < \frac{1}{8}$, the mean heat transfer for the tube is very close to that found when the pressure gradient is neglected. Honda and Fujii [12] have carried out conjugate numerical solutions by using a constant inside coolant coefficient and solving the two-dimensional conduction problem in the tube wall. Rose [13] provided a new interpolation formula for forced-convection condensation on a horizontal surface. Memory et al. [14] present the effects of vapour shear, pressure gradient, and surface tension for the laminar condensation on horizontal elliptical tubes. F. Méndez et al. [15] studied the forced condensation process of a saturated vapour in contact with one upper surface of a thin plate, taking into account the finite thermal conductivity of the plate. From them, few studies, such as [3, 6, 8] used the two-phase boundary layer equations to analyze the heat and mass transfer of the laminar forced film condensation.

However, in previous studies for dealing with the governing partial differential equations of the laminar forced convection and film condensation, the traditional Falkner-Skan type transformation has been universally used. However, with this transformation method it is difficult to conveniently analyze and treat the various important issues, such as velocity field, interfacial physical balance conditions, condensate mass transfer, and the variable physical properties for the laminar forced convection and film condensation. All these difficult treatments caused a difficulty for obtaining heat and mass transfer results. Therefore, so far, there still has been a lack of a complete similarity model developed to provide the rigorous analysis and numerical results of the two-phase flow films equations for the rigorous velocity and temperature distributions, as well as condensate heat mass transfer on the forced film condensation.

In this case, the new similarity analysis method reported in Chap. 4 and successfully applied in Chaps. 5, 6, 7, and 8, respectively, for extensive investigations on laminar forced convection will be used for resolving these difficult points on the study of laminar forced condensation. First, the new similarity analysis method will be applied for development of the corresponding complete similarity dimensionless mathematical model. For this purpose, a series of detailed derivations will be done for similarity transformation under the new similarity analysis system. Meanwhile, with the effective approach on treatment of variable physical properties described in Chaps. 7 and 8, respectively, for laminar forced convection of gases and liquids, the system of the physical properties factor will be similarly transformed to the function of the dimensionless temperature based on the new similarity analysis model. Then, a system of complete governing similarity mathematical model will be obtained through a series of rigorously derivation with consideration of temperature-dependent physical properties both for liquid and for vapour films. On this basis, condensate heat and mass transfer will be deeply investigated based on the theoretical heat and mass transfer analysis and the system of rigorous numerical solutions on two-phase film flows velocity and temperature fields.

9.2 Governing Partial Differential Equations

9.2.1 Physical Model and Coordinate System

The present work is focused on the laminar forced film condensation of vapour on an isothermal horizontal flat plate. The analytical model and coordinate system are shown in Fig. 9.1. An isothermal flat plate is located horizontally in a vapour flow at atmospheric pressure. The plate temperature is t_w, the temperature and velocity of the vapour bulk are t_∞ and $w_{x,\infty}$, and the vapour condensate saturated temperature is t_s. If given condition for the model is $t_w < t_s$, a steady film condensation will occur on the plate. We assume that the laminar flows respectively within the liquid and vapour films are induced by the shear force at the liquid–vapour interface. In addition, we take into account the temperature-dependent physical properties of the condensate liquid film and vapour film. Then, the steady laminar governing partial differential equations for mass, momentum, and energy in the two-phase boundary layers are as follows:

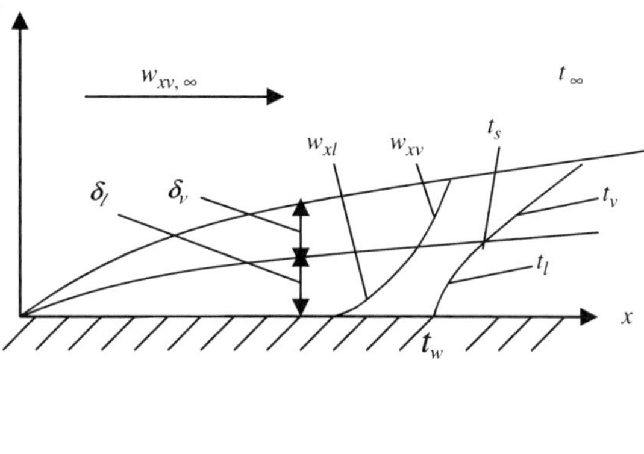

Fig. 9.1 Physical model and coordinate system of the laminar forced film condensation from vapour (δ_l and δ_v denote liquid and vapour film thicknesses, respectively)

9.2.2 Governing Partial Differential Equations

Consulting (7.1), (7.2), and (7.3) for governing partial differential equations of laminar forced convection, the governing partial differential equations for condensate liquid and vapour films of laminar forced film condensation of vapour are expressed as follows with consideration of variable physical properties:

9.2.2.1 For Liquid Film

The governing partial differential equations for condensate liquid film are

$$\frac{\partial}{\partial x}(\rho_l w_{xl}) + \frac{\partial}{\partial y}(\rho_l w_{yl}) = 0, \tag{9.1}$$

$$\rho_l \left(w_{xl} \frac{\partial w_{xl}}{\partial x} + w_{yl} \frac{\partial w_{xl}}{\partial y} \right) = \frac{\partial}{\partial y} \left(\mu_l \frac{\partial w_{xl}}{\partial y} \right), \tag{9.2}$$

$$\rho_l \left[w_{xl} \frac{\partial (c_{p_l} t)}{\partial x} + w_{yl} \frac{\partial (c_{p_l} t)}{\partial y} \right] = \frac{\partial}{\partial y} \left(\lambda_l \frac{\partial t}{\partial y} \right), \tag{9.3}$$

where (9.1), (9.2), and (9.3) are continuity, momentum, and energy conservation equations, respectively. Here the liquid temperature-dependent physical properties, such as density ρ_l, absolute viscosity μ_l, thermal conductivity λ_l, and specific heat c_{p_l} are considered.

9.2.2.2 For Vapour Film

The governing partial differential equations for vapour film are

$$\frac{\partial}{\partial x}(\rho_v w_{xv}) + \frac{\partial}{\partial y}(\rho_v w_{yv}) = 0, \tag{9.4}$$

$$\rho_v \left(w_{xv} \frac{\partial w_{xv}}{\partial x} + w_{yv} \frac{\partial w_{xv}}{\partial y} \right) = \frac{\partial}{\partial y} \left(\mu_v \frac{\partial w_{xv}}{\partial y} \right), \tag{9.5}$$

$$\rho_v \left(w_{xv} \frac{\partial (c_{p_v} t_v)}{\partial x} + w_{yv} \frac{\partial (c_{p_v} t_v)}{\partial y} \right) = \frac{\partial}{\partial y} \left(\lambda_v \frac{\partial t_v}{\partial y} \right), \tag{9.6}$$

where (9.4), (9.5), and (9.6) are continuity, momentum, and energy conservation equations, respectively. Here the vapour temperature-dependent physical properties, such as density ρ_v, absolute viscosity μ_v, thermal conductivity λ_v, and specific heat c_{p_v} are considered.

9.2.2.3 For Boundary Conditions

The boundary conditions of the laminar forced film condensation of vapour are

$$\mathrm{y} = 0: \quad w_{xl} = 0, \quad w_{yl} = 0, \quad t_l = t_w \tag{9.7}$$

$$y = \delta_l: \quad w_{xl,s} = w_{xv,s} \tag{9.8}$$

$$\rho_{l,s} \left(w_{xl} \frac{\partial \delta_l}{\partial x} - w_{yl} \right)_{l,s} = \rho_{v,s} \left(w_{xv} \frac{\partial \delta_v}{\partial x} - w_{yv} \right)_{v,s} = g_x \tag{9.9}$$

$$\mu_{l,s} \left(\frac{\partial w_{xl}}{\partial y} \right)_s = \mu_{v,s} \left(\frac{\partial w_{xv}}{\partial y} \right)_s \tag{9.10}$$

$$\lambda_{l,s}\left(\frac{\partial t_l}{\partial y}\right)_{y=\delta_l} = \lambda_{v,s}\left(\frac{\partial t_v}{\partial y}\right)_{y=\delta_l} + h_{fg}\rho_{v,s}\left(w_{xv}\frac{\partial \delta_v}{\partial x} - w_{yv}\right)_s, \tag{9.11}$$

$$t = t_s \tag{9.12}$$

$$y \to \infty: \quad w_{xv} = w_{xv,\infty},\ t_v = t_\infty, \tag{9.13}$$

where (9.7) expresses the physical conditions at the plate. Equation (9.8), (9.9), (9.10), (9.11), and (9.12) express the physical matching conditions at the liquid–vapour interface. Here, (9.8) expresses the velocity component continuity, (9.9) expresses the mass flow rate continuity, (9.10) expresses the balance of shear force, (9.11) expresses the energy balance, and (9.12) expresses the temperature continuity. Equation (9.13) denotes the bulk physical conditions.

The set of the governing partial differential equations (9.1), (9.2), (9.3), (9.4), (9.5), and (9.6) with their boundary condition equations (9.7), (9.8), (9.9), (9.10), (9.11), (9.12), and (9.13) are suitable for saturated and superheated bulk vapour temperatures, i.e. $\Delta t_\infty = t_\infty - t_s \geq 0$.

9.3 Similarity Variables

9.3.1 For Liquid Film

Consulting the dimensionless similarity variables reported in Chap. 8 for liquid laminar forced convection with consideration of variable physical properties, the following equations are set up for the related dimensionless similarity variables η_l, $\mathrm{Re}_{xl,s}$, θ_l, $W_{xl}(\eta_l)$, and $W_{yl}(\eta_l)$ of the condensate liquid film on laminar forced film condensation of vapour:

$$\eta_l = \frac{y}{x}\left(\frac{1}{2}\mathrm{Re}_{xl,s}\right)^{1/2}, \tag{9.14}$$

$$\mathrm{Re}_{xl,s} = \frac{w_{xv,\infty}x}{\nu_{l,s}}, \tag{9.15}$$

$$\theta_l(\eta_l) = \frac{t_l - t_s}{t_w - t_s}, \tag{9.16}$$

$$w_{xl} = w_{xv,\infty}W_{xl}(\eta_l), \tag{9.17}$$

$$w_{yl} = w_{xv,\infty}\left(\frac{1}{2}\mathrm{Re}_{xl,s}\right)^{-1/2} W_{yl}(\eta_l), \tag{9.18}$$

where η_l is the dimensionless coordinate variable, $\mathrm{Re}_{xl,s}$ is the local Reynolds number, $\theta_l(\eta_l)$ is the dimensionless temperature, and $W_{xl}(\eta_l)$ and $W_{yl}(\eta_l)$ are the dimensionless velocity components of the condensate liquid film in x and y coordinates, respectively. Here, $\nu_{l,s}$ is the condensate liquid kinetic viscosity at the liquid–vapour interface.

9.3.2 For Vapour Film

For vapour film, consulting the dimensionless similarity variables reported in Chap. 7 for gas laminar forced convection with consideration of variable physical properties, the following equations are set up for the related dimensionless similarity variables η_v, $\mathrm{Re}_{xv,\infty}$, θ_v, $W_{xv}(\eta_l)$, and $W_{yv}(\eta_l)$ on laminar forced film condensation of vapour:

$$\eta_v = \frac{y}{x}\left(\frac{1}{2}\mathrm{Re}_{xv,\infty}\right)^{1/2}, \tag{9.19}$$

$$\mathrm{Re}_{xv,\infty} = \frac{w_{xv,\infty}x}{\nu_{v,\infty}}, \tag{9.20}$$

$$\theta_v(\eta_v) = \frac{t_v - t_\infty}{t_s - t_\infty}, \tag{9.21}$$

$$w_{xv} = w_{xv,\infty}W_{xv}(\eta_v), \tag{9.22}$$

$$w_{yv} = w_{xv,\infty}\left(\frac{1}{2}\mathrm{Re}_{xv,s}\right)^{-1/2} W_{yv}(\eta_v), \tag{9.23}$$

where η_v is the dimensionless coordinate variable, $\mathrm{Re}_{xv,\infty}$ is the local Reynolds number, θ_v is the dimensionless temperature, and $W_{xv}(\eta_v)$ and $W_{yv}(\eta_v)$ are the dimensionless velocity components in x and y coordinates of the vapour film, respectively. Here, $\nu_{xv,\infty}$ is vapour kinetic viscosity in the vapour bulk.

9.4 Similarity Transformation of Governing Partial Differential Equations

9.4.1 For Liquid Film

9.4.1.1 Transformation of (9.1)

Equation (9.1) can be changed to

$$w_{xl}\frac{\partial \rho_l}{\partial x} + w_{yl}\frac{\partial \rho_l}{\partial y} + \rho_l\left(\frac{\partial w_{xl}}{\partial x} + \frac{\partial w_{yl}}{\partial y}\right) = 0, \tag{9.1a}$$

where

$$\frac{\partial w_{xl}}{\partial x} = w_{xv,\infty}\frac{\mathrm{d}W_{xl}(\eta_l)}{\mathrm{d}\eta_l}\cdot\frac{\partial \eta_l}{\partial x},$$

$$\frac{\partial \eta_l}{\partial x} = -\frac{1}{2}\eta_l x^{-1}. \tag{a}$$

With (a) we have

$$\frac{\partial w_{xl}}{\partial x} = -\frac{1}{2}\eta_l x^{-1} w_{xv,\infty} \frac{\mathrm{d}W_{xl}(\eta_l)}{\mathrm{d}\eta_l}. \tag{b}$$

In addition,

$$\begin{aligned}
w_{yl} &= w_{xv,\infty}\left(\frac{1}{2}\mathrm{Re}_{xl,s}\right)^{-1/2} W_{yl}(\eta_l),\\
\frac{\partial w_{yl}}{\partial y} &= w_{xv,s\infty}\left(\frac{1}{2}\mathrm{Re}_{xl,s}\right)^{-1/2} \frac{\mathrm{d}W_{yl}(\eta_l)}{\mathrm{d}\eta_l}\cdot\frac{\partial \eta_l}{\partial y},\\
\frac{\partial \eta_l}{\partial y} &= x^{-1}\left(\frac{1}{2}\mathrm{Re}_{xl,s}\right)^{1/2}.
\end{aligned} \tag{c}$$

With (c) we have

$$\frac{\partial w_{yl}}{\partial y} = w_{xv,s\infty}\frac{\mathrm{d}W_{yl}(\eta_l)}{\mathrm{d}\eta_l}\cdot x^{-1}. \tag{d}$$

Furthermore,

$$\frac{\partial \rho_l}{\partial x} = \frac{\mathrm{d}\rho_l}{\mathrm{d}\eta_l}\frac{\partial \eta_l}{\partial x}.$$

With (a), the above equation becomes

$$\begin{aligned}
\frac{\partial \rho_l}{\partial x} &= -\frac{1}{2}\frac{\mathrm{d}\rho_l}{\mathrm{d}\eta_l}\eta_l x^{-1},\\
\frac{\partial \rho_l}{\partial y} &= \frac{\mathrm{d}\rho_l}{\mathrm{d}\eta_l}\frac{\partial \eta_l}{\partial y}.
\end{aligned} \tag{e}$$

With (c) the above equation becomes

$$\frac{\partial \rho_l}{\partial y} = \frac{\mathrm{d}\rho_l}{\mathrm{d}\eta_l}x^{-1}\left(\frac{1}{2}\mathrm{Re}_{xl,s}\right)^{1/2}. \tag{f}$$

With (9.17), (9.18), (b), (d), (e), (f), (9.1a) is changed to

$$\begin{aligned}
&\rho_l\left[-\frac{1}{2}\eta_l x^{-1} w_{xv,\infty}\frac{\mathrm{d}W_{xl}(\eta_l)}{\mathrm{d}\eta_l} + w_{xv,s\infty}\frac{\mathrm{d}W_{yl}(\eta_l)}{\mathrm{d}\eta_l}\cdot x^{-1}\right]\\
&-\frac{1}{2}\frac{\partial \rho_l}{\partial \eta_l}\eta_l x^{-1} w_{xv,\infty} W_{xl}(\eta_l) + \frac{\mathrm{d}\rho_l}{\mathrm{d}\eta_l}x^{-1}\left(\frac{1}{2}\mathrm{Re}_{xl,s}\right)^{1/2} w_{xv,\infty}\left(\frac{1}{2}\mathrm{Re}_{xl,s}\right)^{-1/2} W_{yl}(\eta_l) = 0
\end{aligned}$$

The above equation is divided by $x^{-1}w_{xv,\infty}$ and becomes

$$\rho_l \left[-\frac{1}{2}\eta_l \frac{\mathrm{d}W_{xl}(\eta_l)}{\mathrm{d}\eta_l} + \frac{\mathrm{d}W_{yl}(\eta_l)}{\mathrm{d}\eta_l} \right] - \frac{1}{2}\frac{\mathrm{d}\rho_l}{\mathrm{d}\eta_l}\eta_l W_{xl}(\eta_l) + \frac{\mathrm{d}\rho_l}{\mathrm{d}\eta_l} W_{yl}(\eta_l),$$

i.e

$$-\eta_l \frac{\mathrm{d}W_{xl}(\eta_l)}{\mathrm{d}\eta_l} + 2\frac{\partial W_{yl}(\eta_l)}{\mathrm{d}\eta_l} + \frac{1}{\rho_l}\frac{\mathrm{d}\rho_l}{\mathrm{d}\eta_l}[-\eta_l W_{xl}(\eta_l) + 2W_{yl}(\eta_l)] = 0. \quad (9.24)$$

9.4.1.2 Transformation of (9.2)

Equation (9.2) is changed to

$$\rho_l \left(w_{xl}\frac{\partial w_{xl}}{\partial x} + w_{yl}\frac{\partial w_{xl}}{\partial y} \right) = \mu_l \frac{\partial^2 w_{xl}}{\partial y^2} + \frac{\partial \mu_l}{\partial y}\frac{\partial w_{xl}}{\partial y}, \quad (9.2a)$$

where

$$\frac{\partial w_{xl}}{\partial y} = w_{xv,\infty}\frac{\mathrm{d}W_x(\eta_l)}{\mathrm{d}\eta_l}\frac{\partial \eta_l}{\partial y}.$$

With (c), the above equation becomes

$$\frac{\partial w_{xl}}{\partial y} = w_{xv,\infty}\frac{\mathrm{d}W_x(\eta_l)}{\mathrm{d}\eta_l}x^{-1}\left(\frac{1}{2}\mathrm{Re}_{xl,s}\right)^{1/2}. \quad (g)$$

Then,

$$\frac{\partial^2 w_{xl}}{\partial y^2} = w_{xv,\infty}\frac{\mathrm{d}^2 W_{xl}(\eta_l)}{\mathrm{d}\eta_l^2}x^{-2}\left(\frac{1}{2}\mathrm{Re}_{xl,s}\right). \quad (h)$$

Similar to (f), the following equation can be obtained for $\frac{\partial \mu_l}{\partial y}$

$$\frac{\partial \mu_l}{\partial y} = \frac{\mathrm{d}\mu_l}{\mathrm{d}\eta_l}x^{-1}\left(\frac{1}{2}\mathrm{Re}_{xl,s}\right)^{1/2}. \quad (i)$$

With (9.17), (9.18), (b), and (g) to (i), (9.2a) is changed to

$$\rho_l \left[-w_{xv,\infty} W_{xl}\frac{1}{2}\eta_l x^{-1} w_{xv,\infty}\frac{\mathrm{d}W_{xl}(\eta_l)}{\mathrm{d}\eta_l} \right.$$
$$\left. + w_{xv,\infty}\left(\frac{1}{2}\mathrm{Re}_{xl,s}\right)^{-1/2} W_{yl}(\eta_l)\cdot w_{xv,\infty}\frac{\mathrm{d}W_{xl}(\eta_l)}{\mathrm{d}\eta_l}x^{-1}\left(\frac{1}{2}\mathrm{Re}_{xl,s}\right)^{1/2} \right]$$

$$= \mu_l w_{xv,\infty} \frac{\mathrm{d}^2 W_{xl}(\eta_l)}{\mathrm{d}\eta_l^2} x^{-2} \left(\frac{1}{2}\mathrm{Re}_{xl,s}\right)$$

$$+ \frac{\mathrm{d}\mu_l}{\mathrm{d}\eta_l} x^{-1} \left(\frac{1}{2}\mathrm{Re}_{xl,s}\right)^{1/2} w_{xv,\infty} \frac{\mathrm{d}W_{xl}(\eta_l)}{\mathrm{d}\eta_l} x^{-1} \left(\frac{1}{2}\mathrm{Re}_{xl,s}\right)^{1/2}$$

The above equation is simplified to

$$\rho_l \left[-w_{xv,\infty} W_{xl} \frac{1}{2} \eta_l x^{-1} w_{xv,\infty} \frac{\mathrm{d}W_{xl}(\eta_l)}{\mathrm{d}\eta_l} + w_{xv,\infty} W_{yl}(\eta_l) \cdot w_{xv,\infty} \frac{\mathrm{d}W_{xl}(\eta_l)}{\mathrm{d}\eta_l} x^{-1} \right]$$

$$= \mu_l w_{xv,\infty} \frac{\mathrm{d}^2 W_{xl}(\eta_l)}{\mathrm{d}\eta_l^2} x^{-2} \left(\frac{1}{2}\mathrm{Re}_{xl,s}\right) + \frac{\mathrm{d}\mu_l}{\mathrm{d}\eta_l} x^{-2} \left(\frac{1}{2}\mathrm{Re}_{xl,s}\right) w_{xv,\infty} \frac{\mathrm{d}W_{xl}(\eta_l)}{\mathrm{d}\eta_l}$$

With the definition of $\mathrm{Re}_{xl,s}$, the above equation becomes

$$\rho_l \left[-w_{xv,\infty} W_{xl}(\eta_l) \frac{1}{2} \eta_l x^{-1} w_{xv,\infty} \frac{\mathrm{d}W_{xl}(\eta_l)}{\mathrm{d}\eta_l} + w_{xv,\infty} W_{yl} w_{xv,\infty} \frac{\mathrm{d}W_{xl}(\eta_l)}{\mathrm{d}\eta_l} x^{-1} \right]$$
$$= \mu_l w_{xv,\infty} \frac{\mathrm{d}^2 W_{xl}(\eta_l)}{\mathrm{d}\eta_l^2} x^{-2} \left(\frac{1}{2} \frac{w_{xv,\infty} x}{\nu_{l,s}}\right) + \frac{\mathrm{d}\mu_l}{\mathrm{d}\eta_l} x^{-2} \left(\frac{1}{2} \frac{w_{xv,\infty} x}{\nu_{l,s}}\right) w_{xv,\infty} \frac{\mathrm{d}W_{xl}(\eta_l)}{\mathrm{d}\eta_l}$$

The above equation is divided by $x^{-1} \cdot w_{xv,\infty}{}^2$, and becomes

$$\rho_l \left[-W_{xl}(\eta_l) \frac{1}{2} \eta_l \frac{\mathrm{d}W_{xl}(\eta_l)}{\mathrm{d}\eta_l} + W_{yl} \frac{\mathrm{d}W_{xl}(\eta_l)}{\mathrm{d}\eta_l} \right] = \mu_l \frac{\mathrm{d}^2 W_{xl}(\eta_l)}{\mathrm{d}\eta_l^2} \left(\frac{1}{2} \frac{1}{\nu_{l,s}}\right)$$
$$+ \frac{\mathrm{d}\mu_l}{\mathrm{d}\eta_l} \left(\frac{1}{2} \frac{1}{\nu_{l,s}}\right) \frac{\mathrm{d}W_{xl}(\eta_l)}{\mathrm{d}\eta_l}$$

The above equation is multiplied by $2\frac{\nu_{l,s}}{\mu_l}$, and becomes

$$\frac{\nu_{l,s}}{\nu_l} [-\eta_l W_{xl}(\eta_l) + 2W_{yl}(\eta_l)] \frac{\mathrm{d}W_{xl}(\eta_l)}{\mathrm{d}\eta_l} = \frac{\mathrm{d}^2 W_{xl}(\eta_l)}{\mathrm{d}\eta_l^2} + \frac{1}{\mu_l} \frac{\mathrm{d}\mu_l}{\mathrm{d}\eta_l} \frac{\mathrm{d}W_{xl}(\eta_l)}{\mathrm{d}\eta_l}. \quad (9.25)$$

9.4.1.3 Transformation of (9.3)

Equation (9.3) is changed to

$$\rho_l \left(w_{xl} c_{pl} \frac{\partial t_l}{\partial x} + w_{xl} t_l \frac{\partial c_{pl}}{\partial x} + w_{yl} c_{pl} \frac{\partial t_l}{\partial y} + w_{yl} t_l \frac{\partial c_{pl}}{\partial y} \right) = \lambda_l \frac{\partial^2 t_l}{\partial y^2} + \frac{\partial \lambda_l}{\partial y} \frac{\partial t_l}{\partial y}, \quad (9.3\text{a})$$

where

$$\frac{\partial t_l}{\partial x} = \frac{\mathrm{d}t_l}{\mathrm{d}\eta_l}\frac{\partial \eta_l}{\partial x},$$

while,

$$\frac{\partial t_l}{\partial \eta_l} = (t_w - t_s)\frac{\mathrm{d}\theta_l(\eta_l)}{\mathrm{d}\eta_l}. \tag{j}$$

With (a) and (j), we have

$$\frac{\partial t_l}{\partial x} = -\frac{1}{2}\eta_l x^{-1}(t_w - t_s)\frac{\mathrm{d}\theta_l(\eta_l)}{\mathrm{d}\eta_l}. \tag{k}$$

Additionally,

$$\frac{\partial t_l}{\partial y} = \frac{\mathrm{d}t_l}{\mathrm{d}\eta_l}\frac{\partial \eta_l}{\partial y}.$$

With (c) and (j), the above equation becomes

$$\frac{\partial t_l}{\partial y} = x^{-1}\left(\frac{1}{2}\mathrm{Re}_{xl,s}\right)^{1/2}(t_w - t_s)\frac{\mathrm{d}\theta_l(\eta_l)}{\mathrm{d}\eta_l}. \tag{l}$$

Then

$$\frac{\partial^2 t_l}{\partial y^2} = x^{-2}\left(\frac{1}{2}\mathrm{Re}_{xl,s}\right)(t_w - t_s)\frac{\mathrm{d}^2 t_l}{\mathrm{d}\eta_l^2}. \tag{m}$$

Similar to (i), $\frac{\partial \lambda_l}{\partial y}$ and $\frac{\partial c_{pl}}{\partial y}$ can be expressed as respectively

$$\frac{\partial \lambda_l}{\partial y} = \frac{\mathrm{d}\lambda_l}{\mathrm{d}\eta_l}x^{-1}\left(\frac{1}{2}\mathrm{Re}_{xl,s}\right)^{1/2}, \tag{n}$$

$$\frac{\partial c_{pl}}{\partial y} = \frac{\mathrm{d}c_{pl}}{\mathrm{d}\eta_l}x^{-1}\left(\frac{1}{2}\mathrm{Re}_{xl,s}\right)^{1/2}. \tag{o}$$

Similar to (e), $\frac{\partial c_{pl}}{\partial x}$ is expressed as

$$\frac{\partial c_{pl}}{\partial x} = -\frac{1}{2}\frac{\mathrm{d}c_{pl}}{\mathrm{d}\eta_l}\eta_l x^{-1}. \tag{p}$$

With (9.17), (9.18), and (k) to (p), (9.3a) is changed to

$$\rho_l\left[w_{xv,\infty}W_{xl}(\eta_l)c_{p_l}\left(-\tfrac{1}{2}\eta_l x^{-1}(t_w-t_s)\tfrac{d\theta_l(\eta_l)}{d\eta_l}\right)+w_{xv,\infty}W_{xl}(\eta_l)t_l\left(-\tfrac{1}{2}\eta_l x^{-1}\tfrac{dc_{p_l}}{d\eta_l}\right)\right.$$
$$+w_{xv,\infty}\left(\tfrac{1}{2}\mathrm{Re}_{xl,s}\right)^{-1/2}W_{yl}(\eta_l)c_{p_l}x^{-1}\left(\tfrac{1}{2}\mathrm{Re}_{xl,s}\right)^{1/2}(t_w-t_s)\tfrac{d\theta_l(\eta_l)}{d\eta_l}+w_{xv,\infty}\left(\tfrac{1}{2}\mathrm{Re}_{xl,s}\right)^{-1/2}$$
$$\left.W_{yl}(\eta_l)t_l\tfrac{dc_{p_l}}{d\eta_l}x^{-1}\left(\tfrac{1}{2}\mathrm{Re}_{xl,s}\right)^{1/2}\right]$$
$$=\lambda_l x^{-2}\left(\tfrac{1}{2}\mathrm{Re}_{xl,s}\right)(t_w-t_s)\tfrac{d^2\theta_l(\eta_l)}{d\eta_l^2}+\tfrac{\partial\lambda_l}{\partial\eta_l}x^{-1}\left(\tfrac{1}{2}\mathrm{Re}_{xl,s}\right)^{1/2}x^{-1}\left(\tfrac{1}{2}\mathrm{Re}_{xl,s}\right)^{1/2}(t_w-t_s)\tfrac{d\theta_l(\eta_l)}{d\eta_l}$$

With definition of $\mathrm{Re}_{xl,s}$, the above equation is changed to

$$\rho_l\left[\begin{array}{l}w_{xv,\infty}W_{xl}(\eta_l)c_{p_l}\left(-\tfrac{1}{2}\eta_l x^{-1}(t_w-t_s)\tfrac{d\theta_l(\eta_l)}{d\eta_l}\right)+w_{xv,\infty}W_{xl}(\eta_l)t_l\left(-\tfrac{1}{2}\eta_l x^{-1}\tfrac{dc_{p_l}}{d\eta_l}\right)\\ \quad+w_{xv,\infty}W_{yl}(\eta_l)c_{p_l}x^{-1}(t_w-t_s)\tfrac{d\theta_l(\eta_l)}{d\eta_l}+w_{x,\infty}W_{yl}(\eta_l)t_l\tfrac{dc_{p_l}}{d\eta_l}x^{-1}\end{array}\right]$$
$$=\lambda_l x^{-2}\left(\frac{1}{2}\frac{w_{xv,\infty}x}{\nu_{l,s}}\right)(t_w-t_s)\frac{d^2\theta_l(\eta_l)}{d\eta_l^2}+\frac{d\lambda_l}{d\eta_l}x^{-1}\left(\frac{1}{2}\frac{w_{xv,\infty}x}{\nu_{l,s}}\right)x^{-1}(t_w-t_s)\frac{d\theta_l(\eta_l)}{d\eta_l}$$

The above equation is divided by $\frac{1}{2}(t_w-t_s)\frac{c_{p_l}w_{xv,\infty}\lambda_l}{x\nu_{l,s}}$, and simplified to

$$\frac{\nu_{l,s}\rho_l}{\lambda_l}\left[\begin{array}{l}W_{xl}(\eta_l)\left(-\eta_l\tfrac{d\theta_l(\eta_l)}{d\eta_l}\right)+W_{xl}(\eta_l)\tfrac{t_l}{t_w-t_s}\left(-\eta_l\tfrac{1}{c_{p_l}}\tfrac{dc_{p_l}}{d\eta_l}\right)+2W_{yl}(\eta_l)\tfrac{d\theta_l(\eta_l)}{d\eta_l}\\ \quad+2W_{yl}(\eta_l)\tfrac{t_l}{t_w-t_s}\tfrac{1}{c_{p_l}}\tfrac{dc_{p_l}}{d\eta_l}\end{array}\right]$$
$$=\frac{1}{c_{p_l}}\frac{d^2\theta_l(\eta_l)}{d\eta_l^2}+\frac{1}{c_{p_l}}\frac{1}{\lambda_l}\frac{d\lambda_l}{d\eta_l}\frac{d\theta_l(\eta_l)}{d\eta_l} \tag{9.3b}$$

With (9.16), we have

$$\frac{t_l}{t_w-t_s}=\theta_l(\eta_l)+\frac{t_s}{t_w-t_s}. \tag{q}$$

With (q), (9.3b) is further changed to

$$\frac{c_{p_l}\nu_{l,s}\rho_l}{\lambda_l}\left[\begin{array}{l}W_{xl}(\eta_l)\left(-\eta_l\tfrac{d\theta_l(\eta_l)}{d\eta_l}\right)+W_{xl}(\eta_l)\left[\theta_l(\eta_l)+\tfrac{t_s}{t_w-t_s}\right]\left(-\eta_l\tfrac{1}{c_{p_l}}\tfrac{dc_{p_l}}{d\eta_l}\right)\\ \quad+2W_{yl}(\eta_l)\tfrac{d\theta_l(\eta_l)}{d\eta_l}+2W_{yl}(\eta_l)\left[\theta_l(\eta_l)+\tfrac{t_s}{t_w-t_s}\right]\tfrac{1}{c_{p_l}}\tfrac{dc_{p_l}}{d\eta_l}\end{array}\right]$$
$$=\frac{d^2\theta_l(\eta_l)}{d\eta_l^2}+\frac{1}{\lambda_l}\frac{d\lambda_l}{d\eta_l}\frac{d\theta_l(\eta_l)}{d\eta_l}$$

or

$$\mathrm{Pr}_l\frac{\nu_{l,s}}{\nu_l}\left[\begin{array}{l}-\eta_l\cdot W_{xl}(\eta_l)\cdot\tfrac{d\theta_l(\eta_l)}{d\eta_l}+2W_{yl}(\eta_l)\cdot\tfrac{d\theta_l(\eta_l)}{d\eta_l}-\eta_l W_{xl}(\eta_l)\left[\theta_l(\eta_l)+\tfrac{t_s}{t_w-t_s}\right]\left(\tfrac{1}{c_{p_l}}\tfrac{dc_{p_l}}{d\eta_l}\right)\\ \quad+2W_{yl}(\eta_l)\left[\theta_l(\eta_l)+\tfrac{t_s}{t_w-t_s}\right]\tfrac{1}{c_{p_l}}\tfrac{dc_{p_l}}{d\eta_l}\end{array}\right]$$
$$=\frac{d^2\theta_l(\eta_l)}{d\eta_l^2}+\frac{1}{\lambda_l}\frac{d\lambda_l}{d\eta_l}\frac{d\theta_l(\eta_l)}{d\eta_l}$$

The above equation is further changed to

$$\Pr_l \frac{\nu_{l,s}}{\nu_l}\left\{[-\eta_l \cdot W_{xl}(\eta_l) + 2W_{yl}(\eta_l)]\frac{\mathrm{d}\theta_l(\eta_l)}{\mathrm{d}\eta_l} + [-\eta_l W_{xl}(\eta_l) + 2W_{yl}(\eta_l)]\left[\theta_l(\eta_l) + \frac{t_s}{t_w - t_s}\right]\frac{1}{c_{pl}}\frac{\mathrm{d}c_{pl}}{\mathrm{d}\eta_l}\right\}$$
$$= \frac{\mathrm{d}^2\theta_l(\eta_l)}{\mathrm{d}\eta_l^2} + \frac{1}{\lambda_l}\frac{\mathrm{d}\lambda_l}{\mathrm{d}\eta_l}\frac{\mathrm{d}\theta_l(\eta_l)}{\mathrm{d}\eta_l} \tag{9.26}$$

9.4.2 For Vapour Film

9.4.2.1 Transformation of (9.4)

Equation (9.4) is changed to

$$\rho_v\left[\frac{\partial w_{xv}}{\partial x} + \frac{\partial w_{yv}}{\partial y}\right] + w_{xv}\frac{\partial \rho_v}{\partial x} + w_{yv}\frac{\partial \rho_v}{\partial y} = 0. \tag{9.4a}$$

Consulting (a) to (f) for liquid film, we have the following equations for vapour film:

$$\frac{\partial \eta_v}{\partial x} = -\frac{1}{2}\eta_v x^{-1}, \tag{aa}$$

$$\frac{\partial w_{xv}}{\partial x} = -\frac{1}{2}\eta_v x^{-1} w_{xv,\infty}\frac{\mathrm{d}W_{xv}(\eta_v)}{\mathrm{d}\eta_v}, \tag{bb}$$

$$\frac{\partial \eta_v}{\partial y} = x^{-1}\left(\frac{1}{2}\mathrm{Re}_{xv,\infty}\right)^{1/2}, \tag{cc}$$

$$\frac{\partial w_{yv}}{\partial y} = w_{xv,\infty}\frac{\mathrm{d}W_{yv}(\eta_v)}{\mathrm{d}\eta_v}\cdot x^{-1}, \tag{dd}$$

$$\frac{\partial \rho_v}{\partial x} = -\frac{1}{2}\frac{\mathrm{d}\rho_v}{\mathrm{d}\eta_v}\eta_v x^{-1}, \tag{ee}$$

$$\frac{\partial \rho_v}{\partial y} = \frac{\mathrm{d}\rho_v}{\mathrm{d}\eta_v}x^{-1}\left(\frac{1}{2}\mathrm{Re}_{xv,\infty}\right)^{1/2}. \tag{ff}$$

With (9.22), (9.23), (bb), (dd), (ee) and (ff), (9.4a) is changed to

$$\rho_v\left[-\frac{1}{2}\eta_v x^{-1} w_{xv,\infty}\frac{\mathrm{d}W_{xv}(\eta_v)}{\mathrm{d}\eta_v} + w_{xv,\infty}\frac{\mathrm{d}W_{yv}(\eta_v)}{\mathrm{d}\eta_v}x^{-1}\right]$$
$$-\frac{1}{2}\frac{\mathrm{d}\rho_v}{\mathrm{d}\eta_v}\eta_v x^{-1} w_{xv,\infty}W_{xv}(\eta_v) + \frac{\mathrm{d}\rho_v}{\mathrm{d}\eta_v}x^{-1}\left[\frac{1}{2}\mathrm{Re}_{xv,\infty}\right]^{1/2} w_{xv,\infty}\left(\frac{1}{2}\mathrm{Re}_{xv,\infty}\right)^{-1/2}$$
$$W_{yv}(\eta_v) = 0$$

The above equation is divided by $w_{xv,\infty}$, and simplified to

$$\rho_v\left[-\frac{1}{2}\eta_v x^{-1}\frac{\mathrm{d}W_{xv}(\eta_v)}{\mathrm{d}\eta_v}+\frac{\mathrm{d}W_{yv}(\eta_v)}{\mathrm{d}\eta_v}x^{-1}\right]-\frac{1}{2}\frac{\mathrm{d}\rho_v}{\mathrm{d}\eta_v}\eta_v x^{-1}W_{xv}(\eta_v)$$
$$+\frac{\mathrm{d}\rho_v}{\mathrm{d}\eta_v}x^{-1}W_{yv}(\eta_v)=0.$$

The above equation is divided by $\frac{1}{2}x^{-1}$, and simplified to

$$-\eta_v\frac{\mathrm{d}W_{xv}(\eta_v)}{\mathrm{d}\eta_v}+2\frac{\mathrm{d}W_{yv}(\eta_v)}{\mathrm{d}\eta_v}+\frac{1}{\rho_v}\frac{\mathrm{d}\rho_v}{\mathrm{d}\eta_v}[-\eta_v W_{xv}(\eta_v)+2W_{yv}(\eta_v)]=0. \quad (9.27)$$

9.4.2.2 Transformation of (9.5)

Equation (9.5) is changed to

$$\rho_v\left(w_{xv}\frac{\partial w_{xv}}{\partial x}+w_{yv}\frac{\partial w_{xv}}{\partial y}\right)=\mu_v\frac{\partial^2 w_{xv}}{\partial y^2}+\frac{\partial\mu_v}{\partial y}\frac{\partial w_{xv}}{\partial y}. \quad (9.5a)$$

Consulting (g) to (h) for liquid film, we have the following equations for vapour film:

$$\frac{\partial w_{xv}}{\partial y}=w_{xv,\infty}\frac{\mathrm{d}W_{xv}(\eta_v)}{\mathrm{d}\eta_v}x^{-1}\left(\frac{1}{2}\mathrm{Re}_{xv,\infty}\right)^{1/2}, \quad (gg)$$

$$\frac{\partial^2 w_{xv}}{\partial y^2}=w_{xv,\infty}\frac{\mathrm{d}^2W_{xv}(\eta_v)}{\mathrm{d}\eta_v^2}x^{-2}\left(\frac{1}{2}\mathrm{Re}_{xv,\infty}\right), \quad (hh)$$

$$\frac{\partial\mu_v}{\partial y}=\frac{\mathrm{d}\mu_v}{\mathrm{d}\eta_v}x^{-1}\left(\frac{1}{2}\mathrm{Re}_{xv,\infty}\right)^{1/2}. \quad (ii)$$

With (9.22), (9.23), (bb), (gg), (hh), and (ii), (9.5a) is changed to

$$\rho_v\left[-\frac{1}{2}\eta_v x^{-1}w_{xv,\infty}\frac{\mathrm{d}W_{xv}\left(\eta_v\right)}{\mathrm{d}\eta_v}w_{xv,\infty}W_{xv}\left(\eta_v\right)+w_{xv,\infty}\left(\frac{1}{2}\mathrm{Re}_{xv,s}\right)^{-1/2}\right.$$
$$\left.\times W_{yv}\left(\eta_v\right)w_{xv,\infty}\frac{\mathrm{d}W_{xv}\left(\eta_v\right)}{\mathrm{d}\eta_v}x^{-1}\left(\frac{1}{2}\mathrm{Re}_{xv,\infty}\right)^{1/2}\right]$$
$$=\mu_v w_{xv,\infty}\frac{\mathrm{d}^2W_{xv}\left(\eta_v\right)}{\mathrm{d}\eta_v^2}x^{-2}\left(\frac{1}{2}\mathrm{Re}_{xv,\infty}\right)+\frac{\mathrm{d}\mu_v}{\mathrm{d}\eta_v}x^{-1}\left(\frac{1}{2}\mathrm{Re}_{xv,\infty}\right)^{1/2}$$
$$w_{xv,\infty}\frac{\mathrm{d}W_{xv}\left(\eta_v\right)}{\mathrm{d}\eta_v}x^{-1}\left(\frac{1}{2}\mathrm{Re}_{xv,\infty}\right)^{1/2}$$

i.e.

$$\rho_v \left[-\frac{1}{2}\eta_v x^{-1} w_{xv,\infty} \frac{\mathrm{d}W_{xv}(\eta_v)}{\mathrm{d}\eta_v} w_{xv,\infty} W_{xv}(\eta_v) + w_{xv,\infty} W_{yv}(\eta_v) w_{xv,\infty} \frac{\mathrm{d}W_{xv}(\eta_v)}{\mathrm{d}\eta_v} x^{-1} \right]$$
$$= \mu_v w_{xv,\infty} \frac{\mathrm{d}^2 W_{xv}(\eta_v)}{\mathrm{d}\eta_v^2} x^{-2} \left(\frac{1}{2} \frac{w_{xv,\infty} x}{\nu_{v,\infty}} \right)$$
$$+ \frac{\mathrm{d}\mu_v}{\mathrm{d}\eta_v} x^{-1} \left(\frac{1}{2} \frac{w_{xv,\infty} x}{\nu_{v,\infty}} \right) w_{xv,\infty} \frac{\mathrm{d}W_{xv}(\eta_v)}{\mathrm{d}\eta_v} x^{-1}$$

The above equation divided by $\frac{1}{2} w_{xv,\infty}^2 \cdot x^{-1}$, and becomes

$$\rho_v \left[-\eta_v \mathrm{d} \frac{\mathrm{d}W_{xv}(\eta_v)}{\mathrm{d}\eta_v} W_{xv}(\eta_v) + 2W_{yv}(\eta_v) \frac{\mathrm{d}W_{xv}(\eta_v)}{\mathrm{d}\eta_v} \right]$$
$$= \mu_v \frac{\mathrm{d}^2 W_{xv}(\eta_v)}{\mathrm{d}\eta_v^2} \left(\frac{1}{\nu_{v,\infty}} \right) + \frac{\mathrm{d}\mu_v}{\mathrm{d}\eta_v} \left(\frac{1}{\nu_{v,\infty}} \right) \frac{\mathrm{d}W_{xv}(\eta_v)}{\mathrm{d}\eta_v}$$

The above equation is divided by $\frac{\mu_v}{\nu_{v,\infty}}$, and becomes

$$\frac{\nu_{v,\infty}}{\nu_v} \left[-\eta_v W_{xv}(\eta) + 2W_{yv}(\eta) \right] \frac{\mathrm{d}W_{xv}(\eta_v)}{\mathrm{d}\eta_v} = \frac{\mathrm{d}^2 W_{xv}(\eta_v)}{\mathrm{d}\eta_v^2} + \frac{1}{\mu_v} \frac{\mathrm{d}\mu_v}{\mathrm{d}\eta_v} \frac{\mathrm{d}W_{xv}(\eta_v)}{\mathrm{d}\eta_v} \tag{9.28}$$

9.4.2.3 Transformation of (9.6)

Equation (9.6) is changed to

$$\rho_v \left(w_{xv} c_{p_v} \frac{\partial t_v}{\partial x} + w_{xv} t_v \frac{\partial c_{p_v}}{\partial x} + w_{yv} c_{p_v} \frac{\partial t_v}{\partial y} + w_{yv} t_v \frac{\partial c_{p_v}}{\partial y} \right) = \lambda_v \frac{\partial^2 t_v}{\partial y^2} + \frac{\partial \lambda_v}{\partial y} \frac{\partial t_v}{\partial y}. \tag{9.6a}$$

Consulting (j) to (p) for liquid film, we have the following equations for vapour film:

$$\frac{\partial t_v}{\partial \eta_v} = (t_s - t_\infty) \frac{\mathrm{d}\theta_v(\eta_v)}{\mathrm{d}\eta_v}, \tag{jj}$$

$$\frac{\partial t_v}{\partial x} = -\frac{1}{2}\eta_v x^{-1} (t_s - t_\infty) \frac{\mathrm{d}\theta_v(\eta_v)}{\mathrm{d}\eta_v}, \tag{kk}$$

$$\frac{\partial t_v}{\partial y} = x^{-1} \left(\frac{1}{2} \mathrm{Re}_{xv,\infty} \right)^{1/2} (t_s - t_\infty) \frac{\mathrm{d}\theta_v(\eta_v)}{\mathrm{d}\eta_v}, \tag{ll}$$

$$\frac{\partial^2 t_v}{\partial y^2} = x^{-2} \left(\frac{1}{2} \mathrm{Re}_{xv,\infty} \right) (t_s - t_\infty) \frac{\mathrm{d}^2\theta_v(\eta_v)}{\mathrm{d}\eta_v^2}, \tag{mm}$$

$$\frac{\partial \lambda_v}{\partial y} = \frac{\mathrm{d}\lambda_v}{\mathrm{d}\eta_v} x^{-1}\left(\frac{1}{2}\mathrm{Re}_{xv,\infty}\right)^{1/2}, \quad \text{(nn)}$$

$$\frac{\partial c_{p_v}}{\partial y} = \frac{\mathrm{d}c_{p_v}}{\mathrm{d}\eta_v} x^{-1}\left(\frac{1}{2}\mathrm{Re}_{xv,\infty}\right)^{1/2}, \quad \text{(oo)}$$

$$\frac{\partial c_{p_v}}{\partial x} = -\frac{1}{2}\frac{\mathrm{d}c_{p_v}}{\mathrm{d}\eta_v}\eta_v x^{-1}. \quad \text{(pp)}$$

With (9.22), (9.23), and (kk) to (pp), (9.6a) is changed to

$$\rho_v \left[\begin{array}{l} w_{xv,\infty} W_{xv}(\eta_v) c_{p_v}\left(-\frac{1}{2}\eta_v x^{-1}(t_s - t_\infty)\frac{\mathrm{d}\theta_v(\eta_v)}{\mathrm{d}\eta_v}\right) + w_{xv,\infty} W_{xv}(\eta_v) t_v \left(-\frac{1}{2}\eta_v x^{-1}\frac{\mathrm{d}c_{p_v}}{\mathrm{d}\eta_v}\right) \\ + w_{xv,\infty}\left(\frac{1}{2}\mathrm{Re}_{xv,\infty}\right)^{-1/2} W_{yv}(\eta_v) c_{p_v} x^{-1}\left(\frac{1}{2}\mathrm{Re}_{xv,\infty}\right)^{1/2}(t_s - t_\infty)\frac{\mathrm{d}\theta_v(\eta_v)}{\mathrm{d}\eta_v} \\ + w_{x,\infty}\left(\frac{1}{2}\mathrm{Re}_{xv,\infty}\right)^{-1/2} W_{yv}(\eta_v) t_v \frac{\mathrm{d}c_{p_{vl}}}{\mathrm{d}\eta_v} x^{-1}\left(\frac{1}{2}\mathrm{Re}_{xv,\infty}\right)^{1/2} \end{array}\right]$$

$$= \lambda_l x^{-2}\left(\frac{1}{2}\mathrm{Re}_{xv,\infty}\right)(t_s - t_\infty)\frac{\mathrm{d}^2\theta_v(\eta_v)}{\mathrm{d}\eta_v^2} + \frac{\mathrm{d}\lambda_v}{\mathrm{d}\eta_v} x^{-1}\left(\frac{1}{2}\mathrm{Re}_{xv,\infty}\right)^{1/2} x^{-1}\left(\frac{1}{2}\mathrm{Re}_{xv,\infty}\right)^{1/2} (t_s - t_\infty)\frac{\mathrm{d}\theta_v(\eta_v)}{\mathrm{d}\eta_v}$$

i.e.

$$\rho_v \left[\begin{array}{l} -\frac{1}{2} w_{xv,\infty} W_{xv}(\eta_v) c_{p_v}\eta_v x^{-1}(t_s - t_\infty)\frac{\mathrm{d}\theta_v(\eta_v)}{\mathrm{d}\eta_v} - \frac{1}{2} w_{xv,\infty} W_{xv}(\eta_v) t_v \eta_v x^{-1}\frac{\mathrm{d}c_{p_v}}{\mathrm{d}\eta_v} \\ + w_{xv,\infty} W_{yv}(\eta_v) c_{p_v} x^{-1}(t_s - t_\infty)\frac{\mathrm{d}\theta_v(\eta_v)}{\mathrm{d}\eta_v} + w_{x,\infty} W_{yv}(\eta_v) t_v \frac{\mathrm{d}c_{p_{vl}}}{\mathrm{d}\eta_v} x^{-1} \end{array}\right]$$

$$= \lambda_l x^{-2}\left(\frac{1}{2}\mathrm{Re}_{xv,\infty}\right)(t_s - t_\infty)\frac{\mathrm{d}^2\theta_v(\eta_v)}{\mathrm{d}\eta_v^2} + \frac{\mathrm{d}\lambda_v}{\mathrm{d}\eta_v} x^{-1}\left(\frac{1}{2}\mathrm{Re}_{xv,\infty}\right) x^{-1}(t_s - t_\infty)\frac{\mathrm{d}\theta_v(\eta_v)}{\mathrm{d}\eta_v}$$

With definition of $\mathrm{Re}_{xv,\infty}$, the above equation is changed to

$$\rho_v \left[\begin{array}{l} -\frac{1}{2} w_{xv,\infty} W_{xv}(\eta_v) c_{p_v}\eta_v x^{-1}(t_s - t_\infty)\frac{\mathrm{d}\theta_v(\eta_v)}{\mathrm{d}\eta_v} - \frac{1}{2} w_{xv,\infty} W_{xv}(\eta_v) t_v \eta_v x^{-1}\frac{\mathrm{d}c_{p_v}}{\mathrm{d}\eta_v} \\ + w_{xv,\infty} W_{yv}(\eta_v) c_{p_v} x^{-1}(t_s - t_\infty)\frac{\mathrm{d}\theta_v(\eta_v)}{\mathrm{d}\eta_v}) + w_{x,\infty} W_{yv}(\eta_v) t_v \frac{\mathrm{d}c_{p_{vl}}}{\mathrm{d}\eta_v} x^{-1} \end{array}\right]$$

$$= \lambda_v x^{-2}\left(\frac{1}{2}\frac{w_{xv,\infty}x}{\nu_{v,\infty}}\right)(t_s - t_\infty)\frac{\mathrm{d}^2\theta_v(\eta_v)}{\mathrm{d}\eta_v^2} + \frac{\mathrm{d}\lambda_v}{\mathrm{d}\eta_v} x^{-1}\left(\frac{1}{2}\frac{w_{xv,\infty}x}{\nu_{v,\infty}}\right) x^{-1}(t_s - t_\infty)\frac{\mathrm{d}\theta_v(\eta_v)}{\mathrm{d}\eta_v}$$

The above equation is divided by $\frac{1}{2}(t_s - t_\infty)\frac{c_{p_v} w_{xv,\infty}\lambda_v}{x\nu_{v,\infty}}$, and simplified to

$$\frac{\nu_{v,\infty}\rho_v}{\lambda_v}\left[\begin{array}{l} -W_{xv}(\eta_v)\eta_v\frac{\mathrm{d}\theta_v(\eta_v)}{\mathrm{d}\eta_v} - W_{xv}(\eta_v)\frac{t_v}{t_s - t_\infty}\eta_v\frac{1}{c_{p_v}}\frac{\mathrm{d}c_{p_{vl}}}{\mathrm{d}\eta_v} \\ + 2W_{yv}(\eta_v)\frac{\mathrm{d}\theta_v(\eta_v)}{\mathrm{d}\eta_v}) + 2W_{yv}(\eta_v)\frac{t_v}{t_s - t_\infty}\frac{1}{c_{p_v}}\frac{\mathrm{d}c_{p_v}}{\mathrm{d}\eta_v} \end{array}\right]$$

$$= \frac{1}{c_{p_v}}\frac{\mathrm{d}^2\theta_v(\eta_v)}{\mathrm{d}\eta_v^2} + \frac{1}{c_{p_v}}\frac{1}{\lambda_v}\frac{\mathrm{d}\lambda_v}{\mathrm{dd}\eta_v}\frac{\mathrm{d}\theta_v(\eta_v)}{\mathrm{d}\eta_v} \quad \text{(9.6b)}$$

With (9.16), we have

$$\frac{t_v}{t_s - t_\infty} = \theta_v(\eta_v) + \frac{t_\infty}{t_s - t_\infty}. \tag{r}$$

With (q), (9.6b) is further simplified to

$$\mathrm{Pr}_v \frac{\nu_{v,\infty}}{\nu_v} \left[\begin{array}{l} -W_{xv}(\eta_v)\eta_v \dfrac{\mathrm{d}\theta_v(\eta_v)}{\mathrm{d}\eta_v} - W_{xv}(\eta_v)\left(\theta_v(\eta_v) + \dfrac{t_\infty}{t_s - t_\infty}\right)\eta_v \dfrac{1}{c_{p_v}} \dfrac{\mathrm{d}c_{p_{vl}}}{\mathrm{d}\eta_v} \\ +2W_{yv}(\eta_v)\dfrac{\mathrm{d}\theta_v(\eta_v)}{\mathrm{d}\eta_v} + 2W_{yv}(\eta_v)\left(\theta_v(\eta_v) + \dfrac{t_\infty}{t_s - t_\infty}\right)\dfrac{1}{c_{p_v}}\dfrac{\mathrm{d}c_{p_v}}{\mathrm{d}\eta_v} \end{array} \right]$$
$$= \frac{\mathrm{d}^2\theta_v(\eta_v)}{\mathrm{d}\eta_v^2} + \frac{1}{\lambda_v}\frac{\mathrm{d}\lambda_v}{\mathrm{d}\eta_v}\frac{\mathrm{d}\theta_v(\eta_v)}{\mathrm{d}\eta_v}$$

or

$$\mathrm{Pr}_v \frac{\nu_{v,\infty}}{\nu_v} \left\{ \begin{array}{l} [-\eta_v W_{xv}(\eta_v) + 2W_{yv}(\eta_v)]\dfrac{\mathrm{d}\theta_v(\eta_v)}{\mathrm{d}\eta_v} \\ +[-W_{xv}(\eta_v)\eta_v + 2W_{yv}(\eta_v)]\left(\theta_v(\eta_v) + \dfrac{t_\infty}{t_s - t_\infty}\right)\dfrac{1}{c_{p_v}}\dfrac{\mathrm{d}c_{p_v}}{\mathrm{d}\eta_v} \end{array} \right\}$$
$$= \frac{\mathrm{d}^2\theta_v(\eta_v)}{\mathrm{d}\eta_v^2} + \frac{1}{\lambda_v}\frac{\mathrm{d}\lambda_v}{\mathrm{d}\eta_v}\frac{\mathrm{d}\theta_v(\eta_v)}{\mathrm{d}\eta_v} \tag{9.29}$$

9.4.3 For Boundary Conditions

9.4.3.1 Transformation of (9.7)

With (9.14), (9.16), (9.17) and (9.18), (9.7) can be transformed as easily

$$\eta_l = 0: \quad W_{xl}(\eta_l) = 0, \quad W_{yl}(\eta_l) = 0, \quad \theta_l(\eta_l) = 1 \tag{9.30}$$

9.4.3.2 Transformation of (9.8)

With (9.14) and (9.17), (9.8) is transformed to

$$\begin{array}{l} \eta_l = \eta_{l\delta}(\eta_v = 0): \\ W_{xv,s}(\eta_{v0}) = W_{xl,s}(\eta_{l\delta}) \end{array} \tag{9.31}$$

9.4.3.3 Transformation of (9.9)

With (9.14), (9.15), and (9.17), we have

$$\delta_l = \eta_{l\delta}\left(\frac{1}{2}\frac{w_{xv,\infty}x}{\nu_{l,s}}\right)^{-1/2} x,$$

$$\frac{\partial \delta_l}{\partial x} = \frac{1}{2}\eta_{l\delta}\left(\frac{1}{2}\frac{w_{xv,\infty}x}{\nu_{l,s}}\right)^{-1/2}. \tag{s}$$

Similarly, with (9.19), (9.20), and (9.22), we have

$$\frac{\partial \delta_v}{\partial x} = \frac{1}{2}\eta_{v\delta}\left(\frac{1}{2}\frac{w_{xv,\infty}x}{\nu_{v,\infty}}\right)^{-1/2}. \tag{t}$$

With ((9.14), (9.17), (9.22), (9.23), (s) and (t), (9.9) is changed to

$$\rho_{l,s}\left(w_{xl}\frac{\partial \delta_l}{\partial x} - w_{yl}\right)_{l,s}$$
$$= \rho_{l,s}\left[w_{xv,\infty}W_{xl,s}(\eta_{l\delta})\frac{1}{2}\eta_{l\delta}\left(\frac{1}{2}\frac{w_{xv,\infty}x}{\nu_{l,s}}\right)^{-1/2} - w_{xv,\infty}\left(\frac{1}{2}\frac{w_{xv,\infty}x}{\nu_{l,s}}\right)^{-1/2}W_{yl,s}(\eta_{l\delta})\right],$$
$$\rho_{v,s}\left(w_{xv}\frac{\partial \delta_v}{\partial x} - w_{yv}\right)_{v,s}$$
$$= \rho_{v,s}\left[w_{xv,\infty}W_{xv,s}(\eta_{v0})\frac{1}{2}\eta_{v0}\left(\frac{1}{2}\frac{w_{xv,\infty}x}{\nu_{v,\infty}}\right)^{-1/2} - w_{xv,\infty}\left(\frac{1}{2}\frac{w_{xv,\infty}x}{\nu_{v,\infty}}\right)^{-1/2}W_{yv,s}(\eta_{v0})\right]_{v,s}.$$

Then

$$\rho_{l,s}\left[w_{xv,\infty}W_{xl,s}(\eta_{l\delta})\frac{1}{2}\eta_{l\delta}\left(\frac{1}{2}\frac{w_{xv,\infty}x}{\nu_{l,s}}\right)^{-1/2} - w_{xv,\infty}\left(\frac{1}{2}\frac{w_{xv,\infty}x}{\nu_{l,s}}\right)^{-1/2}W_{yl,s}(\eta_{l\delta})\right]$$
$$= \rho_{v,s}\left[w_{xv,\infty}W_{xv,s}(\eta_{v0})\frac{1}{2}\eta_{v0}\left(\frac{1}{2}\frac{w_{xv,\infty}x}{\nu_{v,\infty}}\right)^{-1/2} - w_{xv,\infty}\left(\frac{1}{2}\frac{w_{xv,\infty}x}{\nu_{v,\infty}}\right)^{-1/2}W_{yv,s}(\eta_{v0})\right]$$

Since $\eta_{v0} = 0$, the above equation becomes

$$\rho_{l,s}\left[w_{xv,\infty}W_{xl,s}(\eta_{l\delta})\frac{1}{2}\eta_{l\delta}\left(\frac{1}{2}\frac{w_{xv,\infty}x}{\nu_{l,s}}\right)^{-1/2} - w_{xv,\infty}\left(\frac{1}{2}\frac{w_{xv,\infty}x}{\nu_{l,s}}\right)^{-1/2}W_{yl,s}(\eta_{l\delta})\right]$$
$$= \rho_{v,s}\left[-w_{xv,\infty}\left(\frac{1}{2}\frac{w_{xv,\infty}x}{\nu_{v,\infty}}\right)^{-1/2}W_{yv,s}(\eta_{v0})\right]$$

Then

$$W_{yv,s}(\eta_{v0}) = -\frac{\rho_{l,s}\left[w_{xv,\infty}W_{xl,s}(\eta_{l\delta})\frac{1}{2}\eta_{l\delta}\left(\frac{1}{2}\frac{w_{xv,\infty}x}{\nu_{l,s}}\right)^{-1/2} - w_{xv,\infty}\left(\frac{1}{2}\frac{w_{xv,\infty}x}{\nu_{l,s}}\right)^{-1/2}W_{yl,s}(\eta_{l\delta})\right]}{\rho_{v,s}w_{xv,\infty}\left(\frac{1}{2}\frac{w_{xv,\infty}x}{\nu_{v,\infty}}\right)^{-1/2}}$$

The above equation is simplified to

$$W_{yv,s}(\eta_{v0}) = -\frac{\left(\frac{\nu_{l,s}}{\nu_{v,\infty}}\right)^{1/2} \rho_{l,s}\left[W_{xl,s}(\eta_{l\delta})\frac{1}{2}\eta_{l\delta} - W_{yl,s}(\eta_{l\delta})\right]}{\rho_{v,s}},$$

i.e.

$$W_{yv,s}(\eta_{v0}) = -\frac{\rho_{l,s}}{\rho_{v,s}}\left(\frac{\nu_{l,s}}{\nu_{v,\infty}}\right)^{1/2}\left[\frac{1}{2}\eta_{l\delta}\cdot W_{xl,s}(\eta_{l\delta}) - W_{yl,s}(\eta_{l\delta})\right] \tag{9.32}$$

9.4.3.4 Transformation of (9.10)

With (g) and (gg), (9.10) is changed to

$$\mu_{l,s}\left[w_{xv,\infty}\frac{\mathrm{d}W_x(\eta_l)}{\mathrm{d}\eta_l}x^{-1}\left(\frac{1}{2}\mathrm{Re}_{xl,s}\right)^{1/2}\right]_s = \mu_{v,s}\left[w_{xv,\infty}\frac{\mathrm{d}W_{xv}(\eta_v)}{\mathrm{d}\eta_v}x^{-1}\left(\frac{1}{2}\mathrm{Re}_{xv,\infty}\right)^{1/2}\right]_s,$$

i.e.

$$\mu_{l,s}w_{xv,\infty}\left(\frac{\mathrm{d}W_x(\eta_l)}{\mathrm{d}\eta_l}\right)_s x^{-1}\left(\frac{1}{2}\frac{w_{xv,\infty}x}{\nu_{l,s}}\right)^{1/2} = \mu_{v,s}w_{xv,\infty}\left(\frac{\mathrm{d}W_{xv}(\eta_v)}{\mathrm{d}\eta_v}\right)_s x^{-1}\left(\frac{1}{2}\frac{w_{x,\infty}x}{\nu_{v,\infty}}\right)^{1/2}.$$

Then

$$\left(\frac{\mathrm{d}W_{xv}(\eta_v)}{\mathrm{d}\eta_v}\right)_s = \frac{\mu_{l,s}}{\mu_{v,s}}\left(\frac{\nu_{v,\infty}}{\nu_{l,s}}\right)^{1/2}\left(\frac{\mathrm{d}W_x(\eta_l)}{\mathrm{d}\eta_l}\right)_s. \tag{9.33}$$

9.4.3.5 Transformation of (9.11)

With (9.22), (9.23), (l), (ll) and (s), (9.11) is changed to

$$\lambda_{l,s}x^{-1}\left[\left(\frac{1}{2}\frac{w_{xv,\infty}x}{\nu_{l,s}}\right)^{1/2}(t_w - t_s)\frac{\mathrm{d}\theta_l(\eta_l)}{\mathrm{d}\eta_l}\right]_s = \lambda_{v,s}\left[x^{-1}\left(\frac{1}{2}\frac{w_{xv,\infty}x}{\nu_{v,\infty}}\right)^{1/2}(t_s - t_\infty)\frac{\mathrm{d}\theta_v(\eta_v)}{\mathrm{d}\eta_v}\right]_s$$

$$+h_{fg}\rho_{v,s}\left[w_{xv,\infty}W_{xv}(\eta_v)\frac{1}{2}\eta_{v0}\left(\frac{1}{2}\frac{w_{xv,\infty}x}{\nu_{v,\infty}}\right)^{-1/2} - w_{xv,\infty}\left(\frac{1}{2}\frac{w_{xv,\infty}x}{\nu_{v,\infty}}\right)^{-1/2}W_{yv}(\eta_v)\right]_s$$

With $\eta_{v0} = 0$, the above equation becomes

$$\lambda_{l,s}x^{-1}\left[\left(\frac{1}{2}\frac{w_{xv,\infty}x}{\nu_{l,s}}\right)^{1/2}(t_w - t_s)\frac{\mathrm{d}\theta_l(\eta_l)}{\mathrm{d}\eta_l}\right]_s$$

$$= \lambda_{v,s}\left[x^{-1}\left(\frac{1}{2}\frac{w_{xv,\infty}x}{\nu_{v,\infty}}\right)^{1/2}(t_s - t_\infty)\frac{\mathrm{d}\theta_v(\eta_v)}{\mathrm{d}\eta_v}\right]_s + h_{fg}\rho_{v,s}\left[-w_{xv,\infty}\left(\frac{1}{2}\frac{w_{xv,\infty}x}{\nu_{v,\infty}}\right)^{-1/2}W_{yv}(\eta_{v0})\right]_s$$

The above equation is divided by $\left(\frac{1}{2}w_{xv,\infty}\right)^{1/2} \cdot \left(\frac{1}{\nu_{v,\infty}}\right)^{1/2} \cdot x^{-1/2}$, and simplified to

$$\lambda_{l,s}\left[\left(\frac{\nu_{v,\infty}}{\nu_{l,s}}\right)^{1/2}(t_w - t_s)\frac{d\theta_l(\eta_l)}{d\eta_l}\right]_s = \lambda_{v,s}\left[(t_s - t_\infty)\frac{d\theta_v(\eta_v)}{d\eta_v}\right]_s + 2h_{fg}\rho_{v,s}[-\nu_{v,\infty}W_{yv}(\eta_{v0})],$$

i.e.

$$\lambda_{v,s}\left[(t_s - t_\infty)\frac{d\theta_v(\eta_v)}{d\eta_v}\right]_s = \lambda_{l,s}\left[\left(\frac{\nu_{v,\infty}}{\nu_{l,s}}\right)^{1/2}(t_w - t_s)\frac{d\theta_l(\eta_l)}{d\eta_l}\right]_s - 2h_{fg}\rho_{v,s}[-\nu_{v,\infty}W_{yv}(\eta_{v0})].$$

Then

$$\left(\frac{d\theta_v(\eta_v)}{d\eta_v}\right)_s = \frac{\lambda_{l,s}\left(\frac{\nu_{v,\infty}}{\nu_{l,s}}\right)^{1/2}(t_w - t_s)\left(\frac{d\theta_l(\eta_l)}{d\eta_l}\right)_s + 2h_{fg}\rho_{v,s}\nu_{v,\infty}W_{yv,s}(\eta_{v0})}{\lambda_{v,s}(t_s - t_\infty)}. \quad (9.34)$$

9.4.3.6 Transformation of (9.12)

With (9.16) and (9.21), (9.12) is changed to

$$\theta_l(\eta_{l\delta}) = 0 \text{ and } \theta_v(\eta_{v0}) = 1 \quad (9.35)$$

9.4.3.7 Transformation of (9.13)

With (9.21) and (9.22), (9.13) becomes

$$\eta_v \to \infty \qquad W_{xv}(\eta_{v,\infty}) = 1, \quad \theta_v(\eta_{v,\infty}) = 0 \quad (9.36)$$

9.5 Remarks

With the new similarity analysis method, the complete dimensionless mathematical model is developed for laminar forced film condensation of vapour. The complete similarity model consists of liquid and vapour film models, connected by five interfacial matching conditions, respectively, for velocity continuity, mass transfer balance, shear force balance, energy balance, and temperature continuity. Meanwhile, both in liquid and vapour films, as well as the interfacial boundary condition equations, the variable physical properties such as temperature-dependent density, thermal conductivity, viscosity, and specific heat exist in the form of the related dimensionless physical property factors based on the new similarity analysis system.

The present developed similarity mathematical model is not only suitable for the laminar forced film condensation of superheated vapour with the superheated temperature $t_\infty - t_s > 0$ but also suitable for that of saturated vapour with the superheated temperature $t_\infty - t_s = 0$. In this case, the condensation of saturated vapour is only regarded as a special case of the superheated vapour.

It is also seen that the new similarity analysis method reported in this book is a better alternative to the traditional Falkner-Skan type transformation for extensive investigation of laminar film condensation with consideration various physical elements including the complicated interfacial matching conditions and variable physical properties.

Exercises

1. If you like, according to Chap. 3 for reviewing the Falkner–Skan-type transformation method, try to transform the governing partial differential equations and their boundary conditions (9.1)– (9.13) into the related similarity governing mathematical model with consideration of variable physical properties.
2. Compare the similarity governing mathematical model obtained in this chapter for laminar forced film condensation of vapour with consideration of variable physical properties based on the present new similarity analysis method with that based on the Falkner–Skan-type transformation method. Which conclusions can you get?
3. If the flat plate and the bulk vapour flow have an angle to the horizontal level for laminar forced film condensation of vapour, please do the following:

   (1) Based on (9.1), (9.2), (9.3), (9.4), (9.5), (9.6), (9.7), (9.8), (9.9), (9.10), (9.11), (9.12), and (9.13) for horizontal laminar film condensation of vapour, provide the related governing partial differential equations for the inclined laminar film condensation of vapour.
   (2) Based on the present new similarity analysis method, provide a new set of the appropriate similarity variables for similarity transformation of the governing partial differential equations for the inclined laminar film condensation of vapour.
   (3) By using the new set of the appropriate similarity variables, transform similarly the governing partial differential equations together with the boundary conditions of the inclined laminar film condensation into the related similarity governing mathematical model.
   (4) Which conclusions do you find from the above work?

References

1. W. Nusselt (1916) Die Oberflächenkondensation des Wasserdampfes. Z. Ver. Dt. Ing. 60:541–546, 569–575
2. R.D. Cess, Laminar film condensation on a flat plate in the absence of a body force, Z. Angew. Math. Pfus. **11**, 426–433 (1960)

3. J.C.Y. Koh, Film condensation in a forced-convection boundary-layer flow. Int J Heat Mass Transfer **5**, 941–954 (1962)
4. I.G. Shekriladze, V.I. Gomelauri, Theoretical study of laminar film condensation of flowing vapor. Int. J. Heat Mass Transfer **9**, 581–591 (1966)
5. H.R. Jacobs, An integral treatment of combined body force and forced convection in laminar film condensation. Int. J. Heat Mass Transfer **9** 637–648 (1966)
6. V.E. Denny, A.F. Mills, Laminar film condensation of a flowingvapor on a horizontal cylinder at normal gravity. ASME J. Heat Transfer **91**, 495–501 (1969)
7. P. Beckett, G. Poots, Laminar film condensation in forced flows. Quart. J. Mech. Appl. Maths **25**(1), 125–152 (1972)
8. T. Fujii, H. Uehara, Laminar filmwise condensation on a vertical surface. Int.I. Heat Mass Transfer **15**, 217–233 (1972)
9. T. Fujii, H. Honda, K. Oda, Condensation of Steam on a Horizontal Tube—The Influence of Oncoming Velocity and Thermal Condition at the Tube Wall, *Condensation Heat Transfer, 18th National Heat Transfer Conference*. San Diego, CA, 1979, pp. 35–43
10. J.W. Rose, Approximate equation for forced-convection condensation in the presence of a non-condensation gas on a flat plate and horizontal tube. Int. J. Heat Mass Transfer **23**, 539–546, (1980)
11. J.W. Rose, Effect of pressurevariation in forced convection film condensation on a horizontal tube. Int. J. Heat Mass Transfer **27**(1), 39–47 (1984)
12. H. Honda, T. Fujii, Condensation of a flowing vapor on a horizontal tube – numerical analysis on a conjugate heat transfer problem. Trans. ASME. J Heat Transfer **106**, 841–848, (1984)
13. J.W. Rose, A new interpolation formula for forced-convection condensation on a horizontal surface. Trans ASME. J Heat Transfer **111**, 818–819 (1989)
14. S.B. Memory, V.H. Adams, Marto, P.J., Free and forced convection laminar film condensation on horizontal elliptical tubes. Int. J. Heat Mass Transfer. **40**(14), 3395–3406 (1997)
15. F. Méndez, C. Treviño, Analysis of a forced laminar film condensation including finite longitudinal heat conduction effects. Heat Mass Transfer **39**, 489–498 (2003)

Chapter 10
Velocity and Temperature Fields on Laminar Forced Film Condensation of Pure Vapour

Abstract The laminar forced film condensation of water vapour is taken as an example for treatment of variable physical properties. The temperature parameter method with the simple power-law equations for gases is used for the treatment of vapour film variable physical properties, and the water polynomial formulations are applied for the treatment of liquid film variable physical properties. Then, an effective numerical procedure is developed for rigorously satisfying a set of the interfacial physical matching conditions. On this basis, a system of the numerical solutions on the velocity and temperature fields of the two-phase film flows is rigorously found out. These numerical solutions demonstrate that increasing the wall subcooled grade causes increase of the condensate liquid film thickness but decrease of vapour film thickness, decrease of the wall temperature gradient but increase of the vapour film temperature gradient, and increase of the interfacial velocity level. Compared with the effect of the wall subcooled grade on these physical variables, the effect of the vapour subcooled grade is very slight.

10.1 Introduction

In Chap. 9, we obtained the complete dimensionless similarity mathematical model for laminar forced film condensation of vapour. Such similarity mathematical model consists of two parts of film flows, the liquid film flow and vapour film flow. The two parts of model are combined by the set of the interfacial matching conditions, i.e. velocity continuity, mass transfer balance, shear force balance, energy balance, and temperature continuity. In the similarity mathematical model on the laminar forced film condensation, the variable thermophysical properties of gas and liquid exist in the form of the related dimensionless physical property factors.

The present work is the successive investigation of Chap. 9, and will take the laminar forced film condensation of water vapour as an example to find out the velocity and temperature fields of the two-phase film flows. First, the treatment of the variable physical properties will be done for liquid and vapour film flows as well as the interfacial matching condition equations. The temperature parameter method with the simple-power law equations reported in [1] and [2], and collected in book [3] is used for the treatment of variable physical properties for vapour film flow,

D.Y. Shang, *Theory of Heat Transfer with Forced Convection Film Flows*,
Heat and Mass Transfer, DOI 10.1007/978-3-642-12581-2_10,

and the polynomial approach [4] are used for treatment of the variable physical properties of condensate liquid film low. Then, the set of physical property factors in the governing similarity mathematical model become the functions of the dimensionless temperature based on the new similarity analysis method. On these bases, the developed complete similarity mathematical model on the laminar forced film condensation of vapour is solved numerically, in which the set of the interfacial matching conditions are rigorously satisfied and the variable physical properties of liquid and vapour films are rigorously treated. Then, a system of solutions with the velocity and temperature fields is given for different wall subcooled and vapour superheated grads, which is a basis on extensive investigation of heat and mass transfer of laminar forced film condensation of vapour.

10.2 Treatment of Temperature-Dependent Physical Properties

As presented in our previous studies such as in [1–6] and Chaps. 7 and 8 of this book, it is clear that fluid variable physical properties have great effects on heat transfer with convection and film flows, and so should be appropriately treated for reliable and rigorous solutions of the related governing equations. In this present work on the two-phase flow condensation problem, the variable physical properties exist respectively in liquid and vapour films and, therefore, should be treated separately. Take the laminar forced film condensation of water vapour as an example, the treatment approach related to the variable physical properties are presented as follows respectively.

10.2.1 For Liquid Film Medium

Consulting Chap. 8 for treatment approach of liquid variable physical properties, as an example, the water variable density, thermal conductivity, absolute viscosity, and specific heat are expressed as follows for condensate water film flow:

$$\rho_l = -4.48 \times 10^{-3} t^2 + 999.9, \tag{10.1}$$

$$\lambda_l = -8.01 \times 10^{-6} t^2 + 1.94 \times 10^{-3} t + 0.563, \tag{10.2}$$

$$\mu_l = \exp\left(-1.6 - \frac{1150}{T} + \left(\frac{690}{T}\right)^2\right) \times 10^{-3}. \tag{10.3}$$

While, the water specific heat can be taken as a constant $c_{p_l} = 4200\,\mathrm{J/(kg\ K)}$ with temperature variation at atmospheric pressure, and then, specific heat factor can be regarded as

$$\frac{1}{c_{p_l}}\frac{\mathrm{d}c_{p_l}}{\mathrm{d}\eta_l} = 0. \tag{10.4}$$

The physical property factors in the governing ordinary differential equations are expressed as respectively

$$\mathrm{Pr}_l = \frac{\mu_l \cdot c_{p_l}}{\lambda_l} = \frac{\left[\exp\left(-1.6 - \frac{1150}{T} + \left(\frac{690}{T}\right)^2\right) \times 10^{-3}\right] \times 4200}{-8.01 \times 10^{-6} t^2 + 1.94 \times 10^{-3} t + 0.563}, \tag{10.5}$$

$$\frac{1}{\rho_l}\frac{\mathrm{d}\rho_l}{\mathrm{d}\eta_l} = \frac{\left[(-2 \times 4.48 \times 10^{-3} t (t_w - t_s)\frac{\mathrm{d}\theta_l(\eta_l)}{\mathrm{d}\eta_l}\right]}{(-4.48 \times 10^{-3} t^2 + 999.9)}, \tag{10.6}$$

$$\frac{1}{\lambda_l}\frac{d\lambda_l}{d\eta_l} = \frac{\left[(-2 \times 8.01 \times 10^{-6} + 1.94 \times 10^{-3})(t_w - t_s)\frac{\mathrm{d}\theta_l(\eta_l)}{\mathrm{d}\eta_l}\right]}{-8.01 \times 10^{-6} t^2 + 1.94 \times 10^{-3} t + 0.563}, \tag{10.7}$$

$$\frac{1}{\mu_l}\frac{\mathrm{d}\mu_l}{\mathrm{d}\eta_l} = \left(\frac{1150}{T^2} - 2 \times \frac{690^2}{T^3}\right)(t_w - t_s)\frac{\mathrm{d}\theta_l(\eta_l)}{\mathrm{d}\eta_l}, \tag{10.8}$$

$$\frac{\nu_{l,\infty}}{\nu_l} = \left(\frac{\mu_{l,\infty}}{\rho_{l,\infty}}\right) / \left(\frac{\mu_l}{\rho_l}\right) = \left(\frac{\mu_{l,\infty}}{\mu_l}\right)\left(\frac{\rho_l}{\rho_{l,\infty}}\right),$$

$$= \frac{\exp\left[-1.6 - \frac{1150}{T_\infty} + \left(\frac{690}{T_\infty}\right)^2\right]}{\exp\left[-1.6 - \frac{1150}{T} + \left(\frac{690}{T}\right)^2\right]} \times \frac{-4.48 \times 10^{-3} t^2 + 999.9}{-4.48 \times 10^{-3} t_\infty^2 + 999.9}, \tag{10.9}$$

where

$$t = (t_w - t_s)\theta(\eta) + t_s \text{ and } T = t + 273$$

Thus, these dependent physical property factors become the function of dimensionless temperature under the new similarity analysis system.

10.2.2 For Vapour Film Medium

Consulting Chap. 7 for treatment approach of gas variable physical properties, the vapour variable density, thermal conductivity, absolute viscosity, and specific heat are expressed as following simple power-aw equations:

$$\frac{\rho_v}{\rho_{v,\infty}} = \left(\frac{T_v}{T_\infty}\right)^{-1}, \tag{10.10}$$

$$\frac{\mu_v}{\mu_{v,\infty}} = \left(\frac{T}{T_\infty}\right)^{n_\mu}, \tag{10.11}$$

$$\frac{\lambda_v}{\lambda_{v,\infty}} = \left(\frac{T}{T_\infty}\right)^{n_\lambda}, \tag{10.12}$$

$$\frac{c_{pv}}{c_{pv,\infty}} = \left(\frac{T}{T_\infty}\right)^{n_{cp}}, \tag{10.13}$$

$$\frac{\nu_{v,\infty}}{\nu_v} = \left(\frac{T}{T_\infty}\right)^{-(n_\mu+1)}. \tag{10.14}$$

While the related physical property factor can be expressed as follows respectively:

$$\frac{1}{\rho_v}\frac{\mathrm{d}\rho_v}{\mathrm{d}\eta_v} = -\frac{(T_s/T_\infty - 1)\mathrm{d}\theta_v(\eta_v)/\mathrm{d}\eta_v}{(T_s/T_\infty - 1)\theta_v(\eta_v) + 1}, \tag{10.15}$$

$$\frac{1}{\mu_v}\frac{\mathrm{d}\mu_v}{\mathrm{d}\eta_v} = \frac{n_\mu(T_s/T_\infty - 1)\mathrm{d}\theta_v(\eta_v)/\mathrm{d}\eta_v}{(T_s/T_\infty - 1)\theta_v(\eta_v) + 1}, \tag{10.16}$$

$$\frac{1}{\lambda_v}\frac{\mathrm{d}\lambda_v}{\mathrm{d}\eta_v} = \frac{n_\lambda(T_s/T_\infty - 1)\mathrm{d}\theta_v(\eta_v)/\mathrm{d}\eta_v}{(T_s/T_\infty - 1)\theta_v(\eta_v) + 1}, \tag{10.17}$$

$$\frac{1}{c_{p_v}}\frac{\mathrm{d}c_{p_v}}{\mathrm{d}\eta_v} = \frac{n_{c_p}(T_s/T_\infty - 1)\mathrm{d}\theta_v(\eta_v)/\mathrm{d}\eta_v}{(T_s/T_\infty - 1)\theta_v(\eta_v) + 1}, \tag{10.18}$$

$$\frac{\nu_{v,\infty}}{\nu_v} = [(T_w/T_\infty - 1)\theta(\eta_v) + 1]^{-(n_\mu+1)}, \tag{10.19}$$

$$\mathrm{Pr}_v = \mathrm{Pr}_{v,\infty}[(T_s/T_\infty - 1)\theta_v(\eta_v) + 1]^{n_\mu - n_\lambda + n_{cp}}, \tag{10.20}$$

where the bulk temperature T_∞ is taken as the reference temperature, and n_μ, n_λ, and n_{c_p} denote viscosity, thermal conductivity, and specific heat parameters of vapour, respectively.

As I mentioned that, according to the study in [1–3], for monatomic and diatomic gases, air, and water vapour, the value of specific heat parameter n_{c_p} is so small that it can be ignored without any obvious deviation on heat transfer result prediction, then the specific heat factor $(1/c_{pv})(\mathrm{d}c_{pv}/\mathrm{d}\eta_v)$ can be regarded as zero.

10.3 Numerical Solutions

10.3.1 Calculation Procedure

Satisfying the whole set of the interfacial matching conditions of the governing equations for the forced film condensation is important for obtaining the reliable solutions. However, it is not easy to realize it actually. Then to date, there has been still lack of the rigorous and reliable numerical solutions demonstrated in previous literatures on this field. For overcoming this difficult point, the following procedural steps of numerical calculations are presented for reliable solution of the forced film condensation.

Step 1: Calculation for liquid film equations
In the first step, the governing equations of liquid film are calculated. The solutions of (9.24), (9.25), and (9.26) are found out by using boundary condition equations (9.30) and (9.35), as well as the supposed initial values of condensate liquid film thickness $\eta_{l\delta}$ and liquid interfacial velocity component $W_{xl,s}$ as the boundary conditions. Then the second step of calculation begins.

Step 2: Calculation for vapour film
In this step of calculation, first, (9.31) and (9.32) are used for evaluation of $W_{xv,s}(\eta_{v,0})$ and $W_{xv,s}(\eta_{v,0})$ respectively by using the solutions of $W_{xl,s}(\eta_{l\delta})$ and $W_{xy,s}(\eta_{v,0})$ obtained from the step 1. Then with equations (9.35), (9.36), as well as the evaluated values of $W_{xv,s}(\eta_{v,0})$ and $W_{xv,s}(\eta_{v,0})$ as the boundary conditions, (9.27), (9.28), and (9.29) are solved.

Step 3: Judgment on the convergence of the numerical calculation
In this step of calculation, first (9.33) and (9.34) are applied as judgment equations for verification of the solution convergence. By means of these judgment equations, the calculation is iterated with appropriate change of the values $\eta_{l\delta}$ and $W_{xl,s}$. In each iteration, calculation on step 1 for liquid film equations (9.24), (9.25), and (9.26) and calculation on step 2 for vapour film equations (9.27), (9.28), and (9.29) are done successively. In each step of calculations, the governing equations are solved by a shooting method with fifth-order Runge-Kutta integration. For solving such very strong nonlinear problem, a variable mesh approach is applied to the numerical calculation programs, in order to obtain accurate numerical solutions.

It should be emphasized that satisfying all interfacial physical matching conditions is important for obtaining reliable solutions for the laminar forced film condensation with two-phase flows. Even one of the interfacial physical matching conditions is ignored, the solutions will be multiple, and it is never possible to obtain the sole set of correct solutions.

10.3.2 Velocity and Temperature Fields of the Two-Phase Film Flows

The forced film condensation of water vapour on a horizontal flat plate is taken as example for the numerical calculation. A system of numerical solutions on momentum and temperature fields of liquid and vapour films is plotted in Figs. 10.1, 10.2, 10.3, 10.4, 10.5, 10.6, 10.7, and 10.8 Meanwhile, Figs. 10.1, 10.2, 10.3, 10.4, 10.5, and 10.6 show the condensate film velocity component $W_{xl}(\eta_l)$, condensate film temperature $\theta_l(\eta_l)$, vapour film velocity component $W_{xv}(\eta_v)$, and vapour film temperature $\theta_v(\eta_v)$ with the wall subcooled grade $\frac{\Delta t_w}{t_s} = \frac{t_s - t_w}{t_s} = 0.2$, 0.6 and 1 and different vapour superheated grade $\frac{\Delta t_\infty}{t_s} = \frac{t_\infty - t_s}{t_s}$. Figures 10.7 and 10.8 show the condensate film velocity component $W_{xl}(\eta_l)$, condensate film temperature $\theta_l(\eta_l)$, vapour film velocity component $W_{xv}(\eta_v)$, and vapour film temperature $\theta_v(\eta_v)$ at water vapour superheated grade $\frac{\Delta t_\infty}{t_s} = \frac{t_\infty - t_s}{t_s} = 2.27$ with variation of the wall subcooled grade $\frac{\Delta t_w}{t_s} = \frac{t_s - t_w}{t_s}$. These results demonstrate the following effects of the wall subcooled grade $\frac{\Delta t_w}{t_s}$ and the vapour superheated grade $\frac{\Delta t_\infty}{t_s}$ on velocity and temperature fields of the laminar forced film condensation.

Increasing the wall subcooled grade $\frac{\Delta t_w}{t_s}$ causes

(i) increasing the liquid film thickness $\eta_{l\delta}$ and interfacial velocity $W_{xl,s}(\eta_{l\delta})$,
(ii) decreasing wall dimensionless temperature gradient,

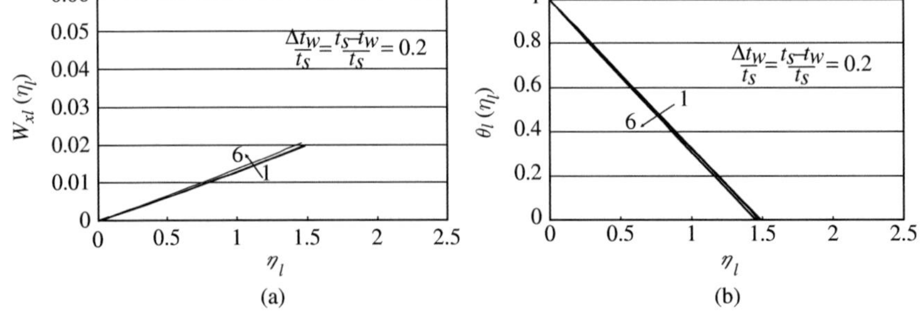

Fig. 10.1 Numerical results on **(a)** condensate film velocity component $W_{xl}(\eta_l)$, and **(b)** condensate film temperature $\theta_l(\eta_l)$ for forced film condensation of water vapour on a horizontal flat plate at $\frac{\Delta t_w}{t_s} = \frac{t_s - t_w}{t_s} = 0.2$ and $t_s = 100°C$ (Lines 1 → 6 denotes $\frac{\Delta t_\infty}{t_s} = \frac{t_\infty - t_s}{t_s} = 0$, 0.27, 1.27, 2.27, 3.27, and 4.27, respectively)

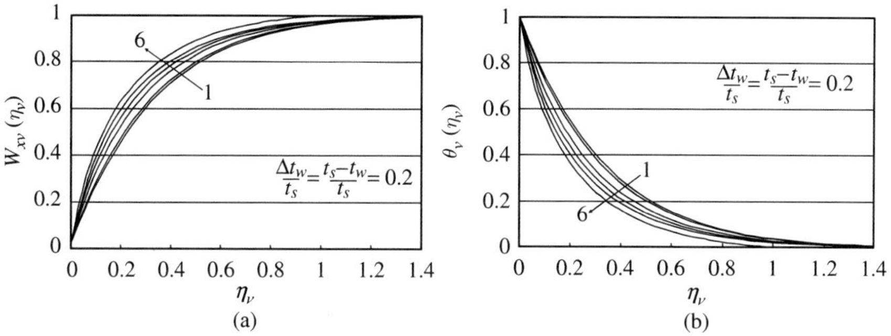

Fig. 10.2 Numerical results on **(a)** vapour film velocity component $W_{xv}(\eta_v)$, and **(b)** vapour film temperature $\theta_v(\eta_v)$ for forced film condensation of water vapour on a horizontal flat plate at $\frac{\Delta t_w}{t_s} = \frac{t_s - t_w}{t_s} = 0.2$ and $t_s = 100°C$ (Lines 1 → 6 denotes $\frac{\Delta t_\infty}{t_s} = \frac{t_\infty - t_s}{t_s} = 0, 0.27, 1.27, 2.27, 3.27$, and 4.27 , respectively)

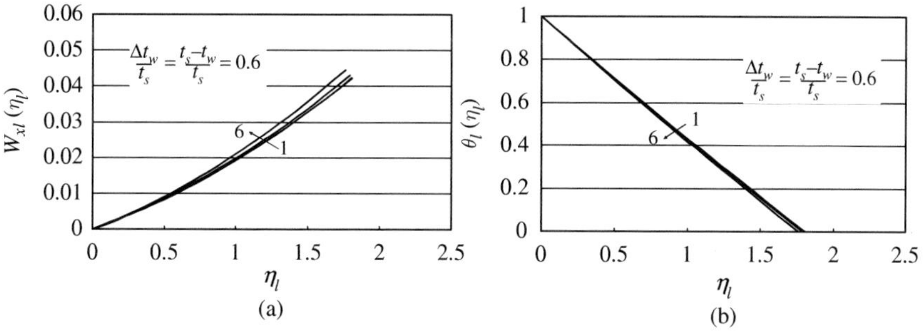

Fig. 10.3 Numerical results on **(a)** condensate film velocity component $W_{xl}(\eta_l)$, and **(b)** condensate film temperature $\theta_l(\eta_l)$ for forced film condensation of water vapour on a horizontal flat plate at $\frac{\Delta t_w}{t_s} = \frac{t_s - t_w}{t_s} = 0.6$ and $t_s = 100°C$ (Lines 1 → 6 denotes $\frac{\Delta t_\infty}{t_s} = \frac{t_\infty - t_s}{t_s} = 0, 0.27, 1.27, 2.27$, 3.27, and 4.27 , respectively)

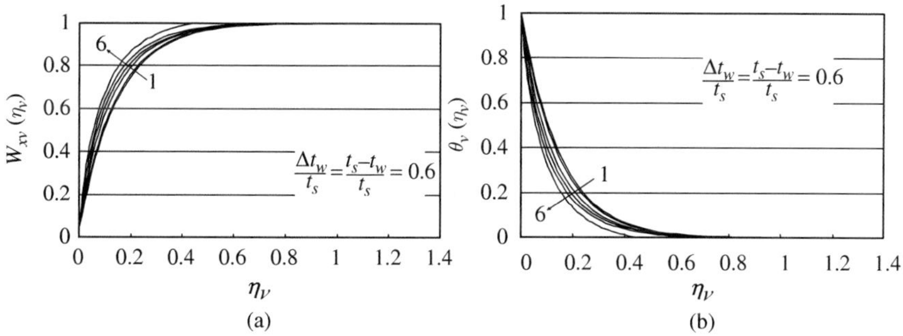

Fig. 10.4 Numerical results on **(a)** vapour film velocity component $W_{xv}(\eta_v)$, and **(b)** vapour film temperature $\theta_v(\eta_v)$ for forced film condensation of water vapour on a horizontal flat plate at $\frac{\Delta t_w}{t_s} = \frac{t_s - t_w}{t_s} = 0.6$ and $t_s = 100°C$ (Lines 1 → 6 denotes $\frac{\Delta t_\infty}{t_s} = \frac{t_\infty - t_s}{t_s} = 0, 0.27, 1.27, 2.27, 3.27$, and 4.27, respectively)

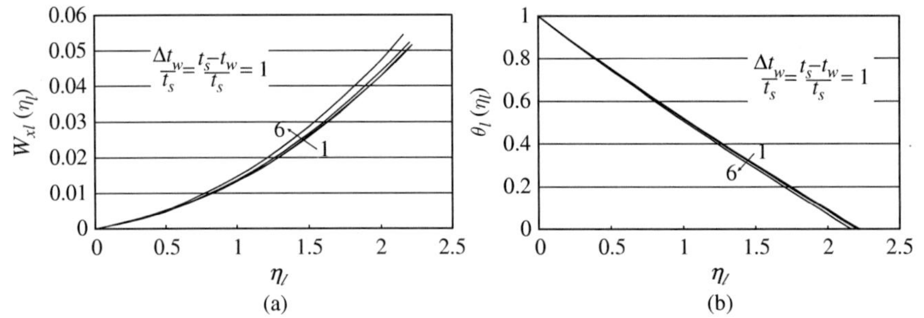

Fig. 10.5 Numerical results on **(a)** condensate film velocity component $W_{xl}(\eta_l)$, **(b)** condensate film temperature $\theta_l(\eta_l)$ for forced film condensation of water vapour on a horizontal flat plate at $\frac{\Delta t_w}{t_s} = \frac{t_s - t_w}{t_s} = 1$ and $t_s = 100°C$ (Lines 1 → 6 denotes $\frac{\Delta t_\infty}{t_s} = \frac{t_\infty - t_s}{t_s} = 0, 0.27, 1.27, 2.27, 3.27,$ and 4.27 , respectively)

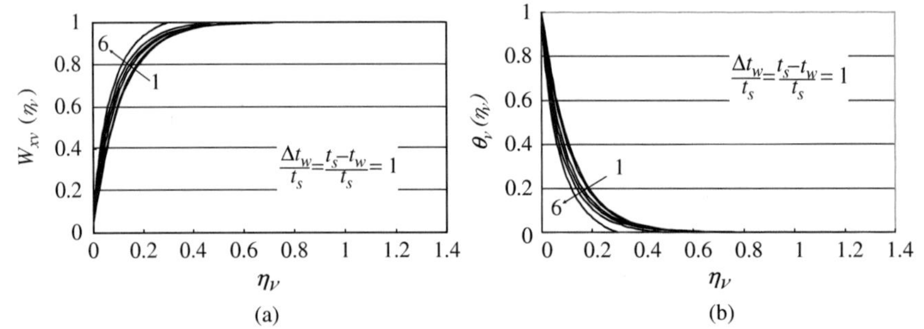

Fig. 10.6 Numerical results on **(a)** vapour film velocity component $W_{xv}(\eta_v)$, and **(b)** vapour film temperature $\theta_v(\eta_v)$ for forced film condensation of water vapour on a horizontal flat plate at $\frac{\Delta t_w}{t_s} = \frac{t_s - t_w}{t_s} = 1$ and $t_s = 100°C$ (Lines 1 → 6 denotes $\frac{\Delta t_\infty}{t_s} = \frac{t_\infty - t_s}{t_s} = 0, 0.27, 1.27, 2.27, 3.27,$ and 4.27, respectively)

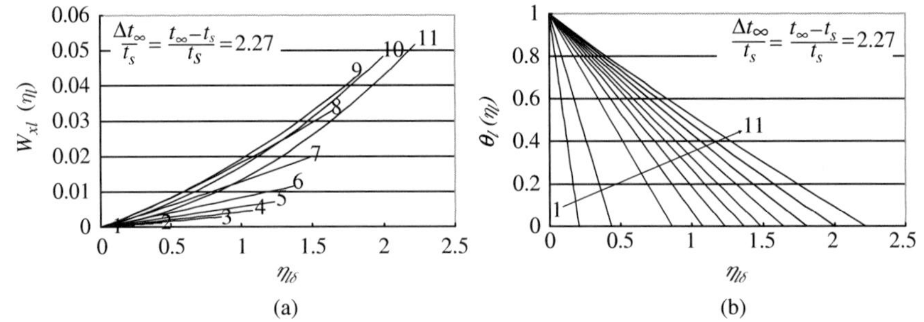

Fig. 10.7 Numerical solutions on **(a)** velocity profiles $W_{xl}(\eta_l)$ and **(b)** temperature profiles $\theta_l(\eta_l)$ of liquid film for laminar film condensation of water vapour on a horizontal flat plate at $\frac{\Delta t_\infty}{t_s} = \frac{t_\infty - t_s}{t_s} = 2.27$ (Lines 1 → 11 denote $\frac{\Delta t_w}{t_s} = \frac{t_s - t_w}{t_s} = 0.0001, 0.001, 0.01, 0.025, 0.05, 0.1, 0.2,$ 0.4, 0.6, 0.8, and 1, respectively)

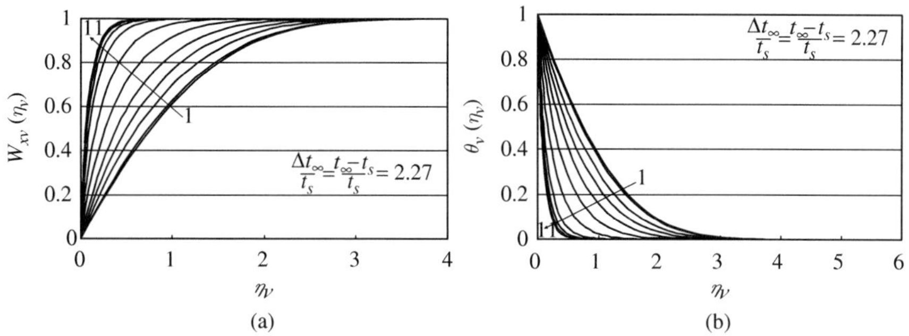

Fig. 10.8 Numerical solutions on **(a)** velocity profiles $W_{xv}(\eta_v)$ and **(b)** temperature profiles $\theta_v(\eta_v)$ of vapour film for laminar film condensation of water vapour on a horizontal flat plate at $\frac{\Delta t_\infty}{t_s} = \frac{t_\infty - t_s}{t_s} = 2.27$ (Lines 1 → 11 denote $\frac{\Delta t_w}{t_s} = \frac{t_s - t_w}{t_s} = 0.0001$, 0.001, 0.01, 0.025, 0.05, 0.1, 0.2, 0.4, 0.6, 0.8, and 1, respectively)

(iii) decreasing the vapour film thickness and increasing the vapour film flow velocity level,
(iv) increasing the vapour film dimensionless temperature gradient.

While increasing the vapour superheated grade $\frac{\Delta t_\infty}{t_s}$ causes

(i) decreasing the liquid film thickness slightly and increasing the liquid film flow velocity slightly;
(ii) increasing the wall dimensionless temperature gradient slightly;
(iii) decreasing the vapour film thickness and increasing vapour film flow velocity level slightly;
(iv) increasing the vapour film dimensionless temperature gradient slightly.

To sum up, compared with the effect of the wall subcooled grade $\frac{\Delta t_w}{t_s}$ on the laminar forced film condensation, the effect of the vapour superheated grade $\frac{\Delta t_\infty}{t_s}$ is very slight. The reason is that compared with the latent heat h_{fg} of the vapour, the vapour heat capacity $c_p \cdot \Delta t_\infty$ is very small generally.

10.4 Remarks

Based on the complete similarity mathematical model proposed in Chap. 9 for laminar forced film condensation with consideration of variable physical properties, the advanced procedure for solution on velocity and temperature fields is developed in this chapter, so that all interfacial physical matching conditions are rigorously satisfied for reliable solution. Together with the careful consideration of the variable physical properties of liquid and vapour film flows, the reliable numerical results on velocity and temperature fields are found out.

Increasing the wall subcooled grade $\frac{\Delta t_w}{t_s}$ causes increasing the liquid film thickness $\eta_{l,\delta}$ and interfacial velocities, decreasing wall dimensionless temperature gradient, decreasing the vapour film thickness, increasing the vapour film flow velocity level, and increasing the vapour film dimensionless temperature gradient.

Increasing the vapour superheated grade $\frac{\Delta t_\infty}{t_s}$ causes decreasing the liquid film thickness slightly, increasing the liquid velocity slightly, increasing the wall dimensionless temperature gradient slightly, decreasing the vapour film thickness, increasing vapour film flow velocity level, and increasing the vapour film dimensionless temperature gradient.

Compared with the effect of the wall subcooled grade $\frac{\Delta t_w}{t_s}$ on the laminar forced film condensation, the effect of the vapour superheated grade $\frac{\Delta t_\infty}{t_s}$ is very slight. The reason is that compared with the latent heat h_{fg} of the vapour, the vapour heat capacity $c_p \cdot \Delta t_\infty$ is very small generally. All these express the coupled effects of the variable physical properties of vapour and liquid on the laminar forced film condensation.

It should be emphasized that satisfying all interfacial physical matching conditions is important for obtaining reliable solutions for the laminar forced film condensation with two-phase flows. Even one of the interfacial physical matching conditions is ignored, the solutions will be multiple, and it is never possible to obtain the sole set of correct solutions.

Exercises

1. From the numerical results on velocity and temperature fields of laminar forced film condensation of vapour reported in this chapter, it is seen the obvious coupled effects of variable physical properties on the laminar forced film condensation. Please point out the correct meaning of “the coupled effects”.
2. Please indicate all liquid–vapour interfacial matching conditions of laminar forced film condensation of vapour and explain why all these interfacial matching conditions should be satisfied for a reliable solution.

References

1. D.Y. Shang, B.X. Wang, Effect of variable thermophysical properties on laminar free convection of gas. Int. J. Heat Mass Transfer **33**(7), 1387–1395 (1990)
2. D.Y. Shang, B.X. Wang, Effect of variable thermophysical properties on laminar free convection of polyatomic gas. Int. J. Heat Mass Transfer **34**(3) 749–755 (1991)
3. D.Y. Shang, *Free Convection Film Flows and Heat Transfer* (Springer, Berlin, Heidelberg and New York, NY, 2006)
4. D.Y. Shang, B.X. Wang, Y. Wang, Y. Quan, Study on liquid laminar free convection with consideration of variable thermophysical properties. Int. J. Heat Mass Transfer **36**(14), 3411–3419 (1993)

5. D.Y. Shang, T. Adamek, Study on laminar film condensation of saturated steam on a vertical flat late for consideration of various physical factors including variable thermophysical properties. Warme- und Stoffubertragung **30**(2), 89–100 (1994)
6. D.Y. Shang, B.X. Wang, An extended study on steady-state laminar film condensation of a superheated vapor on an isothermal vertical plate. Int. J. Heat Mass Transfer **40**(4), 931–941 (1997)

Chapter 11
Heat and Mass Transfer on Laminar Forced Film Condensation of Pure Vapour

Abstract On the basis of Chaps. 9 and 10, heat and mass transfer investigation is done under the new similarity analysis system for laminar forced film condensation of pure vapour. The heat transfer analysis equations are provided where the wall dimensionless temperature gradient becomes only one no-given variable for prediction of condensate heat transfer. By means of the condensate mass transfer analysis, the condensate mass flow rate parameter is induced as the function of the dimensionless condensate liquid film thickness and the interfacial dimensionless condensate liquid velocity components. It is found that the condensate mass flow rate parameter is the only one no-given variable for prediction of condensate mass transfer. Take the laminar forced film condensation of water vapour on a horizontal flat plate as an example, the systems of key numerical solutions on the wall dimensionless temperature gradient and mass flow rate parameter is obtained, and further rigorously formulated by using curve-fitting approach. With the rigorously formulated equations on the wall dimensionless temperature gradient and mass flow rate parameter, the theoretical analysis equations on condensate heat and mass become available for simple and reliable prediction of heat and mass transfer on laminar forced film condensation of saturated and superheated water vapour. Thus, the effects of the wall subcooled grade and vapour superheated grads on the condensate heat and mass transfer are clarified. Although increasing the wall subcooled grade causes decreasing the wall temperature gradient, it will increase the condensate heat transfer. Increasing the wall subcooled grade causes increasing the condensate mass flow rate parameter, and further causes increasing the condensate mass flow rate. Increasing the vapour superheated grade causes decreasing the condensate mass flow rate parameter, and further causes decreasing the condensate mass flow rate. Compared with the effect of the wall subcooled grade on heat and mass transfer of the forced film condensation, the effect of the vapour subcooled grade is so slight that it can be ignored in the special range of the superheated grade. At last, a condensate mass–energy transformation equation is reported in this chapter under the new similarity analysis system for a better clarification of the internal relation of film condensate heat and mass transfer.

D.Y. Shang, *Theory of Heat Transfer with Forced Convection Film Flows*,
Heat and Mass Transfer, DOI 10.1007/978-3-642-12581-2_11,

11.1 Introduction

On the basis of Chaps. 9 and 10, respectively, for complete similarity mathematical model and the numerical results of velocity and temperature fields, in this chapter further investigation will be done for heat and mass transfer of laminar forced film condensation of pure vapour. First, the appropriate heat transfer analysis will be done for providing the appropriate heat transfer theoretical equations in which the key solution, i.e. the wall dimensionless temperature gradient, will dominate condensate heat and mass transfer. Meanwhile, the condensate mass flow rate parameter will be induced by the mass transfer analysis, which will be the function of the dimensionless liquid film thickness, as well as the condensate interfacial velocity components, and will dominate condensate mass transfer. Furthermore, the key solutions on the wall dimensionless temperature gradient and the condensate mass flow rate parameter of the similarity governing mathematical model will be formulated based on the rigorous numerical results, so that the theoretical equations on condensate heat and mass transfer can be conveniently used for simple and reliable prediction. In addition, the mass-energy transformation equation is reported in this chapter for description of the transformation relation between the key solutions, i.e. the wall temperature gradient and the mass flow rate parameter, which dominate the condensate mass and heat transfer, respectively.

11.2 Condensate Heat Transfer Analysis

In the previous chapters, the heat transfer analysis was done for laminar forced convection, while the following heat and mass transfer analysis is for laminar forced film condensation.

The local heat transfer rate $q_{x,w}$ of the film condensation at position x per unit area on the plate can be calculated by applying Fourier's law as $q_x = -\lambda_{l,w}\left(\frac{\partial t}{\partial y}\right)_{y=0}$ with consideration of variable physical properties. With the assumed transformation variable of (9.14), (9.15), and (9.16) for liquid film, $q_{x,w}$ is described as

$$q_{x,w} = \lambda_{l,w}(t_w - t_s)\left(\frac{1}{2}\mathrm{Re}_{xl,s}\right)^{1/2} x^{-1}\left(-\frac{\mathrm{d}\theta_l(\eta_l)}{\mathrm{d}\eta_l}\right)_{\eta_l=0}. \tag{11.1}$$

Then the local heat transfer coefficient, defined as $q_{x,w} = \alpha_{x,w}(t_w - t_s)$, will be

$$\alpha_{x,w} = \lambda_{l,w}\left(\frac{1}{2}\mathrm{Re}_{xl,s}\right)^{1/2} x^{-1}\left(-\frac{\mathrm{d}\theta_l(\eta_l)}{\mathrm{d}\eta_l}\right)_{\eta_l=0}. \tag{11.2}$$

The local Nusselt number, defined as $Nu_{xl,w} = \frac{\alpha_{x,w} x}{\lambda_{l,w}}$, is expressed by

$$Nu_{xl,w} = \left(\frac{1}{2}\mathrm{Re}_{xl,s}\right)^{1/2} \left(-\frac{\mathrm{d}\theta_l(\eta_l)}{\mathrm{d}\eta_l}\right)_{\eta_l=0}. \tag{11.3}$$

The total heat transfer rate for position $x = 0$ to x with width of b on the plate is an integration

$$Q_{x,w} = \iint_A q_{x,w}\mathrm{d}A = \int_0^x q_{x,w} b\,\mathrm{d}x,$$

where the plate condensate heat transfer area is described as $A = b \times x$, where b is the plate width related to heat transfer.

With (11.1), the above equation is changed to

$$Q_{x,w} = \int_0^x b\lambda_{l,w}(t_w - t_s)\left(\frac{1}{2}\mathrm{Re}_{xl,s}\right)^{1/2} x^{-1}\left(-\frac{\mathrm{d}\theta_l(\eta_l)}{\mathrm{d}\eta_l}\right)_{\eta_l=0} \mathrm{d}x.$$

With definition of local Reynolds number $\mathrm{Re}_{xl,s}$ in (9.15), we obtain

$$Q_{x,w} = \sqrt{2}b\lambda_{l,w}(t_w - t_s)(\mathrm{Re}_{xl,s})^{1/2}\left(-\frac{\mathrm{d}\theta_l(\eta_l)}{\mathrm{d}\eta_l}\right)_{\eta_l=0}. \tag{11.4}$$

The average heat transfer coefficient $\overline{\alpha}_{x,w}$ defined as $Q_{x,w} = \overline{\alpha_{x,w}}(t_w - t_s)A$ is expressed as

$$\overline{\alpha}_{x,w} = 2\lambda_{l,w}\left(\frac{1}{4}\mathrm{Re}_{xl,s}\right)^{1/2} x^{-1}\left(-\frac{\mathrm{d}\theta_l(\eta_l)}{\mathrm{d}\eta_l}\right)_{\eta_l=0}. \tag{11.5}$$

The average Nusselt number defined as $\overline{Nu}_{xl,w} = \frac{\overline{\alpha_{x,w}}x}{\lambda_{l,w}}$ will be

$$\overline{Nu}_{xl,w} = \sqrt{2}(\mathrm{Re}_{xl,s})^{1/2}\left(-\frac{\mathrm{d}\theta_l(\eta_l)}{\mathrm{d}\eta_l}\right)_{\eta_l=0}. \tag{11.6}$$

Here, it is necessary to mention the positive and negative signs for heat transfer results of the film condensation, such as those found out in (11.1) and (11.4). In fact, the value of the dimensionless temperature gradient is negative, and the temperature difference $(t_w - t_s)$ is negative. Then, the values of $q_{x,w}$ and $Q_{x,w}$ will be negative always. According to the coordinative system, the positive direction of the heat flux is from the wall to the vapour bulk. Then, the negative values of the heat transfer rate show that the related heat flux direction is just from the vapour bulk to the wall.

11.3 Wall Dimensionless Temperature Gradient

From (11.1), (11.2), (11.3), (11.4), (11.5), and (11.6), it is seen that wall dimensionless temperature gradient $\left(\frac{d\theta_l(\eta_l)}{d\eta_l}\right)_{\eta_l=0}$, as the key solution of the governing equations, is only the un-given variable for evaluation of heat transfer of the film condensation.

Together with the numerical solutions on temperature fields on the laminar forced film condensation of water vapour obtained in Chap. 10, a system of the key numerical solutions on the wall dimensionless temperature gradient $-\left(\frac{d\theta_l(\eta_l)}{d\eta_l}\right)_{\eta_l=0}$ is found out, plotted in Fig. 11.1, and some of them are listed in Table 11.1, with variation of the wall subcooled grade $\frac{\Delta t_w}{t_s}$ and vapour superheated grade $\frac{\Delta t_\infty}{t_s}$. It is seen that decreasing the wall subcooled grade will cause increasing the wall dimensionless temperature gradient, and even increase at an accelerative pace in the range of small wall subcooled grade $\frac{\Delta t_w}{t_s}$. Meanwhile, increasing the vapour superheated grade causes increasing the wall dimensionless temperature gradient very slightly. Compared with the effect of wall subcooled grade on the wall dimensionless temperature gradient, the effect of the vapour superheated grade is so small

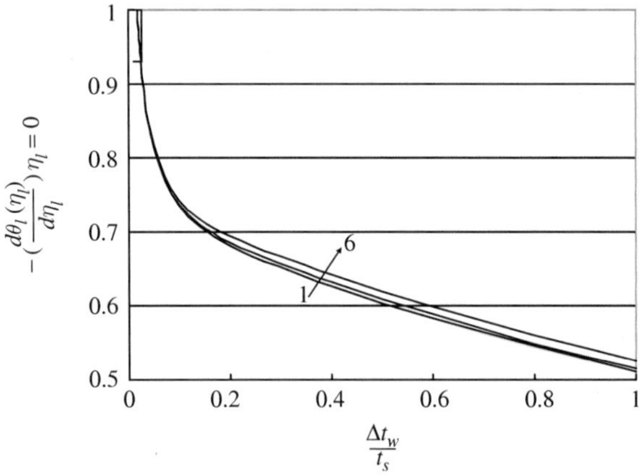

Fig. 11.1 Numerical results $-\left(\frac{d\theta_l(\eta_l)}{d\eta_l}\right)_{\eta_l=0}$ of laminar forced film condensation of saturated and superheated water vapour on a horizontal flat plate

Note: 1. lines 1 → 6 denotes $\left(\frac{\Delta t_\infty}{t_s}\right) = 0, 0.27, 1.27, 2.27, 3.27$, and 4.27, respectively

that it can be ignored within the range of $0 \le \frac{\Delta t_\infty}{t_s} \le 4.27$ without any obvious influence caused on condensate heat transfer prediction. According to such analysis, by using a curve-fitting approach, the system of numerical solutions on the wall dimensionless temperature gradient for laminar forced film condensation of water vapour along a horizontal flat plate is formulated to the following equations:

for $0 \le \dfrac{\Delta t_\infty}{t_s} \le 4.27$

$$-\left(\frac{d\theta_l(\eta_l)}{d\eta_l}\right)_{\eta_l=0} = \frac{0.7421\dfrac{\Delta t_w}{t_s} + 0.3124}{\left(\dfrac{\Delta t_w}{t_s}\right)^{0.28}} \quad \left(0.01 \le \frac{\Delta t_w}{t_s} \le 0.1\right) \tag{11.7}$$

$$-\left(\frac{d\theta_l(\eta_l)}{d\eta_l}\right)_{\eta_l=0} = 0.1466\left(\frac{\Delta t_w}{t_s}\right)^2 - 0.3974\left(\frac{\Delta t_w}{t_s}\right) + 0.7652 \quad \left(0.1 \le \frac{\Delta t_w}{t_s} \le 1\right) \tag{11.8}$$

The results on the wall dimensionless temperature gradient $\left(\frac{d\theta_l(\eta_l)}{d\eta_l}\right)_{\eta_l=0}$ predicted by using (11.7) and (11.8) are listed in Table 11.1. It is seen that the prediction deviation of (11.7) is less than 0.47% for $\frac{\Delta t_\infty}{t_s} = 0$, and less than 0.87% for $0 < \frac{\Delta t_\infty}{t_s} \le 4.27$, while the prediction deviation of (11.8) is less than 1.48% for $\frac{\Delta t_\infty}{t_s} = 0$, and less than 3.32% for $0 < \frac{\Delta t_\infty}{t_s} \le 4.27$. These predicted results are well coincident to the related numerical results for $0 \le \frac{\Delta t_\infty}{t_s} \le 4.27$.

Obviously, (11.7) and (11.8) are suitable for laminar forced film condensation of water vapour on a horizontal flat plate.

11.4 Prediction Equations on Heat Transfer

Combined with the formulated equations (11.7) and (11.8), the theoretical analysis equations (11.1) to (11.6) on heat transfer become available for prediction of the condensate heat transfer results. For example, combined with the formulated equations (11.7) and (11.8), (11.4) and (11.6) are changed to the following equations for prediction of the heat transfer result and average Nusselt number within $\frac{\Delta t_\infty}{t_s} \le 4.27$ for laminar forced film condensation of water vapour on the horizontal flat plate.

Table 11.1 Numerical results $-(\mathrm{d}\theta_l(\eta_l)/\mathrm{d}\eta_l)_{\eta_l=0}$ of laminar forced film condensation of saturated and superheated water vapour on a horizontal flat plate (a denotes numerical solutions, b denotes predicted values by using (11.7) and (11.8), and c denotes predicted deviation)

		$\frac{\Delta t_w}{t_s}$								
		0.01	0.025	0.05	0.1	0.2	0.4	0.6	0.8	1
$\frac{\Delta t_\infty}{t_s}$		$-\left(\frac{\mathrm{d}\theta_l}{\mathrm{d}\eta_l}\right)_{\eta_l=0}$								
0	a	1.159	0.928502	0.81245	0.7353442	0.681475	0.626769	0.583673	0.545086	0.5113987
	b	1.1612	0.929699	0.808615	0.736670	0.691584	0.629696	0.579536	0.541104	0.5144
	c	−0.0019	−0.00129	0.00472	−0.00180	0.014834	0.00467	−0.00709	−0.00731	0.005869
00.27	a	1.159739	0.928717	0.812572	0.7354518	0.681496	0.626772	0.583673	0.545085	0.511484
	b	1.1612	0.929699	0.808615	0.736670	0.689089	0.637459	0.589698	0.545515	0.504642
	c	−0.00126	−0.00106	0.00487	−0.00165	0.014803	0.004665	−0.00709	−0.0073	0.005701
10.27	a	1.160344	0.929191	0.812921	0.7356132	0.681561	0.626772	0.583673	0.545098	0.5114841
	b	1.1612	0.929699	0.808615	0.736670	0.691584	0.629696	0.579536	0.541104	0.5144
	c	−0.00074	−0.00055	0.005297	−0.00143	0.014706	0.004665	−0.00709	−0.00733	0.005701
20.27	a	1.161017	0.92964	0.813231	0.7358825	0.68168	0.626982	0.583899	0.545385	0.5118526
	b	1.1612	0.929699	0.808615	0.736670	0.691584	0.629696	0.579536	0.541104	0.5144
	c	−0.00016	−6.4E–05	0.005676	−0.001070	0.014529	0.004329	−0.00747	−0.00785	0.004977
30.27	a	1.161584	0.930098	0.813694	0.7369621	0.684828	0.631521	0.589009	0.549464	0.5156593
	b	1.1612	0.929699	0.808615	0.736670	0.691584	0.629696	0.579536	0.541104	0.5144
	c	0.000331	0.000428	0.006241	0.000396	0.009865	0.004329	−0.00747	−0.00785	0.004977
40.27	a	1.16219	0.930711	0.815574	0.7430987	0.694397	0.642479	0.599185	0.55966	0.5255988
	b	1.1612	0.929699	0.808615	0.736670	0.691584	0.629696	0.579536	0.541104	0.5144
	c	0.000852	0.001087	0.008533	0.008650	−0.00405	−0.0199	−0.03279	−0.03316	−0.02131

$$Q_{x,w} = \sqrt{2}b\lambda_{l,w}(t_w - t_s){\mathrm{Re}_{xl,s}}^{1/2}\frac{0.7421\dfrac{\Delta t_w}{t_s} + 0.3124}{\left(\dfrac{\Delta t_w}{t_s}\right)^{0.28}} \quad \left(0.01 \le \frac{\Delta t_w}{t_s} \le 0.1\right) \tag{11.9}$$

$$Q_{x,w} = \sqrt{2}b\lambda_{l,w}(t_w - t_s){\mathrm{Re}_{xl,s}}^{1/2}\left[0.1466\left(\frac{\Delta t_w}{t_s}\right)^2 - 0.3974\left(\frac{\Delta t_w}{t_s}\right) + 0.7652\right] \quad \left(0.1 \le \frac{\Delta t_w}{t_s} \le 1\right) \tag{11.10}$$

$$\overline{Nu}_{x,w} = \sqrt{2}\mathrm{Re}_{xl,s}^{1/2}\frac{0.7421\dfrac{\Delta t_w}{t_s} + 0.3124}{\left(\dfrac{\Delta t_w}{t_s}\right)^{0.28}} \quad \left(0.01 \le \frac{\Delta t_w}{t_s} \le 0.1\right) \tag{11.11}$$

$$\overline{Nu}_{x,w} = \sqrt{2}\mathrm{Re}_{xl,s}^{1/2}\left[0.1466\left(\frac{\Delta t_w}{t_s}\right)^2 - 0.3974\left(\frac{\Delta t_w}{t_s}\right) + 0.7652\right] \quad \left(0.1 \le \frac{\Delta t_w}{t_s} \le 1\right) \tag{11.12}$$

11.5 Mass Transfer Analysis

Set g_x to be a local mass flow rate entering the liquid film through the liquid–vapour interface at position x corresponding to unit area of the plate. According to the boundary layer theory of fluid mechanics, g_x is expressed as

$$g_x = \rho_{l,s}\left(w_{xl,s}\frac{d\delta_l}{dx} - w_{yl,s}\right)_s.$$

With (9.15), (9.17) and (9.18) for the assumed dimensionless coordinate variable $\eta_{l\delta}$, local Reynolds number $\mathrm{Re}_{xl,s}$, and dimensionless velocity components $W_{xl,s}(\eta_{l\delta})$ and $W_{yl,s}(\eta_{l\delta})$ of the liquid film, the above equation is changed to

$$g_x = \rho_{l,s}\left[w_{xv,\infty}W_{xl,s}(\eta_{l\delta})\frac{\mathrm{d}\delta_l}{\mathrm{d}x} - w_{xv,\infty}\left(\frac{1}{2}\mathrm{Re}_{xl,s}\right)^{-1/2}W_{yl,s}(\eta_{l\delta})\right]_s \tag{11.13}$$

and

$$\delta_l = \eta_{l\delta} \left(\frac{1}{2} \frac{w_{xv,\infty} x}{\nu_{l,s}} \right)^{-1/2} x,$$

$$\frac{\mathrm{d}\delta_l}{\mathrm{d}x} = \frac{1}{2} \eta_{l\delta} \left(\frac{1}{2} \mathrm{Re}_{xl,s} \right)^{-1/2}. \tag{11.14}$$

With (11.14), (11.13) is changed to

$$g_x = \rho_{l,s} \left[w_{xv,\infty} W_{xl,s}(\eta_{l\delta}) \frac{1}{2} \eta_{l\delta,s} \left(\frac{1}{2} \mathrm{Re}_{xl,s} \right)^{-1/2} - w_{xv,\infty} \left(\frac{1}{2} \mathrm{Re}_{xl,s} \right)^{-1/2} W_{yl,s}(\eta_{l\delta}) \right],$$

i.e.

$$g_x = \frac{1}{2} \rho_{l,s} w_{xv,\infty} \left(\frac{1}{2} \mathrm{Re}_{xl,s} \right)^{-1/2} [\eta_{l\delta,s} W_{xl,s}(\eta_{l\delta}) - 2W_{yl,s}(\eta_{l\delta})].$$

The above equation is rewritten as

$$g_x = \frac{1}{2} \rho_{l,s} w_{xv,\infty} \left(\frac{1}{2} \mathrm{Re}_{xl,s} \right)^{-1/2} \Phi_s, \tag{11.15}$$

where Φ_s can be defined as mass flow rate parameter of laminar forced film condensation with the following expression:

$$\Phi_s = \eta_{l\delta} \cdot W_{xl,s}(\eta_{l\delta}) - 2W_{yl,s}(\eta_{l\delta}). \tag{11.16}$$

Take G_x to express total mass flow rate entering the liquid film through the liquid–vapour interface from position $x = 0$ to x with width of b of the plate, it should be the following integration:

$$G_x = \iint_A g_x \mathrm{d}A = b \int_0^x g_x \mathrm{d}x,$$

where $A = b \cdot x$ is related area of the plate with b as the plate width corresponding to the condensation.

With (11.15), the above equation is changed to

$$G_x = \frac{1}{2} b \int_0^x \rho_{l,s} w_{xv,\infty} \left(\frac{1}{2} \mathrm{Re}_{xl,s} \right)^{-1/2} \Phi_s \mathrm{d}x.$$

With the definition of $\mathrm{Re}_{xl,s}$, the above equation is changed to

$$G_x = \frac{1}{2} b \int_0^x \rho_{l,s} w_{xv,\infty} \left(\frac{1}{2} \frac{w_{xv,\infty} x}{\nu_{l,s}} \right)^{-1/2} \Phi_s \mathrm{d}x.$$

Then

$$G_x = 2 \cdot \frac{1}{2} b \rho_{l,s} w_{xv,\infty} x^{1/2} \left(\frac{1}{2} \frac{w_{xv,\infty}}{\nu_{l,s}} \right)^{-1/2} \Phi_s,$$

i.e.

$$G_x = 2b\rho_{l,s}\nu_{l,s} \left(\frac{1}{2} \frac{w_{xv,\infty} x}{\nu_{l,s}} \right)^{1/2} \Phi_s,$$

or

$$G_x = \sqrt{2} b \mu_{l,s} \mathrm{Re}_{xl,s}{}^{1/2} \Phi_s. \tag{11.17}$$

Obviously, (11.16) and (11.17) are universally suitable for general laminar forced film condensation.

11.6 Mass Flow Rate Parameter

From (11.15) and (11.17), it is seen that the condensate mass flow rate parameter Φ_s, as the key solution of the governing equations, is only the un-given variable for evaluation of the condensate mass flow rate of the laminar forced film condensation. In this case, the evaluation of the condensate mass flow rate parameter is necessary for calculation of the condensate mass flow rate.

Equation (11.16) demonstrates that the condensate mass flow rate parameter Φ_s is the function of the key solutions on dimensionless liquid film thickness $\eta_{l\delta}$, and dimensionless interfacial liquid velocity components $W_{xl,s}(\eta_{l\delta})$ and $W_{yl,s}(\eta_{l\delta})$. Then, first it is necessary to investigate the variation regulations of these key solutions.

The rigorous key numerical solutions for dimensionless condensate film thickness $\eta_{l\delta}$, and the interfacial dimensionless liquid velocity components $W_{xl,s}(\eta_{l\delta})$ and $W_{yl,s}(\eta_{l\delta})$, obtained together with the solutions in Chap. 10 are plotted in Figs. 11.2, 11.3, and 11.4 and some of their values are listed in Tables 11.2, 11.3, and 11.4, respectively. Then, a system of solutions on the condensate mass flow rate parameter Φ_s found out with (11.16) are plotted in Fig. 11.5 and listed in Table 11.5. From these numerical results, the following points should be addressed.

Decreasing the wall subcooled grade causes decreasing the condensate liquid film thickness. In the range of small wall subcooled grade, the condensate liquid film thickness will decrease at an accelerative pace with decreasing the wall subcooled grade. In addition, increasing the wall subcooled grade will causes increasing the interfacial dimensionless condensate liquid velocity component $W_{yl,s}(\eta_{l\delta})$

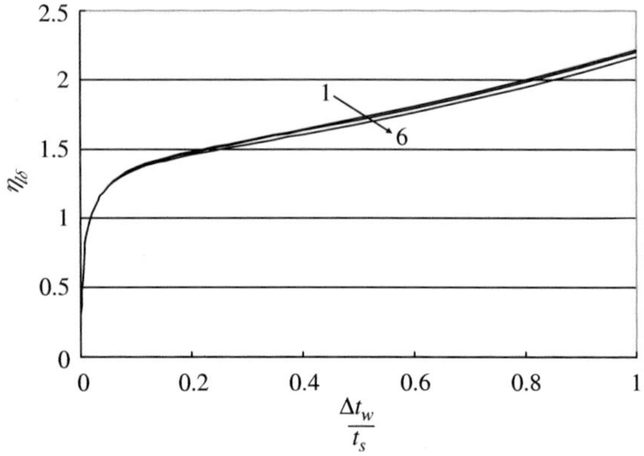

Fig. 11.2 Numerical results of the dimensionless condensate film thickness $\eta_{l\delta}$ of laminar forced film condensation of water vapour on a horizontal flat plate (lines 1 → 6 denotes $\frac{\Delta t_\infty}{t_s} = 0, 0.27,$ 1.27, 2.27, 3.27, and 4.27, respectively)

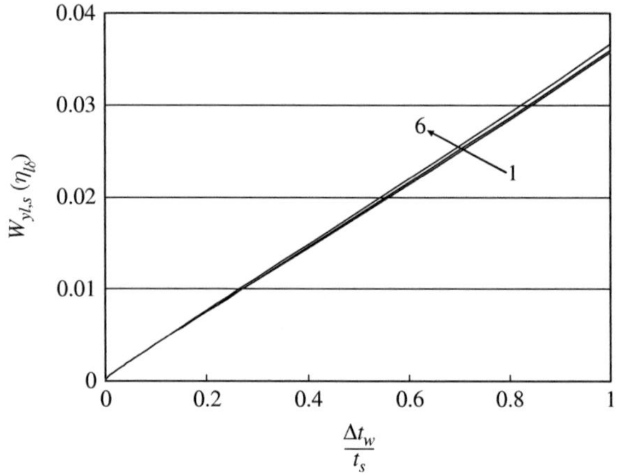

Fig. 11.3 Numerical results of the interfacial dimensionless liquid velocity component $W_{yl,s}(\eta_{l\delta})$ of laminar forced film condensation of superheated water vapour on a horizontal flat plate (lines 1 → 6 denotes $\frac{\Delta t_\infty}{t_s} = 0, 0.27,$ 1.27, 2.27, 3.27, and 4.27, respectively)

proportionally, increasing the interfacial dimensionless condensate liquid velocity component $W_{xl,s}(\eta_{l\delta})$, and increasing the condensate mass flow rate parameter. Compared with the effect of the wall subcooled grade on these key solutions, the effect of the vapour superheated grade is very slight in the range $0 \le \frac{\Delta t_\infty}{t_s} \le 4.27$, and can be ignored for practical prediction of these key solutions without any obvious deviation to the prediction results. On this basis, a curve-fitting approach is used for development related formulated equations, and then the system of these key solutions are formulated respectively to the following equations:

for $0 \le \frac{\Delta t_\infty}{t_s} \le 4.27$

Table 11.2 Numerical results $\eta_{l\delta}$ of laminar forced film condensation of water vapour on a horizontal flat plate

		$\frac{\Delta t_w}{t_s}$								
		0.01	0.025	0.05	0.1	0.2	0.4	0.6	0.8	1
$\frac{\Delta t_\infty}{t_s}$		$\eta_{l\delta}$								
0	a	0.86277	1.07815	1.233585	1.3665	1.484	1.64235	1.8078	2.0001	2.2235
	b	0.878334	1.055665	1.213230	1.390224	1.465073	1.627077	1.806995	2.006808	2.228715
	c	0.0180	−0.0208	−0.01659	0.01736	−0.0127	−0.0093	−0.00044	0.00335	0.00234
0.27	a	0.86262	1.0779	1.2334	1.3663	1.48395	1.64235	1.8078	2.0001	2.2231
	b	0.878334	1.055665	1.213230	1.390224	1.465073	1.627077	1.806995	2.006808	2.228716
	c	0.0182	−0.0206	−0.0163	0.01751	−0.0127	−0.0093	−0.00044	0.00335	0.002526
1.27	a	0.86217	1.07735	1.23287	1.366	1.48381	1.64235	1.8078	2.00005	2.2231
	b	0.878334	1.055665	1.213230	1.390224	1.465073	1.627077	1.806995	2.006808	2.228716
	c	0.01874	−0.02012	−0.0159	0.01773	−0.0126	−0.0093	−0.00044	0.00338	0.002526
2.27	a	0.86167	1.07683	1.2324	1.3655	1.48355	1.6418	1.8071	1.999	2.2215
	b	0.878334	1.055665	1.213230	1.390224	1.465073	1.627077	1.806995	2.006808	2.228716
	c	0.01934	−0.0196	−0.0155	0.0181	−0.01245	−0.0089	−5.785E–05	0.00390	0.003248
3.27	a	0.86125	1.0763	1.2317	1.3635	1.47673	1.63	1.791421	1.98416	2.2051
	b	0.878334	1.055665	1.213230	1.390224	1.465073	1.627077	1.806995	2.006808	2.228716
	c	0.0198	−0.01917	−0.01499	0.01960	−0.00789	−0.00179	0.00869	0.01141	0.010709
4.27	a	0.8608	1.07559	1.22886	1.35224	1.45638	1.6022	1.761	1.94801	2.1634
	b	0.878334	1.055665	1.213230	1.390224	1.465073	1.627077	1.806995	2.006808	2.228716
	c	0.02037	−0.0185	−0.012719	0.02809	0.00597	0.0155	0.0261	0.03018	0.030191

Table 11.3 Numerical results $W_{yl,s}(\eta_{l\delta})$ of laminar forced film condensation of superheated water vapour on a horizontal flat plate

$\frac{\Delta t_\infty}{t_s}$		$\frac{\Delta t_w}{t_s}$: 0.01	0.025	0.05	0.1	0.2	0.4	0.6	0.8	1
		$W_{yl,s}(\eta_{l\delta})$								
0	a	0.00062	0.001245	0.002187	0.003994	0.007547	0.0145	0.021417	0.028425	0.035695
	b	0.000643	0.0012025	0.002135	0.003993	0.007536	0.014622	0.021708	0.028794	0.03588
	c	0.0371	−0.0341	−0.0238	−0.00025	−0.00146	0.00841	0.0136	0.013	0.00518
0.27	a	0.00062	0.001245	0.002187	0.0039937	0.007543	0.014514	0.021419	0.028419	0.035649
	b	0.000643	0.0012025	0.002135	0.003993	0.007536	0.014622	0.021708	0.028794	0.03588
	c	0.0371	−0.0341	−0.0238	−0.00018	−0.00093	0.0074	0.0135	0.0132	0.00648
1.27	a	0.000621	0.001245	0.002188	0.003995	0.007544	0.014514	0.021419	0.02842	0.035650
	b	0.000643	0.0012025	0.002135	0.003993	0.007536	0.014622	0.021708	0.028794	0.03588
	c	0.0354	−0.0341	−0.0242	−0.0005	−0.00106	0.00744	0.01349	0.01316	0.00645
2.27	a	0.000621	0.001246	0.002189	0.003996	0.007545	0.014519	0.021428	0.028435	0.035676
	b	0.000643	0.0012025	0.002135	0.003993	0.007536	0.014622	0.021708	0.028794	0.03588
	c	0.0354	−0.0349	−0.0246	−0.00075	−0.00119	0.00709	0.013	0.0126	0.005718
3.27	a	0.000621	0.001247	0.00219	0.004002	0.00758	0.014624	0.021615	0.028648	0.035941
	b	0.000643	0.0012025	0.002135	0.003993	0.007536	0.014622	0.021708	0.028794	0.03588
	c	0.0354	−0.0357	−0.0251	−0.00225	−0.0058	−0.00014	0.0043	0.005096	−0.0017
4.27	a	0.000622	0.001247	0.002195	0.004035	0.007686	0.014878	0.021989	0.02918	0.036634
	b	0.000643	0.0012025	0.002135	0.003993	0.007536	0.014622	0.021708	0.028794	0.03588
	c	0.03376	−0.0357	−0.0273	−0.0104	−0.0195	−0.0172	−0.0128	−0.0132	−0.0206

Table 11.4 Numerical results $W_{xl,s}(\eta_{l\delta})$ of laminar forced film condensation of superheated water vapour on a horizontal flat plate

$\frac{\Delta t_\infty}{t_s}$		$\frac{\Delta t_w}{t_s}$: 0.01	0.025	0.05	0.1	0.2	0.4	0.6	0.8	1
		$W_{xl,s}(\eta_{l\delta})$								
0	a	0.002871	0.004603	0.007042	0.011522	0.019724	0.03295	0.04225	0.048175	0.0516
	b	0.002991	0.004418	0.006795	0.01226	0.0199	0.03272	0.04226	0.04852	0.0515
	c	0.0418	−0.0403	−0.0351	0.06405	0.0089	−0.00698	0.000237	0.00716	−0.00194
0.27	a	0.00287	0.004603	0.007044	0.011524	0.019714	0.03298	0.042253	0.048165	0.051545
	b	0.002991	0.004418	0.006795	0.01226	0.0199	0.03272	0.04226	0.04852	0.0515
	c	0.04216	−0.0403	−0.03535	0.06389	0.00943	−0.00788	0.000165	0.00737	−0.00087
1.27	a	0.002875	0.004608	0.00705	0.011529	0.019718	0.032981	0.042254	0.048169	0.051547
	b	0.002991	0.004418	0.006795	0.01226	0.0199	0.03272	0.04226	0.04852	0.0515
	c	0.0403	−0.0413	−0.03617	0.0634	0.00923	−0.00791	0.000142	0.007287	−0.00091
2.27	a	0.002879	0.004613	0.007055	0.011538	0.019725	0.033003	0.042287	0.048219	0.051621
	b	0.002991	0.004418	0.006795	0.01226	0.0199	0.03272	0.04226	0.04852	0.0515
	c	0.0389	−0.04238	−0.03685	0.0626	0.00887	−0.00857	−0.00065	0.00624	−0.00234
3.27	a	0.002881	0.004617	0.007063	0.011572	0.019908	0.033483	0.043031	0.048943	0.052392
	b	0.002991	0.004418	0.006795	0.01226	0.0199	0.03272	0.04226	0.04852	0.0515
	c	0.03818	−0.0432	−0.03794	0.05949	−0.0004	−0.02279	−0.0179	−0.00864	−0.017
4.27	a	0.002884	0.004623	0.007096	0.011765	0.020468	0.034655	0.04453	0.050777	0.054431
	b	0.002991	0.004418	0.006795	0.01226	0.0199	0.03272	0.04226	0.04852	0.0515
	c	0.0371	−0.04445	−0.0424	0.04207	−0.02775	−0.0558	−0.05098	−0.04445	−0.05385

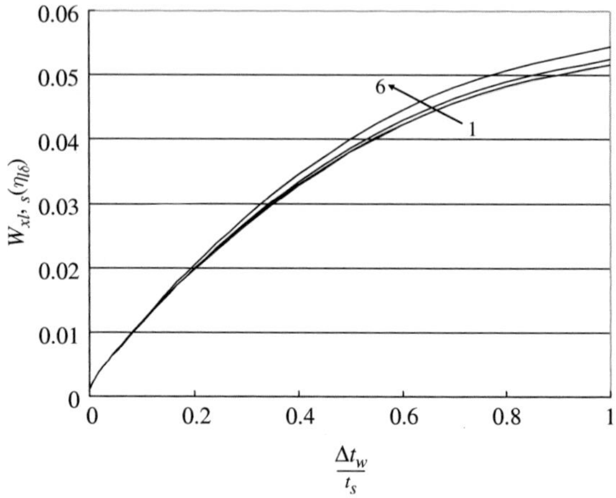

Fig. 11.4 Numerical results of the interfacial dimensionless liquid velocity component $W_{xl,s}(\eta_{l\delta})$ of laminar forced film condensation of superheated water vapour on a horizontal flat plate (lines 1 → 6 denotes $\frac{\Delta t_\infty}{t_s} = 0, 0.27,$ 1.27, 2.27, 3.27, and 4.27, respectively)

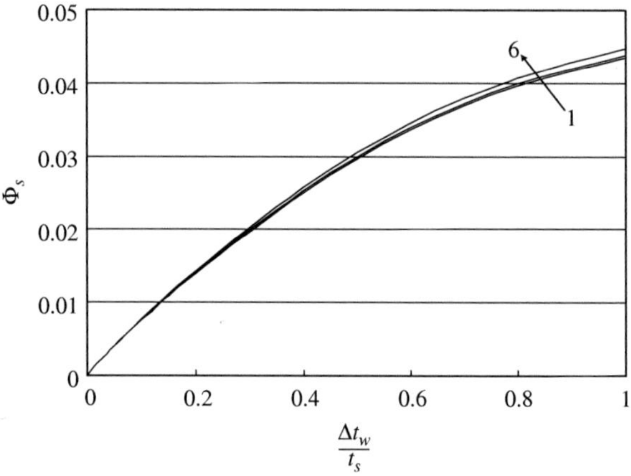

Fig. 11.5 Numerical results of condensate mass flow rate parameter Φ_s of laminar forced film condensation of superheated water vapour on a horizontal flat plate
Note: 1. lines 1 → 6 denotes $\frac{\Delta t_\infty}{t_s} = 0, 0.27, 1.27, 2.27,$ 3.27, and 4.27, respectively

$$\eta_{l\delta} = 2.2134 \left(\frac{\Delta t_w}{t_s}\right)^{0.2007} \quad \left(0.01 \le \frac{\Delta t_w}{t_s} \le 0.1\right) \tag{11.18}$$

$$\eta_{l\delta} = 1.3192 \exp\left(0.5244\left(\frac{\Delta t_w}{t_s}\right)\right) \quad \left(0.1 \le \frac{\Delta t_w}{t_s} \le 1\right) \tag{11.19}$$

$$W_{yl,s}(\eta_{l\delta}) = 0.0373\left(\frac{\Delta t_w}{t_s}\right) + 0.00027 \quad \left(0.01 \le \frac{\Delta t_w}{t_s} \le 0.1\right) \tag{11.20}$$

$$W_{yl,s}(\eta_{l\delta}) = 0.03543\left(\frac{\Delta t_w}{t_s}\right) + 0.00045 \quad \left(0.1 \le \frac{\Delta t_w}{t_s} \le 1\right) \tag{11.21}$$

$$W_{xl,s}(\eta_{l\delta}) = 0.0951\left(\frac{\Delta t_w}{t_s}\right) + 0.00204 \quad \left(0.01 \le \frac{\Delta t_w}{t_s} \le 0.1\right) \tag{11.22}$$

Table 11.5 Numerical results Φ_s of laminar forced film condensation of superheated water vapour on a horizontal flat plate

$\frac{\Delta t_\infty}{t_s}$		$\frac{\Delta t_w}{t_s}$ = 0.01	0.025	0.05	0.1	0.2	0.4	0.6	0.8	1
		Φ_s								
0	a	0.001237	0.002473	0.004313	0.0077575	0.014176	0.025115	0.033545	0.039506	0.043342
	b	0.001289	0.0023675	0.004165	0.007749	0.014096	0.024984	0.033464	0.039536	0.0432
	c	0.042	−0.04266	−0.0343	−0.0011	−0.00564	−0.0052	−0.00241	0.000759	−0.00328
0.27	a	0.001237	0.002472	0.004313	0.007758	0.014168	0.025138	0.033547	0.039497	0.043292
	b	0.001289	0.0023675	0.004165	0.007749	0.014096	0.024984	0.033464	0.039536	0.0432
	c	0.042	−0.04227	−0.0343	−0.0011	−0.00508	−0.00613	−0.00247	0.000987	−0.00213
1.27	a	0.001238	0.002474	0.004315	0.007760	0.01417	0.025138	0.033548	0.039499	0.043293
	b	0.001289	0.0023675	0.004165	0.007749	0.014096	0.024984	0.033464	0.039536	0.0432
	c	0.0412	−0.043	−0.03476	−0.00135	−0.00522	−0.00613	−0.0025	0.000937	−0.00215
2.27	a	0.001239	0.002475	0.004317	0.007762	0.014173	0.025147	0.033561	0.03952	0.043324
	b	0.001289	0.0023675	0.004165	0.007749	0.014096	0.024984	0.033464	0.039536	0.0432
	c	0.04036	−0.0434	−0.03521	−0.00167	−0.0054	−0.00648	−0.00289	0.000405	−0.00286
3.27	a	0.001239	0.002476	0.004319	0.007774	0.014238	0.025329	0.033855	0.039816	0.043647
	b	0.001289	0.0023675	0.004165	0.007749	0.014096	0.024984	0.033464	0.039536	0.0432
	c	0.04036	−0.0438	−0.0356	−0.00319	−0.00997	−0.0136	−0.01155	−0.00703	−0.0102
4.27	a	0.00124	0.002478	0.004329	0.007839	0.014437	0.025769	0.03444	0.040555	0.044488
	b	0.001289	0.0023675	0.004165	0.007749	0.014096	0.024984	0.033464	0.039536	0.0432
	c	0.0395	−0.0446	−0.0379	−0.0114	−0.02362	−0.0305	−0.0283	−0.0251	−0.0289

$$W_{xl,s}(\eta_{l\delta}) = -0.041\left(\frac{\Delta t_w}{t_s}\right)^2 + 0.0887\left(\frac{\Delta t_w}{t_s}\right) + 0.0038 \left(0.1 \le \frac{\Delta t_w}{t_s} \le 1\right) \tag{11.23}$$

$$\Phi_s = 0.0719\left(\frac{\Delta t_w}{t_s}\right) + 0.00057 \left(0.01 \le \frac{\Delta t_w}{t_s} \le 0.1\right) \tag{11.24}$$

$$\Phi_s = -0.0301\left(\frac{\Delta t_w}{t_s}\right)^2 + 0.0725\left(\frac{\Delta t_w}{t_s}\right) + 0.0008 \left(0.1 \le \frac{\Delta t_w}{t_s} \le 1\right) \tag{11.25}$$

The results on the dimensionless condensate film thickness $\eta_{l\delta}$, the interfacial dimensionless condensate liquid velocity components $W_{xl,s}(\eta_{l\delta})$ and $W_{yl,s}(\eta_{l\delta})$, and the condensate mass flow rate parameter Φ_s, evaluated by using (11.18), (11.19), (11.20), (11.21), (11.22), (11.23), (11.24), and (11.25), are listed in Tables 11.2, 11.3, 11.4, and 11.5, respectively. It is seen that these evaluated results are well coincident to the related numerical solutions. Meanwhile, for the condensate mass flow rate parameter Φ_s, the key solution of the laminar forced film condensation of water vapour, the deviations predicted by using (11.24) and (11.25), are respectively less than 1.46% and 0.56% for $\frac{\Delta t_\infty}{t_s} = 0$, and respectively less than 1.67% and 3.05% for $0 < \frac{\Delta t_\infty}{t_s} \le 4.27$.

Obviously, (11.18), (11.19), (11.20), (11.21), (11.22), (11.23), (11.24), and (11.25) are suitable for laminar forced film condensation of water vapour.

11.7 Prediction Equations on Condensate Mass Transfer

Combined with the reliable formulated equations (11.24) and (11.25) on the condensate mass flow rate parameter, the mass transfer theoretical expression (11.17) will become the following simple and reliable equation for prediction of condensate mass transfer on laminar forced condensation of saturated and superheated water vapour.

for $0 \le \dfrac{\Delta t_\infty}{t_s} \le 4.27$,

$$G_x = 1.414 b \mu_{l,s} \,\mathrm{Re}_{xl,s}{}^{1/2} \left[0.0719\left(\frac{\Delta t_w}{t_s}\right) + 0.00057\right] \left(0.01 \le \frac{\Delta t_w}{t_s} \le 0.1\right) \tag{11.26}$$

$$G_x = 1.414 b \mu_{l,s} \,\mathrm{Re}_{xl,s}{}^{1/2} \left[-0.0301\left(\frac{\Delta t_w}{t_s}\right)^2 + 0.0725\left(\frac{\Delta t_w}{t_s}\right) + 0.0008\right] \left(0.1 \le \frac{\Delta t_w}{t_s} \le 1\right) \tag{11.27}$$

Obviously, the reliability of (11.26) and (11.27) is same as that of (11.24) and (11.25).

11.8 Condensate Mass–Energy Transformation Equation

11.8.1 Derivation on Condensate Mass–Energy Transformation Equation

The diagram of heat transfer balance with the two-phase flow forced film condensation is expressed in Fig. 11.6. At the liquid–vapour interface, the liquid heat conduction $(q_x)_l$ is balanced by the latent heat of vapour condensation, $(q_x)_h = h_{fg}g_x$, and the vapour heat conduction $(q_x)_v$, i.e.

$$-(q_x)_l = h_{fg}g_x - (q_x)_v. \tag{11.28}$$

It is indicated that the negative signs of the heat transfer rate in (11.28) express that the direction of the related heat fluxes is negative to the coordinative direction. Additionally, the interfacial liquid heat conductivity rate $(q_x)_l$ is divided to two parts, and shown as below:

$$-(q_x)_l = -q_{x,w} + q_{x,l}, \tag{11.29}$$

where $q_{x,w}$ is the defined condensation heat transfer rate on the wall and $q_{x,l}$ is the heat transfer rate brought out by the liquid film flow.

From (11.28) and (11.29), we have the following equivalent equation on heat transfer at liquid-vapour interface for the forced two-phase flow film condensation:

$$h_{fg}g_x - (q_x)_v \equiv -q_{x,w} + q_{x,l}. \tag{11.30}$$

Now, let us analyse the amount of latent heat $h_{fg}g_x$ of vapour condensation and vapour heat conduction rate $(q_x)_v$ at the liquid–vapour interface. As we know, the latent heat of vapour condensation per unit vapour mass h_{fg} is very larger, but the related vapour thermal conductivity λ is very small. For example, for water

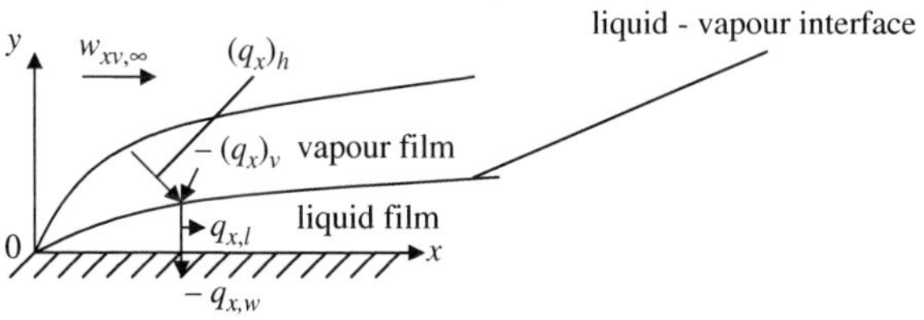

Fig. 11.6 Diagram of equivalent heat and mass transfer of the forced film condensation with the two-phase flow

vapour, h_{fg} = 2257100 J/kg, but λ = 0.00245 W/(m K). Therefore, the latent heat of vapour condensation $h_{fg}g_x$ will be far larger than the vapour heat conduction rate $(q_x)_v$ at the liquid–vapour interface, and the vapour heat conduction $(q_x)_v$ can be ignored compared with the latent heat of the vapour condensation. Then (11.30) can be changed to the following equation for laminar forced film condensation of vapour at a broad range of the vapour superheated grade:

$$h_{fg}g_x = -q_{x,w} + q_{x,l}. \tag{11.30a}$$

In addition, compared with the defined condensation heat transfer rate on the wall, $q_{x,w}$, the heat transfer rate brought out by the liquid film flow, $q_{x,l}$, is very slight. Then (11.30a) is simplified to

$$h_{fg}g_x = -q_{x,w}. \tag{11.31}$$

With (11.1) and (11.15), the above equation is changed to

$$h_{fg}\frac{1}{2}\rho_{l,s}w_{xv,\infty}\left(\frac{1}{2}\mathrm{Re}_{xl,s}\right)^{-1/2}\Phi_s = -\lambda_{l,w}(t_w - t_s)\left(\frac{1}{2}\mathrm{Re}_{xl,s}\right)^{1/2}x^{-1}\left(-\frac{\mathrm{d}\theta_l(\eta_l)}{\mathrm{d}\eta_l}\right)_{\eta_l=0}.$$

With the definition of $\mathrm{Re}_{xl,s}$, the above equation is simplified to

$$\left(-\frac{\mathrm{d}\theta_l(\eta_l)}{\mathrm{d}\eta_l}\right)_{\eta_l=0} = -\frac{h_{fg}\frac{1}{2}\rho_{l,s}w_{xv,\infty}}{\lambda_{l,w}(t_w - t_s)\left(\frac{1}{2}\frac{w_{xv,\infty}x}{\nu_{l,s}}\right)x^{-1}}\Phi_s,$$

i.e.

$$\left(-\frac{\mathrm{d}\theta_l(\eta_l)}{\mathrm{d}\eta_l}\right)_{\eta_l=0} = C_{mh}\Phi_s, \tag{11.32}$$

where

$$C_{mh} = \frac{\mu_{l,s}h_{fg}}{\lambda_{l,w}(t_s - t_w)}. \tag{11.33}$$

Since the condensate mass flow rate parameter Φ_s and the wall temperature gradient $\left(-\frac{d\theta_l(\eta_l)}{d\eta_l}\right)_{\eta_l=0}$ dominate the condensate mass and heat transfer respectively, (11.32) expresses the internal transformation relationship between the condensate mass and heat transfer, and can be regarded as the condensate mass–energy transformation equation of the laminar forced film condensation, where the dimensional factor C_{mh} can be regarded as the mass–energy transformation coefficient. The present condensate mass–energy transformation equation (11.32) with (11.33) will be used for a convenient transformation between condensate heat and mass transfer, as well as for judgement of the reliability of the solutions of the forced film condensation.

From the derivation it is concluded that, the condensate mass–energy transformation equation (11.32) with (11.33) is universally suitable for any type of laminar forced film condensation.

11.8.2 Mass–Energy Transformation Coefficient

Equation (11.33) shows that the condensate mass–energy transformation coefficient is independent of the geometric conditions and only dependent on the vapour physical properties, as well as the temperature boundary conditions of laminar forced film condensation.

Now, let us analyse the physical significance of the condensate mass–energy transformation coefficient C_{mh}. The mass–energy transformation coefficient in (11.33) can be rewritten as

$$C_{mh} = \mathrm{Pr}_{l,s} \frac{\lambda_{l,s}}{\mu_{l,s} c_{p_{l,s}}} \frac{\mu_{l,s} h_{fg}}{\lambda_{l,w}(t_s - t_w)}.$$

Then the above equation becomes

$$C_{mh} = \mathrm{Pr}_{l,s} \frac{\lambda_{l,s}}{\lambda_{l,w}} \cdot \frac{h_{fg}}{c_{p_{l,s}}(t_s - t_w)}. \tag{11.34}$$

The physical factor $\frac{h_{fg}}{c_{p_{l,s}}(t_s - t_w)}$ expresses the ratio of the condensate latent heat of vapour to the condensate liquid heat capacity. With the physical factor $\frac{h_{fg}}{c_{p_{l,s}}(t_s - t_w)}$, (11.34) shows a clear physical significance to the mass–energy transformation coefficient C_{mh}. It follows that the mass–energy transformation coefficient is proportional to the physical factor $\frac{h_{fg}}{c_{p_{l,s}}(t_s - t_w)}$, interfacial liquid Prandtl number $\mathrm{Pr}_{l,s}$, and the liquid thermal conductivity ratio $\frac{\lambda_{l,s}}{\lambda_{l,w}}$. Obviously, values of mass–energy transformation coefficient C_{mh} only depends on the physical properties of the related condensate liquid and its vapour.

The values of mass–energy transformation coefficient C_{mh} are found out by using (11.33) with different wall subcooled temperature $t_s - t_w$ of laminar forced film condensation of water vapour, and listed in Table 11.6. Meanwhile, the values of mass–energy transformation coefficient C_{mh} are obtained by using (11.32) $C_{mh} = \frac{\left(-\frac{\mathrm{d}\theta_l(\eta_l)}{\mathrm{d}\eta_l}\right)_{\eta_l=0}}{\Phi_s}$ with the related numerical results of the wall dimensionless temperature gradient $\left(-\frac{\mathrm{d}\theta_l(\eta_l)}{\mathrm{d}\eta_l}\right)_{\eta_l=0}$ and condensate mass flow rate parameter Φ_s for laminar forced film condensation of water vapour on a horizontal flat plate, and

Table 11.6 Evaluated results on C_{mh} based on (11.33), and (11.32) with numerical solutions Φ_s and $\left(-\frac{d\theta_l(\eta_l)}{d\eta_l}\right)_{\eta_l=0}$ for laminar forced condensation of water vapour on a horizontal flat plate

	$\frac{\Delta t_w}{t_s}$								
	0.01	0.025	0.05	0.1	0.2	0.4	0.6	0.8	1
	$t_s - t_w$, °C								
	1	2.5	5	10	20	40	60	80	100
	t_w, °C								
	99	97.5	95	90	80	60	40	20	0
	$\lambda_{l,w}$, W/(m K)								
	0.6766	0.6760	0.6750	0.6727	0.6669	0.6506	0.6278	0.5986	0.563
$\frac{\Delta t_\infty}{t_s}$	$\left(-\frac{d\theta_l(\eta_l)}{d\eta_l}\right)_{\eta_l=0}$ (numerical solutions)								
0	1.159	0.928502	0.81245	0.735344	0.681475	0.626769	0.583673	0.545086	0.511399
0.27	1.159739	0.928717	0.812572	0.735452	0.681496	0.626772	0.583673	0.545085	0.511484
1.27	1.160344	0.929191	0.812921	0.735613	0.681561	0.626772	0.583673	0.545098	0.511484
2.27	1.161017	0.92964	0.813231	0.735883	0.68168	0.626982	0.583899	0.545385	0.511853
3.27	1.161584	0.930098	0.813694	0.736962	0.684828	0.631521	0.589009	0.549464	0.515659
4.27	1.16219	0.930711	0.815574	0.743099	0.694397	0.642479	0.599185	0.55966	0.525599
	Φ_s (numerical solutions)								
0	0.001237	0.002473	0.004313	0.007757	0.014176	0.025115	0.033545	0.039506	0.043342
0.27	0.001237	0.002472	0.004313	0.007758	0.014168	0.025138	0.033547	0.039497	0.043292
1.27	0.001238	0.002474	0.004315	0.007760	0.01417	0.025138	0.033548	0.039499	0.043293
2.27	0.001239	0.002475	0.004317	0.007762	0.014173	0.025147	0.033561	0.03952	0.043324
3.27	0.001239	0.002476	0.004319	0.007774	0.014238	0.025329	0.033855	0.039816	0.043647
4.27	0.00124	0.002478	0.004329	0.007839	0.014437	0.025769	0.03444	0.040555	0.044488

Table 11.6 (continued)

	$C_{mh}=\left(-\frac{d\theta_l(\eta_l)}{d\eta_l}\right)_{\eta_l=0}\Big/\Phi_s$ (numerical results)								
0	936.9442	375.4557	188.3723	94.7975	48.0724	24.95596	17.3997	13.7975	11.7992
0.27	937.5416	375.6945	188.4006	94.7992	48.1011	24.93325	17.3986	13.8006	11.8147
1.27	937.273	375.5824	188.3942	94.7955	48.0989	24.93324	17.3981	13.8003	11.8145
2.27	937.0597	375.6121	188.3787	94.8058	48.0971	24.93267	17.3981	13.8002	11.8145
3.27	937.5174	375.64540	188.3987	94.7983	48.0986	24.93272	17.3979	13.8001	11.8143
4.27	937.25	375.58960	188.39780	94.7951	48.0984	24.93224	17.3979	13.8000	11.8144
	predicted results C_{mh} with $C_{mh}=\frac{\mu_{l,s}h_{fg}}{\lambda_{l,w}(t_s-t_w)}$ ($\mu_{l,s}=282.5\times10^{-6}$Kg/(m s), $h_{fg}=2257.1$ kJ/kg)								
0	942.4684	377.2935	188.9249	94.7841	47.8030	24.5030	16.9281	13.3151	11.3256
0.27	942.4684	377.2935	188.9249	94.7841	47.8030	24.50300	16.9281	13.3151	11.3256
1.27	942.4684	377.2935	188.9249	94.7841	47.8030	24.5030	16.9281	13.3151	11.3256
2.27	942.4684	377.2935	188.9249	94.7841	47.8030	24.5030	16.9280	13.3151	11.3256
3.27	942.4684	377.2935	188.9249	94.7841	47.8030	24.5030	16.9281	13.3151	11.3256
4.27	942.4684	377.2935	188.9249	94.7841	47.8030	24.5030	16.9281	13.3151	11.3256
	relative deviation of C_{mh} predicted by equations $C_{mh}=\left(-\frac{d\theta_l(\eta_l)}{d\eta_l}\right)_{\eta_l=0}\Big/\Phi_s$ with numerical results								
0	0.0059	0.00489	0.00293	−0.0001	−0.0056	−0.018	−0.027	−0.0350	−0.040
0.27	0.00525	0.00426	0.00278	−0.0001	−0.0062	−0.0173	−0.0270	−0.035	−0.0414
1.27	0.00554	0.00455	0.00282	−0.0001	−0.0061	−0.0173	−0.0270	−0.0351	−0.0414
2.27	0.00577	0.00448	0.0029	−0.0002	−0.0061	−0.0172	−0.0270	−0.0351	−0.0414
3.27	0.00528	0.00439	0.00279	−0.0001	−0.0061	−0.0172	−0.0270	−0.0351	−0.0414
4.27	0.00557	0.00454	0.00280	−0.0001	−0.0061	−0.0172	−0.0270	−0.0351	−0.0414
$\frac{\Delta t_\infty}{t_s}$	$\frac{\Delta t_w}{t_s}$								
	0.01	0.025	0.05	0.1	0.2	0.4	0.6	0.8	1

Note: the present values of physical properties $\lambda_{l,w}$, $\mu_{l,s}$, and h_{fg} are corresponding to the numerical program listed in Table 11.6 also. It is seen that the values of mass–energy transformation coefficient C_{mh} obtained by the different approaches are well coincident. Then, it follows that the present calculated results on the key solutions $\left(-\frac{d\theta_l(\eta_l)}{d\eta_l}\right)_{\eta_l=0}$ and Φ_s are reliable, and the condensate mass–heat transfer transformation equation (11.32) with (11.33) is valid.

11.9 Summary

Now it is necessary to summarize in Table 11.7 the governing partial differential equations, the similarity variables, the similarity governing mathematical models, physical property factor equations, and the condensate heat and mass transfer equations on the two-phase flows laminar forced film condensation of vapour, including consideration of a series of interfacial physical matching conditions.

11.10 Remarks

The theoretical equations on condensate heat and mass transfer reported in this chapter based on the new similarity analysis method demonstrate that the wall dimensionless temperature gradient and condensate mass flow rate parameter are key solutions of the governing similarity mathematical model for dominating the condensate heat and mass transfer results. The laminar forced film condensation of water vapour on a horizontal flat plate is taken as an example, and a system of numerical results on the wall dimensionless temperature gradient and condensate mass flow rate parameter is found out rigorously. Meanwhile, the condensate mass flow rate parameter depends on the dimensionless condensate film thickness and interfacial condensate velocity components. Thus, the above theoretical equations on condensate heat and mass transfer become available for reliable prediction of condensate heat and mass transfer results for laminar forced film condensation of water vapour on a horizontal flat plate.

The variation regulations of the series of physical variables on laminar forced film condensation are clarified. Decreasing the wall subcooled grade will cause increasing the wall dimensionless temperature gradient; and in the region of lower vapour superheated grade $\frac{\Delta t_\infty}{t_s}$, the wall dimensionless temperature gradient will increase at an accelerative pace with decreasing the vapour superheated grade $\frac{\Delta t_\infty}{t_s}$. Decreasing the wall subcooled grade causes decreasing the condensate liquid film thickness. In the range of lower wall subcooled grade, the condensate liquid film thickness will decrease at an accelerative pace with decreasing the condensate liquid film thickness. In addition, increasing the wall subcooled grade will causes increasing the interfacial liquid dimensionless velocity component $W_{yl,s}(\eta_{l\delta})$ proportionally, increasing the interfacial liquid dimensionless velocity component $W_{xl,s}(\eta_{l\delta})$, and increasing the condensate mass flow rate parameter. All these effects demonstrate the influence of variable physical properties on condensate heat and

Table 11.7 Governing partial differential equations, the similarity variables, the similarity governing mathematical models, physical property factor equations, and the condensate heat and mass transfer equations on the two-phase flows laminar forced film condensation of vapour, including consideration of a series of interfacial physical matching conditions

Governing partial differential equations for liquid film

Mass equation
$$\frac{\partial}{\partial x}(\rho_l w_{xl}) + \frac{\partial}{\partial y}(\rho_l w_{yl}) = 0$$

Momentum equation
$$\rho_l\left(w_{xl}\frac{\partial w_{xl}}{\partial x} + w_{yl}\frac{\partial w_{xl}}{\partial y}\right) = \frac{\partial}{\partial y}\left(\mu_l\frac{\partial w_{xl}}{\partial y}\right)$$

Energy equation
$$\rho_l\left[w_x\frac{\partial(c_{p_l}t)}{\partial x} + w_y\frac{\partial(c_{p_l}t)}{\partial y}\right] = \frac{\partial}{\partial y}\left(\lambda_l\frac{\partial t}{\partial y}\right)$$

Governing partial differential equations for vapour film

Mass equation
$$\frac{\partial}{\partial x}(\rho_v w_{xv}) + \frac{\partial}{\partial y}(\rho_{v,} w_{yv}) = 0$$

Momentum equation
$$\rho_v\left(w_{xv}\frac{\partial w_{xv}}{\partial x} + w_{yv}\frac{\partial w_{xv}}{\partial y}\right) = \frac{\partial}{\partial y}\left(\mu_v\frac{\partial w_{xv}}{\partial y}\right)$$

Energy equation
$$\rho_v\left(w_{xv}\frac{\partial(c_{p_v}t_v)}{\partial x} + w_{yv}\frac{\partial(c_{p_v}t_v)}{\partial y}\right) = \frac{\partial}{\partial y}\left(\lambda_v\frac{\partial t_v}{\partial y}\right)$$

Boundary conditions of governing partial differential equations

y = 0
$$w_{xl} = 0, \quad w_{yl} = 0, \quad t_l = t_w$$

$y = \delta_l$:
$$w_{xl,s} = w_{xv,s}$$
$$\rho_{l,s}\left(w_{xl}\frac{\partial\delta_l}{\partial x} - w_{yl}\right)_{l,s} = \rho_{v,s}\left(w_{xv}\frac{\partial\delta_v}{\partial x} - w_{yv}\right)_{v,s}$$
$$\mu_{l,s}\left(\frac{\partial w_{xl}}{\partial y}\right)_s = \mu_{v,s}\left(\frac{\partial w_{xv}}{\partial y}\right)_s$$
$$\lambda_{l,s}\left(\frac{\partial t_l}{\partial y}\right)_{y=\delta_l} = \lambda_{v,s}\left(\frac{\partial t_v}{\partial y}\right)_{y=\delta_l} + h_{fg}\rho_{v,s}\left(w_{xv}\frac{\partial\delta_v}{\partial x} - w_{yv}\right)_s$$
$$t = t_s$$

y → ∞
$$w_{xv} = w_{xv,\infty}, \quad t_v = t_\infty$$

Table 11.7 (continued)

Equations of similarity variables

For liquid film

$$\eta_l = \frac{y}{x}\left(\frac{1}{2}\mathrm{Re}_{xl,s}\right)^{1/2}$$

$$\mathrm{Re}_{xl,s} = \frac{w_{xv,\infty}x}{\nu_{l,s}}$$

$$\theta_l = \frac{t_l - t_s}{t_w - t_s}$$

$$w_{xl} = w_{xv,\infty}W_{xl}(\eta_l)$$

$$w_{yl} = w_{xv,\infty}\left(\frac{1}{2}\mathrm{Re}_{xl,s}\right)^{-1/2} W_{yl}(\eta_l)$$

For vapour film

$$\eta_v = \frac{y}{x}\left(\frac{1}{2}\mathrm{Re}_{xv,\infty}\right)^{1/2}$$

$$\mathrm{Re}_{xv,\infty} = \frac{w_{xv,\infty}x}{\nu_{v,\infty}}$$

$$\theta_v = \frac{t_v - t_\infty}{t_s - t_\infty}$$

$$w_{xv} = w_{xv,\infty}W_{xv}(\eta_v)$$

$$w_{yv} = w_{xv,\infty}\left(\frac{1}{2}\mathrm{Re}_{xv,s}\right)^{-1/2} W_{yv}(\eta_v)$$

Governing ordinary differential equations for liquid film

Mass equation

$$-\eta_l\frac{dW_{xl}(\eta_l)}{d\eta_l} + 2\frac{dW_{yl}(\eta_l)}{d\eta_l} + \frac{1}{\rho_l}\frac{d\rho_l}{d\eta_l}[-\eta_l W_{xl}(\eta_l) + 2W_{yl}(\eta_l)] = 0$$

Momentum equation

$$\frac{\nu_{l,s}}{\nu_l}[-\eta_l W_{xl}(\eta_l) + 2W_{yl}(\eta_l)]\frac{dW_{xl}(\eta_l)}{d\eta_l} = \frac{d^2W_{xl}(\eta_l)}{d\eta_l^2} + \frac{1}{\mu_l}\frac{d\mu_l}{d\eta_l}\frac{dW_{xl}(\eta_l)}{d\eta_l}$$

Energy equation

$$\mathrm{Pr}_l\frac{\nu_{l,s}}{\nu_l}\left\{\begin{array}{l}[-\eta_l \cdot W_{xl}(\eta_l) + 2W_{yl}(\eta_l)]\dfrac{d\theta_l(\eta_l)}{d\eta_l} \\ +[-\eta_l W_{xl}(\eta_l) + 2W_{yl}(\eta_l)]\left[\theta_l(\eta_l) + \dfrac{t_s}{t_w - t_s}\right]\dfrac{1}{c_{p_l}}\dfrac{dc_{p_l}}{d\eta_l}\end{array}\right\}$$

$$= \frac{d^2\theta_l(\eta_l)}{d\eta_l^2} + \frac{1}{\lambda_l}\frac{d\lambda_l}{d\eta_l}\frac{d\theta_l(\eta_l)}{d\eta_l}$$

Table 11.7 (continued)

Governing ordinary differential equations for vapour film

Mass equation

$$-\eta_v \frac{\partial W_{xv}(\eta_v)}{\partial \eta_v} + 2\frac{\partial W_{yv}(\eta_v)}{\partial \eta_v} - \frac{1}{\rho_v}\frac{d\rho_v}{d\eta_v}[\eta_v W_{xv}(\eta_v) + 2W_{yv}(\eta_v)] = 0$$

Momentum equation

$$\frac{\nu_{v,\infty}}{\nu_v}[-\eta_v W_{xv} + 2W_{yv}]\frac{dW_{xv}(\eta_v)}{d\eta_v} = \frac{d^2 W_{xv}(\eta_v)}{d\eta_v^2} + \frac{1}{\mu_v}\frac{d\mu_v}{d\eta_v}\frac{dW_{xv}(\eta_v)}{d\eta_v}$$

Energy equation

$$\Pr_v \frac{\nu_{v,\infty}}{\nu_v}\left\{\begin{array}{l}\left[-W_{xv}(\eta_v)\eta_v + 2W_{yv}(\eta_v)\right]\dfrac{d\theta_v(\eta_v)}{d\eta_v} \\ -[W_{xv}(\eta_v)\eta_v + 2W_{yv}(\eta_v)]\left(\theta_v(\eta_v) + \dfrac{t_\infty}{t_s - t_\infty}\right)\dfrac{1}{c_{p_v}}\dfrac{dc_{p_v}}{d\eta_v}\end{array}\right\}$$

$$= \frac{d^2\theta_v(\eta_v)}{d\eta_v^2} + \frac{1}{\lambda_v}\frac{d\lambda_v}{d\eta_v}\frac{d\theta_v(\eta_v)}{d\eta_v}$$

Boundary conditions of governing ordinary differential equations

$\eta_l = 0:$ $\quad W_{xl}(\eta_l) = 0,\ W_{yl}(\eta_l) = 0,\ \theta(\eta_l) = 1$

$\eta_l = \eta_{l\delta}$ $\quad W_{xv,s}(\eta_{v0}) = W_{xl,s}(\eta_{l\delta})$

$$W_{yv,s}(\eta_{v,s}) = -\frac{\rho_{l,s}}{\rho_{v,s}}\left(\frac{\nu_{l,s}}{\nu_{v,s}}\right)^{1/2}\left[\frac{1}{2}\eta_{l\delta} W_{xl,s}(\eta_{l\delta}) - W_{yl,s}(\eta_{l\delta})\right]$$

$$\left(\frac{dW_{xv}(\eta_v)}{d\eta_v}\right)_s = \frac{\mu_{l,s}}{\mu_{v,s}}\left(\frac{\nu_{v,\infty}}{\nu_{l,s}}\right)^{1/2}\left(\frac{dW_x(\eta_l)}{d\eta_l}\right)_s$$

$$\left(\frac{d\theta_v(\eta_v)}{d\eta_v}\right)_s = \frac{\lambda_{l,s}\left(\dfrac{\nu_{v,\infty}}{\nu_{l,s}}\right)^{1/2}(t_w - t_s)\left(\dfrac{d\theta_l(\eta_l)}{d\eta_l}\right)_s + 2h_{fg}\rho_{v,s}\nu_{v,\infty}W_{yv,s}(\eta_{v,0})}{\lambda_{v,s}(t_s - t_\infty)}$$

$\theta_l(\eta_{l\delta}) = 0$ and $\theta_v(\eta_{v,0} = 0) = 1$

$\eta_v \to \infty:$ $\quad W_{xv}(\eta_{v,\infty}) = 1, \qquad \theta_v(\eta_{v,\infty}) = 0$

Table 11.7 (continued)

Physical property factor equations on condensate liquid film	
For condensate water film medium	$\frac{1}{c_{pl}}\frac{dc_{pl}}{d\eta_l}=0$
	$\Pr_l=\frac{\mu_l\cdot c_{pl}}{\lambda_l}=\frac{\left[\exp\left(-1.6-\frac{1150}{T}+\left(\frac{690}{T}\right)^2\right)\times 10^{-3}\right]\times 4200}{-8.01\times 10^{-6}t^2+1.94\times 10^{-3}t+0.563}$
	$\frac{1}{\rho_l}\frac{d\rho_l}{d\eta_l}=\frac{\left[-2\times 4.48\times 10^{-3}t(t_w-t_s)\frac{d\theta_l}{d\eta_l}\right]}{(-4.48\times 10^{-3}t^2+999.9)^{-1}}$
	$\frac{1}{\lambda_l}\frac{d\lambda_l}{d\eta_l}=\frac{\left[(-2\times 8.01\times 10^{-6}+1.94\times 10^{-3})(t_w-t_s)\frac{d\theta_l(\eta_l)}{d\eta_l}\right]}{-8.01\times 10^{-6}t^2+1.94\times 10^{-3}t+0.563}$
	$\frac{1}{\mu_l}\frac{d\mu_l}{d\eta_l}=\left(\frac{1150}{T^2}-2\times\frac{690^2}{T^3}\right)(t_w-t_s)\frac{d\theta_l}{d\eta_l}$
	$\frac{\nu_{l,\infty}}{\nu_l}=\frac{\exp\left[-1.6-\frac{1150}{T_\infty}+\left(\frac{690}{T_\infty}\right)^2\right]}{\exp\left[-1.6-\frac{1150}{T}+\left(\frac{690}{T}\right)^2\right]}\times\frac{-4.48\times 10^{-3}t^2+999.9}{-4.48\times 10^{-3}t_\infty^2+999.9}$
Physical property factor equations on vapour film	
For general vapour	$\frac{1}{\rho_v}\frac{d\rho_v}{d\eta_v}=-\frac{(T_s/T_\infty-1)d\theta_v(\eta_v)/d\eta_v}{(T_s/T_\infty-1)\theta_v(\eta_v)+1}$
	$\frac{1}{\mu_v}\frac{d\mu_v}{d\eta_v}=\frac{n_\mu(T_s/T_\infty-1)d\theta_v(\eta_v)/d\eta_v}{(T_s/T_\infty-1)\theta_v(\eta_v)+1}$
	$\frac{1}{\lambda_v}\frac{d\lambda_v}{d\eta_v}=\frac{n_\lambda(T_s/T_\infty-1)d\theta_v(\eta_v)/d\eta_v}{(T_s/T_\infty-1)\theta_v(\eta_v)+1}$
	$\frac{1}{c_{pv}}\frac{dc_{pv}}{d\eta_v}=\frac{n_{c_p}(T_s/T_\infty-1)d\theta_v(\eta_v)+1}{(T_s/T_\infty-1)\theta_v(\eta_v)+1}$
	$\frac{\nu_\infty}{\nu_v}=[(T_w/T_\infty-1)\theta_v(\eta_v)+1]^{-(n_\mu+1)}$
	$\Pr_v=\Pr_{v,\infty}[(T_s/T_\infty-1)\theta_v(\eta_v)+1]^{n_\mu-n_\lambda+n_{cp}}$

Table 11.7 (continued)

Condensate heat transfer	
Wall dimensionless temperature gradient $-\left(\frac{d\theta_l(\eta_l)}{d\eta_l}\right)_{\eta_l=0}$	For laminar forced film condensation of water vapour on a horizontal plate for $0 \le \frac{\Delta t_\infty}{t_s} \le 4.27$ $-\left(\frac{d\theta_l(\eta_l)}{d\eta_l}\right)_{\eta_l=0} = \frac{0.7421\frac{\Delta t_w}{t_s}+0.3124}{\left(\frac{\Delta t_w}{t_s}\right)^{0.28}} \left(0.01 \le \frac{\Delta t_w}{t_s} \le 0.1\right)$ $-\left(\frac{d\theta_l(\eta_l)}{d\eta_l}\right)_{\eta_l=0} = 0.1466\left(\frac{\Delta t_w}{t_s}\right)^2 - 0.3974\left(\frac{\Delta t_w}{t_s}\right) + 0.7652$ $\left(0.1 \le \frac{\Delta t_w}{t_s} \le 1\right)$
Local values on heat transfer $Nu_{xl,w} = \frac{\alpha_{x,w}x}{\lambda_{l,w}}$	$q_{x,w} = \lambda_{l,w}(t_w - t_s)\left(\frac{1}{2}\mathrm{Re}_{xl,s}\right)^{1/2} x^{-1}\left(-\frac{d\theta_l(\eta_l)}{d\eta_l}\right)_{\eta_l=0}$ $\alpha_{x,w} = \lambda_{l,w}\left(\frac{1}{2}\mathrm{Re}_{xl,s}\right)^{1/2} x^{-1}\left(-\frac{d\theta_l(\eta_l)}{d\eta_l}\right)_{\eta_l=0}$ $Nu_{xl,w} = \left(\frac{1}{2}\mathrm{Re}_{xl,s}\right)^{1/2}\left(-\frac{d\theta_l(\eta_l)}{d\eta_l}\right)_{\eta_l=0}$
Average values on heat transfer $\overline{Nu}_{xl,w} = \frac{\overline{\alpha}_{x,w}x}{\lambda_{l,w}}$	$\overline{Q}_{x,w} = \sqrt{2}x^{-1}\lambda_{l,w}(t_w - t_s)(\mathrm{Re}_{xl,s})^{1/2}\left(-\frac{d\theta_l(\eta_l)}{d\eta_l}\right)_{\eta_l=0}$ $\overline{\alpha}_{x,w} = \sqrt{2}\lambda_{l,w}\mathrm{Re}_{xl,s}{}^{1/2}x^{-1}\left(-\frac{d\theta_l(\eta_l)}{d\eta_l}\right)_{\eta_l=0}$ $\overline{Nu}_{xl,w} = \sqrt{2}(\mathrm{Re}_{xl,s})^{1/2}\left(-\frac{d\theta_l(\eta_l)}{d\eta_l}\right)_{\eta_l=0}$
Total value on heat transfer	$Q_{x,w} = \sqrt{2}b\lambda_{l,w}(t_w - t_s)(Re_{xl,s})^{1/2}\left(-\frac{d\theta_l(\eta_l)}{d\eta_l}\right)_{\eta_l=0}$

Table 11.7 (continued)

Condensate mass flow rate	
Mass flow rate Parameter Φ_s	For laminar forced film condensation of water vapour on a horizontal plate for $0 \le \frac{\Delta t_\infty}{t_s} \le 4.27$ $\Phi_s = 0.0719\left(\frac{\Delta t_w}{t_s}\right) + 0.00057 \quad \left(0.01 \le \frac{\Delta t_w}{t_s} \le 0.1\right)$ $\Phi_s = -0.0301\left(\frac{\Delta t_w}{t_s}\right)^2 + 0.0725\left(\frac{\Delta t_w}{t_s}\right) + 0.0008 \quad \left(0.1 \le \frac{\Delta t_w}{t_s} \le 1\right)$
Mass flow rate G_x	$G_x = \sqrt{2}b\mu_{l,s}{\mathrm{Re}_{xl,s}}^{1/2}\Phi_s$
Condensate mass–energy transformation equation	
	$\left(-\frac{d\theta_l(\eta_l)}{d\eta_l}\right)_{\eta_l=0} = C_{mh}\Phi_s,$
Condensate mass–energy transformation coefficient	
	$C_{mh} = \frac{\mu_{l,s}h_{fg}}{\lambda_{l,w}(t_s - t_w)}$

mass transfer. Compared with the effect of the wall subcooled grade on these key physical variables, the effect of the vapour superheated grade is very slight in the broad range of the vapour superheated grade, and can be ignored for practical prediction of these key physical variables without any obvious deviation to the prediction results.

On the basis of the above analysis, by using the curve-fitting approach, the system of these rigorous key numerical solutions for laminar forced film condensation of water vapour on a horizontal flat plate is formulated with variation of the wall subcooled grade, and these formulated equations are coincident very well to the related key solutions. With the formulated equations on the wall dimensionless temperature gradient and condensate mass flow rate parameter, the above theoretical equations on heat and mass transfer become available for simple and reliable prediction of condensate heat and mass transfer results.

The condensate mass–energy transformation equation provided in this chapter demonstrates the internal relation between the condensate heat and mass transfer, and is the general equation suitable for all laminar forced film condensation. The condensate mass–energy transformation coefficient is proportional to the physical factor $\frac{h_{fg}}{c_{pl,s}(t_s - t_w)}$, interfacial liquid Prandtl number $\mathrm{Pr}_{l,s}$, and the liquid thermal conductivity ratio $\frac{\lambda_{l,s}}{\lambda_{l,w}}$ in the broad range of the vapour superheated grade.

Table 11.8 The related physical properties of water and water vapour

Materials	Water				
	t_w, °C	t_s, °C			t_f, °C
Temperature	85	100			92.5
Physical properties	$\lambda_{l,w}$, W/(m°C)	$\nu_{l,s}$ m^2/s	$\mu_{l,s}$ Kg/(m s)	h_{fg} kJ/kg	$\nu_{l,f}$ m^2/s
	0.67	0.2911×10^{-6}	279×10^{-6}	2257.1	0.31465×10^{-6}

The new similarity mathematical model transformed by means of the new similarity method with consideration of variable thermophysical properties can yield rigorous and reliable heat and mass transfer results on laminar forced film condensation.

11.11 Calculation Example

Example 1 A flat plate with 0.3 m in width and 0.3 m in length is located horizontally in the flowing water vapour with $w_{x,\infty} = 2.9$ m/s. The wall temperature is $t_w = 85$°C, and the vapour bulk temperatures are respectively $t_\infty = 100$ and 400°C. Please solve the following questions:

(1) calculate the condensate film thicknesses at $x = 0.05, 0.1, 0.15, 0.2$, and 0.3 m, respectively;
(2) calculate heat transfer rate on the plate;
(3) evaluate the condensate mass flow rate,

Solution 1: The related physical properties of water are obtained, and listed in Table 11.8.

First, the local Reynolds number for bulk flow will be calculated as

$\mathrm{Re}_{x_v,f} = \dfrac{w_{xv,\infty}x}{\nu_{v,f}} = \dfrac{2.9 \times 0.3}{26.12 \times 10^{-6}} = 33307.8 < 5 \times 10^5$, then the flow can be regarded as laminar.

With $(t_\infty - t_s)/t_s = (400-100)/100 = 3$ and $\dfrac{\Delta t_w}{t_s} = \dfrac{t_s - t_w}{t_s} = \dfrac{100-85}{100} = 0.15$, (11.19) is used for evaluation of the condensate film thickness, i.e.

$$\eta_{l\delta} = 1.3192 \exp\left(0.5244\left(\frac{\Delta t_w}{t_s}\right)\right) \text{ for } (0 \le (t_\infty - t_s)/t_s \le 4.27).$$

Then

$$\eta_{l\delta} = 1.3192 \times \exp(0.5244 \times 0.15) = 1.4271586.$$

With (9.14)

$$\eta_l = \frac{y}{x}\left(\frac{1}{2}\mathrm{Re}_{xl,s}\right)^{1/2},$$

we have

$$\eta_{l\delta} = \frac{\delta_l}{x}\left(\frac{1}{2}\frac{w_{x,\infty}x}{\nu_{l,s}}\right)^{1/2}.$$

Then,

$$\delta_l = \sqrt{2}x^{1/2}\eta_{l\delta}\left(\frac{\nu_{l,s}}{w_{x,\infty}}\right)^{1/2}.$$

Then the condensate liquid film thickness δ_l is evaluated as

$$\begin{aligned}\delta_l &= \sqrt{2}\times 1.4271586\left(\frac{0.2911\times 10^{-6}}{2.9}\right)^{1/2} x^{1/2},\\ &= 0.000639x^{1/2},\end{aligned}$$

at $x = 0.05\,\mathrm{m}$, $\delta_l = 0.000143\,\mathrm{m}$
at $x = 0.1\,\mathrm{m}$, $\delta_l = 0.000202\,\mathrm{m}$
at $x = 0.15\,\mathrm{m}$, $\delta_l = 0.000247\,\mathrm{m}$
at $x = 0.2\,\mathrm{m}$, $\delta_l = 0.000286\,\mathrm{m}$
at $x = 0.3\,\mathrm{m}$, $\delta_l = 0.00035\,\mathrm{m}$

Solution 2: With $\frac{\Delta t_w}{t_s} = 0.15$, (11.8) is used to calculate the wall dimensionless temperature gradient, i.e.

$$-\left(\frac{d\theta_l(\eta_l)}{d\eta_l}\right)_{\eta_l=0} = 0.1466\left(\frac{\Delta t_w}{t_s}\right)^2 - 0.3974\left(\frac{\Delta t_w}{t_s}\right) + 0.7652. \quad \left(0.1 \le \frac{\Delta t_w}{t_s} \le 1\right) \tag{11.8}$$

Then

$$-\left(\frac{d\theta_l(\eta_l)}{d\eta_l}\right)_{\eta_l=0} = 0.1466\times 0.15^2 - 0.3974\times 0.15 + 0.7652 = 0.708889.$$

Also,

$$\mathrm{Re}_{xl,s} = \frac{w_{xv,\infty}x}{\nu_{l,s}} = \frac{2.9\times 0.3}{0.2911\times 10^{-6}} = 2988664.$$

With (11.4), the heat transfer rate on the plate is calculated by

$$Q_{x,w} = \sqrt{2}b\lambda_{l,w}(t_w - t_s)\mathrm{Re}_{xl,s}{}^{1/2}\left(-\frac{d\theta_l(\eta_l)}{d\eta_l}\right)_{\eta_l=0}. \tag{11.4}$$

Then the plate condensate heat transfer rate is

$$Q_{x,w} = \sqrt{2} \times 0.3 \times 0.67 \times (85 - 100) \times 2988664^{1/2} \times 0.708889 = -5225.4\,\text{W}.$$

The negative result of the evaluated total heat transfer rate show that the heat flux direction is from the vapour bulk to the plate wall.

Solution 3: With $\frac{\Delta t_w}{t_s} = 0.15$, (11.25) is used to calculate the condensate mass flow rate parameter for laminar forced film flows of water vapour on a horizontal flat plate, i.e.

$$\Phi_s = -0.0301\left(\frac{\Delta t_w}{t_s}\right)^2 + 0.0725\left(\frac{\Delta t_w}{t_s}\right) + 0.0008 \qquad \left(0.1 \leq \frac{\Delta t_w}{t_s} \leq 1\right) \tag{11.25}$$

Then

$$\Phi_s = -0.0301 \times 0.15^2 + 0.0725 \times 0.15 + 0.0008 = 0.010998.$$

Equation (11.17) can be used for evaluation of condensate heat transfer rate on the plate, i.e.

$$G_x = \sqrt{2} b \mu_{l,s}\, \text{Re}_{xl,s}{}^{1/2}\, \Phi_s. \tag{11.17}$$

Therefore,

$$G_x = \sqrt{2} \times 0.3 \times 279 \times 10^{-6} \times 2988664^{1/2} \times 0.010998 = 0.002251\,\text{kg/s} = 8.1\,\text{kg/h}.$$

Exercises

1. Point out the suitable scope of the condensate mass-energy transformation equation reported in this chapter.
2. Continued to Example 1, suppose the wall temperature t_w = 65°C, but keep all other conditions. Please solve the following:

   (1) calculate the condensate film thicknesses at x = 0.05, 0.1, 0.15, 0.2, and 0.3 m;
   (2) calculate heat transfer rate on the plate;
   (3) evaluate the condensate mass flow rate.

3. Suppose that the wall dimensionless temperature gradient is 0.68, and keep all other related conditions of Example 1. Please use two different approaches to evaluate the total condensate mass flow rate.

Chapter 12
Complete Similarity Mathematical Models on Laminar Forced Film Condensation of Vapour–Gas Mixture

Abstract The complete similarity mathematical model based on the new similarity analysis method is developed for laminar forced film condensation from vapour–gas mixture. The governing similarity mathematical models for the liquid and vapour film flows are coupled by the seven interfacial matching conditions respectively for the two-dimensional velocity component balances, shear force balance, mass flow rate balance, temperature balance, heat transfer balance, concentration condition, as well as the balance between the condensate liquid mass flow and vapour mass diffusion. The work of this chapter is the successful application of the present new similarity analysis method to the extensive investigation of the laminar forced film condensation of vapour–gas mixture, and the developed mathematical model is a complete dimensionless similarity model. In the complete similarity model both for liquid and vapour–gas mixture films, the concentration and temperature-dependent physical properties, such as variable density, thermal conductivity, viscosity, and specific heat, are taken into account. These variable physical properties exist in the transformed governing similarity mathematical model as the related dimensionless physical factors. The investigation of this chapter proves that the present new similarity analysis method developed in this book is a better alternative to the traditional Falkner-Skan type transformation for extensive investigation of laminar film condensation.

12.1 Introduction

Film condensation phenomenon exists in many industrial departments, and film condensation from vapour in presence of noncondensable gas plays an important role in the design of heat exchangers, and in the heat transfer process of the chemical and power industry including nuclear power plants. Then numerous theoretical and experimental studies have been done for its successive investigations, such as those in [1–23], and some detailed reviews of condensation heat and mass transfer can be found in [24–28]. These studies demonstrate that the bulk concentration of the noncondensable gas could have a decisive effect on the condensate heat transfer rate, and their results indicated greater reductions in heat transfer. This is because of the fact that the presence of non-condensable gas lowers the partial pressure of

D.Y. Shang, *Theory of Heat Transfer with Forced Convection Film Flows*, Heat and Mass Transfer, DOI 10.1007/978-3-642-12581-2_12,

the vapour, and then reduces the condensate saturation temperature at liquid–vapour interface. Due to its complexity, so far, the research on this issue is still a big challenging.

In [29], we reported a complete similarity mathematical model on extensive investigation of laminar free film condensation of vapour–gas mixture based on a novel similarity analysis method. Then a system of rigorous numerical solutions on velocity, temperature and concentration fields related to the two-phase film flows is obtained. With the system of the rigorous key solutions, the theoretical equations on condensate heat and mass transfer become available for practical prediction. In Chap. 9, a complete similarity mathematical model was presented on laminar forced film condensation of pure vapour. On these bases, in this chapter the complete similarity mathematical model on laminar forced condensation of vapour–gas mixture will be set up based on the present new similarity analysis method for the deep investigation on heat and mass transfer of this challenging subject.

12.2 Governing Partial Differential Equations

12.2.1 Physical Model and Coordinate System

The present work focuses on the laminar forced film condensation of vapour–gas mixture on an isothermal horizontal flat plate. The analytical model and coordinate system used for the film condensation on the horizontal plate is shown in Fig. 12.1. An isothermal flat plate is located horizontally in a vapor–gas mixture flow at atmospheric pressure. The plate is parallel to the vapour–gas mixture flow. The plate temperature is t_w, the temperature of the vapor–gas mixture bulk is t_∞, and the vapor condensate saturated temperature between the liquid film and the vapour–gas mixture film is $t_{s,\text{int}}$ (for short, interfacial vapour saturation temperature). If given condition for the model is $t_w < t_{s,\text{int}}$, a steady film condensation will occur on the plate. We assume that the laminar flows within the liquid and vapor–gas mixture films are induced by the shear force at the liquid–vapor interface, and that the mass flow rate of vapor is balanced to the vapor mass diffusion at the liquid–vapor interface in the steady state of the laminar forced film condensation. Then, there is never an additional gas boundary layer near the interface except the induced concentration boundary layer of the vapor–gas mixture. In addition, we take into account

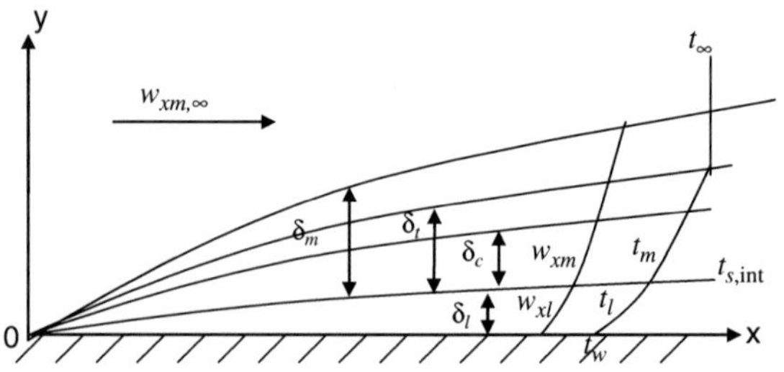

Fig. 12.1 Physical model and coordinate system of the laminar forced film condensation from vapour–gas mixture

the temperature-dependent physical properties of the condensate liquid film and the temperature and concentration-dependent physical properties of the induced vapor–gas mixture film. Then the steady laminar governing equations for mass, momentum, energy and concentration conservations in the two-phase boundary layer are as follows.

12.2.2 Governing Partial Differential Equations

12.2.2.1 For Liquid Film

The governing partial differential equations for condensate liquid film are

$$\frac{\partial}{\partial x}(\rho_l w_{xl}) + \frac{\partial}{\partial y}(\rho_l w_{yl}) = 0, \tag{12.1}$$

$$\rho_l \left(w_{xl} \frac{\partial w_{xl}}{\partial x} + w_{yl} \frac{\partial w_{xl}}{\partial y} \right) = \frac{\partial}{\partial y} \left(\mu_l \frac{\partial w_{xl}}{\partial y} \right), \tag{12.2}$$

$$\rho_v \left[w_{xl} \frac{\partial (c_{p_v} t_l)}{\partial x} + w_{yl} \frac{\partial (c_{p_v} t_l)}{\partial y} \right] = \frac{\partial}{\partial y} \left(\lambda_l \frac{\partial t_l}{\partial y} \right), \tag{12.3}$$

where (12.1), (12.2), and (12.3) are continuity, momentum, and energy conservation equations, respectively. Here the liquid temperature-dependent physical properties, such as density ρ_l, absolute viscosity μ_l, thermal conductivity λ_l, and specific heat c_{p_l} are considered. While, w_{xl} and w_{yl} are liquid film velocity components respectively in x and y coordinates.

12.2.2.2 For Vapour–Gas Mixture Film

The governing partial differential equations for vapour film are,

$$\frac{\partial}{\partial x}(\rho_m w_{xm}) + \frac{\partial}{\partial y}(\rho_m w_{ym}) = 0 \tag{12.4}$$

$$\rho_m \left(w_{xm} \frac{\partial w_{xm}}{\partial x} + w_{ym} \frac{\partial w_{xm}}{\partial y} \right) = \frac{\partial}{\partial y} \left(\mu_m \frac{\partial w_{xm}}{\partial y} \right), \tag{12.5}$$

$$\rho_m \left[w_{xm} \frac{\partial (c_{p_m} t_m)}{\partial x} + w_{ym} \frac{\partial (c_{p_m} t_m)}{\partial y} \right] = \frac{\partial}{\partial y} \left(\lambda_m \frac{\partial t_m}{\partial y} \right) + \frac{\partial}{\partial y} \left[\rho_m D_v (c_{p_v} - c_{p_a}) \frac{\partial C_{mv}}{\partial y} t_m \right], \tag{12.6}$$

$$\frac{\partial (w_{xm} \rho_m C_{mv})}{\partial x} + \frac{\partial (w_{ym} \rho_m C_{mv})}{\partial y} = \frac{\partial}{\partial y} \left(D_v \rho_m \frac{\partial C_{mv}}{\partial y} \right), \tag{12.7}$$

where (12.4), (12.5), and (12.6) are continuity, momentum, and energy conservation equations, while (12.7) is the species conservation equation with mass diffusion. Here, C_{mv} is vapour mass fraction in vapour–gas mixture, ρ_m, μ_m, λ_m, and c_{p_m} are density, absolute viscosity, thermal conductivity, and specific heat of the vapour–gas mixture, respectively, and D_v denotes vapor mass diffusion coefficient in the non-condensable gas, while w_{xm} and w_{ym} are vapour–gas mixture medium velocity components respectively in x and y coordinates.

12.2.2.3 For Boundary Conditions

The boundary conditions are

$$y = 0: \quad w_{xl} = 0, \quad w_{yl} = 0, \quad t_l = t_w \tag{12.8}$$

$$y = \delta_l:$$

$$w_{xl,s} = w_{xm,s} \tag{12.9}$$

$$\rho_{l,s}\left(w_{xl}\frac{\partial \delta_l}{\partial x} - w_{yl}\right)_{l,s} = \rho_{m,s}C_{mv,s}\left(w_{xm}\frac{\partial \delta_m}{\partial x} - w_{ym}\right)_s = g_x, \tag{12.10}$$

$$\mu_{l,s}\left(\frac{\partial w_{xl}}{\partial y}\right)_s = \mu_{m,s}\left(\frac{\partial w_{xm}}{\partial y}\right)_s, \tag{12.11}$$

$$\lambda_{l,s}\left(\frac{\partial t_l}{\partial y}\right)_{s_l} = \lambda_{m,s}\left(\frac{\partial t_m}{\partial y}\right)_s + h_{fg}\rho_{m,s}C_{mv,s}\left(w_{xm}\frac{\partial \delta_m}{\partial x} - w_{ym}\right)_s, \tag{12.12}$$

$$t = t_{s,\text{int}}, \tag{12.13}$$

$$C_{mv} = C_{mv,s}, \tag{12.14}$$

$$\rho_{m,s}C_{mv,s}\left(w_{xm}\frac{\partial \delta_m}{\partial x} - w_{ym}\right)_s = D_v\rho_{m,s}\left(\frac{\partial C_{mv}}{\partial y}\right)_s, \tag{12.15}$$

$$y \to \infty:$$

$$w_{xm} = w_{xm,\infty}, \quad t_m = t_\infty, \quad C_{mv} = C_{mv,\infty}, \tag{12.16}$$

where (11.8) expresses the physical conditions on the plate. Equations (11.9), (11.10), (11.11), (11.12), (11.13), (11.14), and (11.15) express the physical matching conditions at the liquid–vapour interface, in which (11.9) expresses the velocity component continuity, (11.10) expresses the mass flow rate continuity, (11.11) expresses the balance of the shear force, (11.12) expresses the energy balance, (11.13) expresses the temperature continuity, (11.14) expresses the concentration condition, and (11.15) expresses the vapour mass flow rate is balanced to the mass flow rate caused by the vapour mass diffusion. Furthermore, (11.16) expresses the physical conditions in the vapour–gas mixture bulk. In addition, t_s denotes the vapour condensate saturated temperature under atmospheric pressure, $t_{s,\text{int}}$ is the interfacial vapour saturation temperature dependent the interfacial vapour partial pressure, t_∞ denotes the temperature in the vapour–gas mixture bulk, and $C_{mv,s}$ and $C_{mv,\infty}$ denote the vapor mass fraction at the liquid–vapor interface and in the vapor–gas mixture bulk, respectively.

It is necessary to explain (15) for the condition of interfacial mass diffusion balance. Originally, it should be

$$\left[\rho_l\left(w_x\frac{\partial\delta_l}{\partial x}-w_y\right)\right]_{l,s}=\left[\rho_{m,s}C_{mv,s}\left(w_{xm}\frac{\partial\delta_m}{\partial x}-w_{ym}\right)\right]_{v,s}\equiv D_v\rho_{m,s}\left(\frac{\partial C_{mv}}{\partial y}\right)_s. \tag{15 orig.}$$

In order to conveniently transform the balance condition equation to the dimensionless form, we take (15) to express the interfacial mass diffusion balance condition. This form of expression has never caused any change of the calculation results.

12.3 Similarity Variables

Consulting the similarity analysis model developed in Chap. 9 for laminar forced film condensation of pure vapour, we assume the following similarity dimensionless variables for equivalent transformation of the governing partial differential equations on the laminar forced film condensation from vapour–gas mixture.

12.3.1 For Liquid Film

For liquid film the dimensionless coordinate variables η_l and the local Reynolds number $\text{Re}_{xl,s}$ are set up as follows respectively.

$$\eta_l=\frac{y}{x}\left(\frac{1}{2}\text{Re}_{xl,s}\right)^{1/2}, \tag{12.17}$$

$$\text{Re}_{xl,s}=\frac{w_{xm,\infty}x}{\nu_{l,s}}, \tag{12.18}$$

where $\nu_{l,s}$ is condensate liquid kinetic viscosity at the liquid–vapor interface. Here, the velocity in local Reynolds number expression is bulk vapour–gas mixture velocity.

Dimensionless temperature $\theta_l(\eta_l)$ is assumed as

$$\theta_l(\eta_l)=\frac{t_l-t_{s,\text{int}}}{t_w-t_{s,\text{int}}}. \tag{12.19}$$

While the dimensionless velocity components $W_{xl}(\eta_l)$ and $W_{yl}(\eta_l)$ in x - and y - directions are given as, respectively,

$$w_{xl} = w_{xm,\infty} W_{xl}(\eta_l), \tag{12.20}$$

$$w_{yl} = w_{xm,\infty} \left(\frac{1}{2}\mathrm{Re}_{xl,s}\right)^{-1/2} W_{yl}(\eta_l). \tag{12.21}$$

It is noted that here the velocity in equations of the local Reynolds number $\mathrm{Re}_{xl,s}$ and the dimensionless velocity components $W_{xl}(\eta_l)$ and $W_{yl}(\eta_l)$ is the vapour–gas mixture bulk velocity $w_{xm,\infty}$.

12.3.2 For Vapor–Gas Mixture Film

For vapor–gas mixture film, the similarity dimensionless coordinate variable η_m and the local Reynolds number $\mathrm{Re}_{xm,\infty}$ are assumed as respectively

$$\eta_m = \frac{y}{x}\left(\frac{1}{2}\mathrm{Re}_{xm,\infty}\right)^{1/2}, \tag{12.22}$$

$$\mathrm{Re}_{xm,\infty} = \frac{w_{xm,\infty} x}{\nu_{m,\infty}}. \tag{12.23}$$

The dimensionless temperature θ_m is defined as

$$\theta_m(\eta_m) = \frac{t_m - t_\infty}{t_{s,int} - t_\infty}, \tag{12.24}$$

where $t_{s,\mathrm{int}}$ is the interfacial vapour saturation temperature, dependent on the interfacial vapour mass fraction $C_{mv,s}$.

The similarity dimensionless velocity components $W_{xm}(\eta_m)$ and $W_{ym}(\eta_m)$ are assumed as respectively in x and y directions

$$w_{xm} = w_{xm,\infty} W_{xm}(\eta_m), \tag{12.25}$$

$$w_{ym} = w_{xm,\infty} \left(\frac{1}{2}\mathrm{Re}_{xm,\infty}\right)^{-1/2} W_{ym}(\eta_m). \tag{12.26}$$

It is noted that here the velocity in equations of the local Reynolds number $\mathrm{Re}_{xm,\infty}$ and the dimensionless velocity components $W_{xm}(\eta_m)$ and $W_{ym}(\eta_m)$ is the vapour–gas mixture bulk velocity $w_{xm,\infty}$.

The vapor relative mass fraction $\Gamma_{mv}(\eta_m)$ is defined as

$$\Gamma_{mv}(\eta_m) = \frac{C_{mv} - C_{mv,\infty}}{C_{mv,s} - C_{mv,\infty}}, \tag{12.27}$$

where $C_{mv,s}$ and $C_{mv,\infty}$ are vapour mass fraction, respectively, in the liquid–vapour interface and vapour–gas mixture bulk.

12.4 Similarity Transformation of Governing Partial Differential Equations

12.4.1 For Liquid Film

12.4.1.1 Transformation of (12.1)

$$w_{xl}\frac{\partial \rho_l}{\partial x} + w_{yl}\frac{\partial \rho_l}{\partial y} + \rho_l\left(\frac{\partial w_{xl}}{\partial x} + \frac{\partial w_{yl}}{\partial y}\right) = 0, \tag{12.1a}$$

where

$$\frac{\partial w_{xl}}{\partial x} = w_{xm,\infty}\frac{\mathrm{d}W_{xl}(\eta_l)}{\mathrm{d}\eta_l}\frac{\partial \eta_l}{\partial x}.$$

While with (11.17) and (11.18), we have

$$\frac{\partial \eta_l}{\partial x} = -\frac{1}{2}\eta_l x^{-1}. \tag{a}$$

Then

$$\frac{\partial w_{xl}}{\partial x} = -\frac{1}{2}\eta_l x^{-1} w_{xm,\infty}\frac{\mathrm{d}W_{xl}(\eta_l)}{\mathrm{d}\eta_l}. \tag{b}$$

Additionally,

$$\frac{\partial w_{yl}}{\partial y} = w_{xm,\infty}\left(\frac{1}{2}\mathrm{Re}_{xl,s}\right)^{-1/2}\frac{\mathrm{d}W_{yl}(\eta_l)}{\mathrm{d}\eta_l}\frac{\partial \eta_l}{\partial y},$$

where

$$\frac{\partial \eta_l}{\partial y} = x^{-1}\left(\frac{1}{2}\mathrm{Re}_{xl,s}\right)^{1/2}. \tag{c}$$

Then

$$\frac{\partial w_{yl}}{\partial y} = w_{xm,\infty} \frac{\mathrm{d}W_{yl}(\eta_l)}{\mathrm{d}\eta_l} x^{-1}. \tag{d}$$

Furthermore,

$$\frac{\partial \rho_l}{\partial x} = \frac{\mathrm{d}\rho_l}{\mathrm{d}\eta_l}\frac{\partial \eta_l}{\partial x} = -\frac{1}{2}\frac{\mathrm{d}\rho_l}{\mathrm{d}\eta_l}\eta_l x^{-1}, \tag{e}$$

$$\frac{\partial \rho_l}{\partial y} = \frac{\mathrm{d}\rho_l}{\mathrm{d}\eta_l}\frac{\partial \eta_l}{\partial y} = \frac{\mathrm{d}\rho_l}{\mathrm{d}\eta_l} x^{-1}\left(\frac{1}{2}\mathrm{Re}_{xl,s}\right)^{1/2}. \tag{f}$$

With (11.20), (11.21), (b), (d) to (f), (12.1a) is changed to

$$\rho_l\left[-\frac{1}{2}\eta_l x^{-1} w_{xv,\infty}\frac{\mathrm{d}W_{xl}(\eta_l)}{\mathrm{d}\eta_l} + w_{xv,\infty}\frac{\mathrm{d}W_{yl}(\eta_l)}{\mathrm{d}\eta_l}x^{-1}\right]$$
$$-\frac{1}{2}\frac{\mathrm{d}\rho_l}{\mathrm{d}\eta_l}\eta_l x^{-1} w_{xv,\infty} W_{xl}(\eta_l) + \frac{\mathrm{d}\rho_l}{\mathrm{d}\eta_l}x^{-1}\left(\frac{1}{2}\mathrm{Re}_{xl,s}\right)^{1/2} w_{xv,\infty}\left(\frac{1}{2}\mathrm{Re}_{xl,s}\right)^{-1/2}$$
$$W_{yl}(\eta_l) = 0$$

The above equation is divided by $x^{-1}w_{xm,\infty}$ and becomes

$$\rho_l\left[-\frac{1}{2}\eta_l\frac{\mathrm{d}W_{xl}(\eta_l)}{\mathrm{d}\eta_l} + \frac{\mathrm{d}W_{yl}(\eta_l)}{\mathrm{d}\eta_l}\right] - \frac{1}{2}\frac{\mathrm{d}\rho_l}{\mathrm{d}\eta_l}\eta_l W_{xl}(\eta_l) + \frac{\mathrm{d}\rho_l}{\mathrm{d}\eta_l}W_{yl}(\eta_l).$$

The above equation is further divided by $\frac{\rho_l}{2}$ and becomes

$$-\eta_l\frac{\mathrm{d}W_{xl}(\eta_l)}{\mathrm{d}\eta_l} + 2\frac{\mathrm{d}W_{yl}(\eta_l)}{\mathrm{d}\eta_l} + \frac{1}{\rho_l}\frac{\mathrm{d}\rho_l}{\mathrm{d}\eta_l}[-\eta_l W_{xl}(\eta_l) + 2W_{yl}(\eta_l)] = 0. \tag{12.28}$$

12.4.1.2 Transformation of (12.2)

Equation (12.2) is changed to

$$\rho_l\left(w_{xl}\frac{\partial w_{xl}}{\partial x} + w_{yl}\frac{\partial w_{xl}}{\partial y}\right) = \mu_l\frac{\partial^2 w_{xv}}{\partial y^2} + \frac{\partial \mu_l}{\partial y}\frac{\partial w_{xv}}{\partial y}, \tag{12.2a}$$

where

$$\frac{\partial w_{xl}}{\partial y} = w_{xm,\infty}\frac{\mathrm{d}W_{xl}(\eta_l)}{\mathrm{d}\eta_l}\frac{\partial \eta_l}{\partial y}.$$

With (12.d), the above equation is changed to

$$\frac{\partial w_{xl}}{\partial y} = w_{xm,\infty} \frac{\mathrm{d}W_{xl}(\eta_l)}{\mathrm{d}\eta_l} x^{-1} \left(\frac{1}{2}\mathrm{Re}_{xl,s}\right)^{1/2}, \tag{g}$$

or

$$\begin{aligned}\frac{\partial^2 w_{xl}}{\partial y^2} &= w_{xm,\infty} \frac{\mathrm{d}^2 W_{xl}(\eta_l)}{\mathrm{d}\eta_l^2} x^{-1} \left(\frac{1}{2}\mathrm{Re}_{xl,s}\right)^{1/2} \frac{\partial \eta_l}{\partial y} \\ &= w_{xm,\infty} \frac{\mathrm{d}^2 W_{xl}(\eta_l)}{\mathrm{d}\eta_l^2} x^{-2} \left(\frac{1}{2}\mathrm{Re}_{xl,s}\right).\end{aligned} \tag{h}$$

Similar to (f), we have

$$\frac{\partial \mu_l}{\partial y} = \frac{\mathrm{d}\mu_l}{\mathrm{d}\eta_l} x^{-1} \left(\frac{1}{2}\mathrm{Re}_{xl,s}\right)^{1/2}. \tag{i}$$

With (12.20), (12.21), (b), (g) to (i), (12.2a) is changed to

$$\rho_l \left[\begin{aligned} & -w_{xm,\infty} W_{xl}(\eta_l) \frac{1}{2} \eta_l x^{-1} w_{xm,\infty} \frac{\mathrm{d}W_{xl}(\eta_l)}{\mathrm{d}\eta_l} \\ & + w_{xm,\infty} \left(\frac{1}{2}\mathrm{Re}_{xl,s}\right)^{-1/2} W_{yl}(\eta_l) \cdot w_{xm,\infty} \frac{\mathrm{d}W_{xl}(\eta_l)}{\mathrm{d}\eta_l} x^{-1} \left(\frac{1}{2}\mathrm{Re}_{xl,s}\right)^{1/2} \end{aligned} \right]$$

$$= \mu_l w_{xm,\infty} \frac{\mathrm{d}^2 W_{xl}(\eta_l)}{\mathrm{d}\eta_l^2} x^{-2} \left(\frac{1}{2}\mathrm{Re}_{xl,s}\right)$$

$$+ \frac{\mathrm{d}\mu_l}{\mathrm{d}\eta_l} x^{-1} \left(\frac{1}{2}\mathrm{Re}_{xl,s}\right)^{1/2} w_{xm,\infty} \frac{\mathrm{d}W_{xl}(\eta_l)}{\mathrm{d}\eta_l} x^{-1} \left(\frac{1}{2}\mathrm{Re}_{xl,s}\right)^{1/2}.$$

With the definition of $\mathrm{Re}_{xl,s}$, the above equation becomes

$$\rho_l \left[-w_{xm,\infty} W_{xl}(\eta_l) \frac{1}{2} \eta_l x^{-1} w_{xm,\infty} \frac{\mathrm{d}W_{xl}(\eta_l)}{\mathrm{d}\eta_l} + w_{xm,\infty} (W_{yl}(\eta_l) \cdot w_{xm,\infty} \frac{\mathrm{d}W_{xl}(\eta_l)}{\mathrm{d}\eta_l} x^{-1} \right]$$

$$= \mu_l w_{xm,\infty} \frac{\mathrm{d}^2 W_{xl}(\eta_l)}{\mathrm{d}\eta_l^2} x^{-2} \left(\frac{1}{2} \frac{w_{xm,\infty} x}{\nu_{l,s}}\right) + \frac{\mathrm{d}\mu_l}{\mathrm{d}\eta_l} x^{-1} \left(\frac{1}{2} \frac{w_{xm,\infty} x}{\nu_{l,s}}\right) w_{xm,\infty} \frac{\mathrm{d}W_{xl}(\eta_l)}{\mathrm{d}\eta_l} x^{-1}.$$

The above equation is divided by $\frac{1}{2}x^{-1}\cdot w_{xm,\infty}^2$, and becomes

$$\rho_l\left[-W_{xl}(\eta_l)\eta_l\frac{\mathrm{d}W_{xl}(\eta_l)}{\mathrm{d}\eta_l}+2(W_{yl}(\eta_l)\cdot\frac{\mathrm{d}W_{xl}(\eta_l)}{\mathrm{d}\eta_l}\right]=\mu_l\frac{\mathrm{d}^2W_{xl}(\eta_l)}{\mathrm{d}\eta_l^2}\left(\frac{1}{\nu_{l,s}}\right)+\frac{\mathrm{d}\mu_l}{\mathrm{d}\eta_l}\left(\frac{1}{\nu_{l,s}}\right)\frac{\mathrm{d}W_{xl}(\eta_l)}{\mathrm{d}\eta_l}.$$

The above equation is divided by $\frac{\mu_l}{\nu_{l,s}}$, and becomes

$$\frac{\nu_{l,s}}{\nu_l}[-\eta_l W_{xl}(\eta_l)+2(W_{yl}(\eta_l)]\frac{\mathrm{d}W_{xl}(\eta_l)}{\mathrm{d}\eta_l}=\frac{\mathrm{d}^2W_{xl}(\eta_l)}{\mathrm{d}\eta_l^2}+\frac{1}{\mu_l}\frac{\mathrm{d}\mu_l}{\mathrm{d}\eta_l}\frac{\mathrm{d}W_{xl}(\eta_l)}{\mathrm{d}\eta_l}. \quad (12.29)$$

12.4.1.3 Transformation of (12.3)

Equation (12.3) is changed to

$$\rho_l\left(w_{xl}c_{p_l}\frac{\partial t_l}{\partial x}+w_{xl}t_l\frac{\partial c_{p_l}}{\partial x}+w_{yl}c_{p_l}\frac{\partial t_l}{\partial y}+w_{yl}t_l\frac{\partial c_{p_l}}{\partial y}\right)=\lambda_l\frac{\partial^2 t_l}{\partial y^2}+\frac{\partial\lambda_l}{\partial y}\frac{\partial t_l}{\partial y}, \quad (12.3a)$$

where

$$\frac{\partial t_l}{\partial x}=\frac{dt_l}{d\eta_l}\frac{\partial\eta_l}{\partial x}.$$

While

$$\frac{\mathrm{d}t_l}{\mathrm{d}\eta_l}=(t_w-t_{s,\mathrm{int}})\frac{\mathrm{d}\theta_l(\eta_l)}{\mathrm{d}\eta_l}. \quad (j)$$

With (b) and (k), we have

$$\frac{\partial t_l}{\partial x}=-\frac{1}{2}\eta_l x^{-1}(t_w-t_{s,\mathrm{int}})\frac{\mathrm{d}\theta_l(\eta_l)}{\mathrm{d}\eta_l}. \quad (k)$$

Additionally,

$$\frac{\partial t_l}{\partial y}=\frac{dt_l}{d\eta_l}\frac{\partial\eta_l}{\partial y}.$$

With (d) and (k), we have

$$\frac{\partial t_l}{\partial y}=x^{-1}\left(\frac{1}{2}\mathrm{Re}_{xl,s}\right)^{1/2}(t_w-t_{s,\mathrm{int}})\frac{\mathrm{d}\theta_l(\eta_l)}{\mathrm{d}\eta_l}, \quad (l)$$

or

$$\frac{\partial^2 t_l}{\partial y^2} = x^{-2}\left(\frac{1}{2}\mathrm{Re}_{xl,s}\right)(t_w - t_{s,\mathrm{int}})\frac{\mathrm{d}^2\theta_l(\eta_l)}{\mathrm{d}^2\eta_l}. \tag{m}$$

Similar to (f), we have

$$\frac{\partial \lambda_l}{\partial y} = \frac{\mathrm{d}\lambda_l}{\mathrm{d}\eta_l}x^{-1}\left(\frac{1}{2}\mathrm{Re}_{xl,s}\right)^{1/2}, \tag{n}$$

$$\frac{\partial c_{p_l}}{\partial y} = \frac{\mathrm{d}c_{p_l}}{\mathrm{d}\eta_l}x^{-1}\left(\frac{1}{2}\mathrm{Re}_{xl,s}\right)^{1/2}. \tag{o}$$

Similar to (e),

$$\frac{\partial c_{pl}}{\partial x} = -\frac{1}{2}\frac{\mathrm{d}c_{pl}}{\mathrm{d}\eta_l}\eta_l x^{-1}. \tag{p}$$

With (12.20), (12.21), (k) to (p), (12.3a) is changed to

$$\rho_l\left[\begin{array}{l} w_{xm,\infty}W_{xl}(\eta_l)c_{p_l}\left(-\frac{1}{2}\eta_l x^{-1}(t_w - t_{s,\mathrm{int}})\frac{\mathrm{d}\theta_l(\eta_l)}{\mathrm{d}\eta_l}\right) + w_{xm,\infty}W_{xl}(\eta_l)t_l\left(-\frac{1}{2}\eta_l x^{-1}\frac{\mathrm{d}c_{p_l}}{\mathrm{d}\eta_l}\right) \\ +w_{xm,\infty}\left(\frac{1}{2}\mathrm{Re}_{xl,\infty}\right)^{-1/2}W_{yl}(\eta_l)c_{p_l}x^{-1}\left(\frac{1}{2}\mathrm{Re}_{xl,s}\right)^{1/2}(t_w - t_{s,\mathrm{int}})\frac{\mathrm{d}\theta_l(\eta_l)}{\mathrm{d}\eta_l} \\ +w_{xm,\infty}\left(\frac{1}{2}\mathrm{Re}_{xl,s}\right)^{-1/2}W_{yl}(\eta_l)t_l\frac{\mathrm{d}c_{p_l}}{\mathrm{d}\eta_l}x^{-1}\left(\frac{1}{2}\mathrm{Re}_{xl,\infty}\right)^{1/2} \end{array}\right]$$

$$= \lambda_l x^{-2}\left(\frac{1}{2}\mathrm{Re}_{xl,s}\right)(t_w - t_{s,\mathrm{int}})\frac{\mathrm{d}^2\theta_l(\eta_l)}{\mathrm{d}\eta_l^2} + \frac{\partial\lambda_l}{\partial\eta_l}x^{-1}\left(\frac{1}{2}\mathrm{Re}_{xl,s})\right)^{1/2}x^{-1}\left(\frac{1}{2}\mathrm{Re}_{xl,s}\right)^{1/2}$$

$$(t_w - t_{s,\mathrm{int}})\frac{\mathrm{d}\theta_l(\eta_l)}{\mathrm{d}\eta_l},$$

or

$$\rho_l\left[\begin{array}{l} w_{xm,\infty}W_{xl}(\eta_l)c_{p_l}\left(-\frac{1}{2}\eta_l x^{-1}(t_w - t_{s,\mathrm{int}})\frac{\mathrm{d}\theta_l(\eta_l)}{\mathrm{d}\eta_l}\right) + w_{xm,\infty}W_{xl}(\eta_l)t_l\left(-\frac{1}{2}\eta_l x^{-1}\frac{\mathrm{d}c_{p_l}}{\mathrm{d}\eta_l}\right) \\ +w_{xm,\infty}W_{yl}(\eta_l)c_{p_l}x^{-1}(t_w - t_{s,\mathrm{int}})\frac{\mathrm{d}\theta_l(\eta_l)}{\mathrm{d}\eta_l} + w_{xm,\infty}W_{yl}(\eta_l)t_l\frac{\mathrm{d}c_{p_l}}{\mathrm{d}\eta_l}x^{-1} \end{array}\right]$$

$$= \lambda_l x^{-2}\left(\frac{1}{2}\mathrm{Re}_{xl,s}\right)(t_w - t_{s,\mathrm{int}})\frac{\mathrm{d}^2\theta_l(\eta_l)}{\mathrm{d}\eta_l^2} + \frac{\partial\lambda_l}{\partial\eta_l}x^{-1}\left(\frac{1}{2}\mathrm{Re}_{xl,s}\right)x^{-1}(t_w - t_{s,\mathrm{int}})\frac{\mathrm{d}\theta_l(\eta_l)}{\mathrm{d}\eta_l}.$$

With definition of $\mathrm{Re}_{xl,s}$, the above equation is changed to

$$\rho_l \begin{bmatrix} w_{xm,\infty} W_{xl}(\eta_l) c_{p_l} \left(-\frac{1}{2}\eta_l x^{-1}(t_w - t_{s,\mathrm{int}})\frac{\mathrm{d}\theta_l(\eta_l)}{\mathrm{d}\eta_l}\right) + w_{xm,\infty} W_{xl}(\eta_l) t_l \left(-\frac{1}{2}\eta_l x^{-1}\frac{\mathrm{d}c_{p_l}}{\mathrm{d}\eta_l}\right) \\ + w_{xm,\infty} W_{yl}(\eta_l) c_{p_l} x^{-1}(t_w - t_{s,\mathrm{int}})\frac{\mathrm{d}\theta_l(\eta_l)}{\mathrm{d}\eta_l} + w_{xm,\infty} W_{yl}(\eta_l) t_l \frac{\mathrm{d}c_{p_l}}{\mathrm{d}\eta_l} x^{-1} \end{bmatrix}$$

$$= \lambda_l x^{-2}\left(\frac{1}{2}\frac{w_{xm,\infty}x}{\nu_{l,s}}\right)(t_w - t_{s,\mathrm{int}})\frac{\mathrm{d}^2\theta_l(\eta_l)}{\mathrm{d}\eta_l^2} + \frac{\partial \lambda_l}{\partial \eta_l} x^{-1}\left(\frac{1}{2}\frac{w_{xm,\infty}x}{\nu_{l,s}}\right) x^{-1}(t_w - t_{s,\mathrm{int}})\frac{\mathrm{d}\theta_l(\eta_l)}{\mathrm{d}\eta_l}.$$

The above equation is divided by $x^{-1} w_{xm,\infty}(t_w - t_{s,\mathrm{int}})$, and becomes

$$\rho_l c_{p_l} \begin{bmatrix} W_{xl}(\eta_l)\left(-\frac{1}{2}\eta_l \frac{\mathrm{d}\theta_l(\eta_l)}{\mathrm{d}\eta_l}\right) + W_{xl}(\eta_l)\frac{t_l}{t_w - t_{s,\mathrm{int}}}\left(-\frac{1}{2}\eta_l \frac{1}{c_{p_l}}\frac{\mathrm{d}c_{p_l}}{\mathrm{d}\eta_l}\right) \\ + W_{yl}(\eta_l)\frac{\mathrm{d}\theta_l(\eta_l)}{\mathrm{d}\eta_l} + W_{yl}(\eta_l)\frac{t_l}{t_w - t_{s,\mathrm{int}}}\frac{1}{c_{p_l}}\frac{\mathrm{d}c_{p_l}}{\mathrm{d}\eta_l} \end{bmatrix}$$

$$= \lambda_l\left(\frac{1}{2}\frac{1}{\nu_{l,s}}\right)\frac{\mathrm{d}^2\theta_l(\eta_l)}{\mathrm{d}\eta_l^2} + \frac{\partial \lambda_l}{\partial \eta_l}\left(\frac{1}{2}\frac{1}{\nu_{l,s}}\right)\frac{\mathrm{d}\theta_l(\eta_l)}{\mathrm{d}\eta_l}.$$

The above equation is divided by $\frac{\lambda_l}{2\nu_{l,s}}$, and further simplified to

$$\frac{\nu_{l,s}}{\lambda_l}\rho_l c_{p_l} \begin{bmatrix} W_{xl}(\eta_l)\left(-\eta_l \frac{\mathrm{d}\theta_l(\eta_l)}{\mathrm{d}\eta_l}\right) + W_{xl}(\eta_l)\frac{t_l}{t_w - t_{s,\mathrm{int}}}\left(-\eta_l \frac{1}{c_{p_l}}\frac{\mathrm{d}c_{p_l}}{\mathrm{d}\eta_l}\right) \\ + 2W_{yl}(\eta_l)\frac{\mathrm{d}\theta_l(\eta_l)}{\mathrm{d}\eta_l} + 2W_{yl}(\eta_l)\frac{t_l}{t_w - t_{s,\mathrm{int}}}\frac{1}{c_{p_l}}\frac{\mathrm{d}c_{p_l}}{\mathrm{d}\eta_l} \end{bmatrix}$$

$$= \frac{\mathrm{d}^2\theta_l(\eta_l)}{\mathrm{d}\eta_l^2} + \frac{1}{\lambda_l}\frac{\partial \lambda_l}{\partial \eta_l}\frac{\mathrm{d}\theta_l(\eta_l)}{\mathrm{d}\eta_l},$$

i.e.

$$\mathrm{Pr}_l \frac{\nu_{l,s}}{\nu_l}\left([-\eta_l W_{xl}(\eta_l) + 2W_{yl}(\eta_l)]\frac{\mathrm{d}\theta_l(\eta_l)}{\mathrm{d}\eta_l} \right.$$
$$\left. + [-\eta_l W_{xl}(\eta_l) + 2W_{yl}(\eta_l)]\frac{t_l}{t_w - t_{s,\mathrm{int}}}\frac{1}{c_{p_l}}\frac{\mathrm{d}c_{p_l}}{\mathrm{d}\eta_l}\right) = \frac{\mathrm{d}^2\theta_l(\eta_l)}{\mathrm{d}\eta_l^2} + \frac{1}{\lambda_l}\frac{\partial \lambda_l}{\partial \eta_l}\frac{\mathrm{d}\theta_l(\eta_l)}{\mathrm{d}\eta_l}.$$

With $\frac{t_l}{t_w - t_{s,\mathrm{int}}} = \theta_l(\eta_l) + \frac{t_{s,\mathrm{int}}}{t_w - t_{s,\mathrm{int}}}$, the above equation is changed to

$$\mathrm{Pr}_l \frac{\nu_{l,s}}{\nu_l} \left(\begin{array}{l} [-\eta_l W_{xl}(\eta_l) + 2W_{yl}(\eta_l)] \dfrac{\mathrm{d}\theta_l(\eta_l)}{\mathrm{d}\eta_l} \\ + [-\eta_l W_{xl}(\eta_l) + 2W_{yl}(\eta_l)] \left(\theta_l(\eta_l) + \dfrac{t_{s,\mathrm{int}}}{t_w - t_{s,\mathrm{int}}} \right) \dfrac{1}{c_{p_l}} \dfrac{\mathrm{d}c_{p_l}}{\mathrm{d}\eta_l} \end{array} \right) \tag{12.30}$$

$$= \frac{\mathrm{d}^2\theta_l(\eta_l)}{\mathrm{d}\eta_l^2} + \frac{1}{\lambda_l} \frac{\partial \lambda_l}{\partial \eta_l} \frac{\mathrm{d}\theta_l(\eta_l)}{\mathrm{d}\eta_l}.$$

In fact, the derived (12.30) is same as (9.26) although some differences exist in the expressions of the related similarities variables.

12.4.2 For Vapour–Gas Mixture Film

With the similar derivation to that in Sect. 12.4.1, we have the following equations:

$$\frac{\partial \eta_m}{\partial x} = -\frac{1}{2}\eta_m x^{-1}, \tag{aa}$$

$$\frac{\partial w_{xm}}{\partial x} = -\frac{1}{2}\eta_m x^{-1} w_{xm,\infty} \frac{\mathrm{d}W_{xm}(\eta_m)}{\mathrm{d}\eta_m}, \tag{bb}$$

$$\frac{\partial \eta_m}{\partial y} = x^{-1}\left(\frac{1}{2}\mathrm{Re}_{xm,\infty}\right)^{1/2}, \tag{cc}$$

$$\frac{\partial w_{ym}}{\partial y} = w_{xm,\infty} \frac{\mathrm{d}W_{ym}(\eta_m)}{\mathrm{d}\eta_m} x^{-1}, \tag{dd}$$

$$\frac{\partial \rho_m}{\partial x} = -\frac{1}{2}\frac{\mathrm{d}\rho_m}{\mathrm{d}\eta_m}\eta_m x^{-1}, \tag{ee}$$

$$\frac{\partial \rho_m}{\partial y} = \frac{\mathrm{d}\rho_m}{\mathrm{d}\eta_m} x^{-1}\left(\frac{1}{2}\mathrm{Re}_{xm,\infty}\right)^{1/2}, \tag{ff}$$

$$\frac{\partial w_{xm}}{\partial y} = w_{xm,\infty} \frac{\mathrm{d}W_{xm}(\eta_m)}{\mathrm{d}\eta_m} x^{-1}\left(\frac{1}{2}\mathrm{Re}_{xm,\infty}\right)^{1/2}, \tag{gg}$$

$$\frac{\partial^2 w_{xm}}{\partial y^2} = w_{xm,\infty} \frac{\mathrm{d}^2 W_{xm}(\eta_m)}{\mathrm{d}\eta_m^2} x^{-2}\left(\frac{1}{2}\mathrm{Re}_{xm,\infty}\right), \tag{hh}$$

$$\frac{\partial \mu_m}{\partial y} = \frac{\mathrm{d}\mu_m}{\mathrm{d}\eta_m} x^{-1}\left(\frac{1}{2}\mathrm{Re}_{xm,\infty}\right)^{1/2}, \tag{ii}$$

$$\frac{\mathrm{d}t_m}{\mathrm{d}\eta_m} = (t_{s,\mathrm{int}} - t_\infty)\frac{\mathrm{d}\theta_m(\eta_m)}{\mathrm{d}\eta_m}, \tag{jj}$$

$$\frac{\partial t_m}{\partial x} = -\frac{1}{2}\eta_m x^{-1}(t_{s,\mathrm{int}} - t_\infty)\frac{\mathrm{d}\theta_m(\eta_m)}{\mathrm{d}\eta_m}, \tag{kk}$$

$$\frac{\partial t_m}{\partial y} = x^{-1}\left(\frac{1}{2}\mathrm{Re}_{xm,\infty}\right)^{1/2}(t_{s,\mathrm{int}} - t_\infty)\frac{\mathrm{d}\theta_m(\eta_m)}{\mathrm{d}\eta_m}, \tag{ll}$$

$$\frac{\partial^2 t_m}{\partial y^2} = x^{-2}\left(\frac{1}{2}\mathrm{Re}_{xm,\infty}\right)(t_{s,\mathrm{int}} - t_\infty)\frac{\mathrm{d}^2\theta_m(\eta_m)}{\mathrm{d}^2\eta_m}, \tag{mm}$$

$$\frac{\partial \lambda_m}{\partial y} = \frac{\mathrm{d}\lambda_m}{\mathrm{d}\eta_m}x^{-1}\left(\frac{1}{2}\mathrm{Re}_{xm,\infty}\right)^{1/2}, \tag{nn}$$

$$\frac{\partial c_{p_m}}{\partial y} = \frac{\mathrm{d}c_{p_m}}{\mathrm{d}\eta_m}x^{-1}\left(\frac{1}{2}\mathrm{Re}_{xm,\infty}\right)^{1/2}. \tag{oo}$$

Similar to (e),

$$\frac{\partial c_{p_m}}{\partial x} = -\frac{1}{2}\frac{\mathrm{d}c_{p_m}}{\mathrm{d}\eta_m}\eta_m x^{-1}. \tag{pp}$$

Additionally, with (11.27), (aa), and (cc), we have

$$\frac{\partial C_{mv}}{\partial x} = -\frac{1}{2}\eta_m x^{-1}(C_{mv,s} - C_{mv,\infty})\frac{\mathrm{d}\Gamma_{mv}(\eta_m)}{\mathrm{d}\eta_m}, \tag{q}$$

$$\frac{\partial C_{mv}}{\partial y} = (C_{mv,s} - C_{mv,\infty})\frac{\mathrm{d}\Gamma_{mv}(\eta_m)}{\mathrm{d}\eta_m}x^{-1}\left(\frac{1}{2}\mathrm{Re}_{xm,\infty}\right)^{1/2} \tag{r}$$

and

$$\frac{\partial^2 C_{mv}}{\partial y^2} = (C_{mv,s} - C_{mv,\infty})\frac{\mathrm{d}^2\Gamma_{mv}(\eta_m)}{\mathrm{d}\eta_m^2}x^{-2}\left(\frac{1}{2}\mathrm{Re}_{xm,\infty}\right). \tag{s}$$

12.4.2.1 Transformation of (12.4)

Equation (12.4) is changed to

$$w_{xm}\frac{\partial \rho_m}{\partial x} + \rho_m\frac{\partial w_{xm}}{\partial x} + w_{yv}\frac{\partial \rho_m}{\partial y} + \rho_m\frac{\partial w_{ym}}{\partial y} = 0. \tag{12.4a}$$

With (12.25), (12.26), (bb), (gg), (hh), and (ff), (12.4a) is changed to

$$\rho_m \left[-\frac{1}{2}\eta_m x^{-1} w_{xm,\infty} \frac{\mathrm{d}W_{xm}(\eta_m)}{\mathrm{d}\eta_m} + w_{xm,\infty} \frac{\mathrm{d}W_{ym}(\eta_m)}{\mathrm{d}\eta_m} x^{-1} \right]$$

$$+ w_{xm,\infty} W_{xm}(\eta_m) \left(-\frac{1}{2} \frac{\mathrm{d}\rho_m}{\mathrm{d}\eta_m} \eta_m x^{-1} \right)$$

$$+ w_{xm,\infty} \left(\frac{1}{2} \mathrm{Re}_{xm,\infty} \right)^{-1/2} W_{ym}(\eta_m) \frac{\mathrm{d}\rho_m}{\mathrm{d}\eta_m} x^{-1} \left(\frac{1}{2} \mathrm{Re}_{xm,\infty} \right)^{1/2} = 0,$$

i.e.

$$\rho_m \left[-\frac{1}{2}\eta_m x^{-1} w_{xm,\infty} \frac{\mathrm{d}W_{xm}(\eta_m)}{\mathrm{d}\eta_m} + w_{xm,\infty} \frac{\mathrm{d}W_{ym}(\eta_m)}{\mathrm{d}\eta_m} x^{-1} \right]$$

$$+ w_{xm,\infty} W_{xm}(\eta_m) \left(-\frac{1}{2} \frac{\mathrm{d}\rho_m}{\mathrm{d}\eta_m} \eta_m x^{-1} \right) + w_{xm,\infty} W_{ym}(\eta_m) \frac{\mathrm{d}\rho_m}{\mathrm{d}\eta_m} x^{-1} = 0.$$

The above equation is divided by $\frac{1}{2} w_{xv,\infty} x^{-1}$, and simplified to

$$\rho_m \left[-\eta_m \frac{\mathrm{d}W_{xm}(\eta_m)}{\mathrm{d}\eta_m} + 2 \frac{\mathrm{d}W_{ym}(\eta_m)}{\mathrm{d}\eta_m} \right] + W_{xm}(\eta_m) \left(-\frac{\mathrm{d}\rho_m}{\mathrm{d}\eta_m} \eta_m \right) + 2W_{ym}(\eta_m) \frac{\mathrm{d}\rho_m}{\mathrm{d}\eta_m} = 0,$$

i.e.

$$-\eta_m \frac{\mathrm{d}W_{xm}(\eta_m)}{\mathrm{d}\eta_m} + 2 \frac{\mathrm{d}W_{ym}(\eta_m)}{\mathrm{d}\eta_m} + \frac{1}{\rho_m} \frac{\mathrm{d}\rho_m}{\mathrm{d}\eta_m} [-\eta_m \cdot W_{xm}(\eta_m) + 2W_{ym}(\eta_m)] = 0. \tag{12.31}$$

12.4.2.2 Transformation of (12.5)

Equation (12.5) is changed to

$$\rho_m \left(w_{xm} \frac{\partial w_{xm}}{\partial x} + w_{ym} \frac{\partial w_{xm}}{\partial y} \right) = \mu_m \frac{\partial^2 w_{xm}}{\partial y^2} + \frac{\partial \mu_m}{\partial y} \frac{\partial w_{xm}}{\partial y}. \tag{12.5a}$$

With (12.25), (12.26), (bb), (gg), (hh), and (ii), Eg. (12.5a) is changed to

$$\rho_m \left[\begin{array}{l} w_{xm,\infty} W_{xm}(\eta_m) \left(-\frac{1}{2}\eta_m x^{-1} w_{xm,\infty} \frac{\mathrm{d}W_{xm}(\eta_m)}{\mathrm{d}\eta_m} \right) \\ + w_{xm,\infty} \left(\frac{1}{2} \mathrm{Re}_{xm,\infty} \right)^{-1/2} W_{ym}(\eta_m) w_{xm,\infty} \frac{\mathrm{d}W_{xm}(\eta_m)}{\mathrm{d}\eta_m} x^{-1} \left(\frac{1}{2} \mathrm{Re}_{xm,\infty} \right)^{1/2} \end{array} \right]$$

$$= \mu_m w_{xm,\infty} \frac{\mathrm{d}^2 W_{xm}(\eta_m)}{\mathrm{d}\eta_m^2} x^{-2} \left(\frac{1}{2}\mathrm{Re}_{xm,\infty}\right)$$

$$+ \frac{\mathrm{d}\mu_m}{\mathrm{d}\eta_m} x^{-1} \left(\frac{1}{2}\mathrm{Re}_{xm,\infty}\right)^{1/2} w_{xm,\infty} \frac{\mathrm{d}W_{xm}(\eta_m)}{\mathrm{d}\eta_m} x^{-1} \left(\frac{1}{2}\mathrm{Re}_{xm,\infty}\right)^{1/2},$$

i.e.

$$\rho_m \left[w_{xm,\infty} W_{xm}(\eta_m) \left(-\frac{1}{2}\eta_m x^{-1} w_{xm,\infty} \frac{\mathrm{d}W_{xm}(\eta_m)}{\mathrm{d}\eta_m} \right) + w_{xm,\infty} W_{ym}(\eta_m) w_{xm,\infty} \frac{\mathrm{d}W_{xm}(\eta_m)}{\mathrm{d}\eta_m} x^{-1} \right]$$

$$= \mu_m w_{xm,\infty} \frac{\mathrm{d}^2 W_{xm}(\eta_m)}{\mathrm{d}\eta_m^2} x^{-2} \left(\frac{1}{2}\mathrm{Re}_{xm,\infty}\right) + \frac{\mathrm{d}\mu_m}{\mathrm{d}\eta_m} x^{-1} \left(\frac{1}{2}\mathrm{Re}_{xm,\infty}\right) w_{xm,\infty} \frac{\mathrm{d}W_{xm}(\eta_m)}{\mathrm{d}\eta_m} x^{-1}.$$

With definition of $\mathrm{Re}_{xm,\infty}$, the above equation is changed to

$$\rho_m \left[w_{xm,\infty} W_{xm}(\eta_m) \left(-\frac{1}{2}\eta_m x^{-1} w_{xm,\infty} \frac{\mathrm{d}W_{xm}(\eta_m)}{\mathrm{d}\eta_m} \right) + w_{xm,\infty} W_{ym}(\eta_m) w_{xm,\infty} \frac{\mathrm{d}W_{xm}(\eta_m)}{\mathrm{d}\eta_m} x^{-1} \right]$$

$$= \mu_m w_{xm,\infty} \frac{\mathrm{d}^2 W_{xm}(\eta_m)}{\mathrm{d}\eta_m^2} x^{-2} \left(\frac{1}{2}\frac{w_{x,\infty} x}{\nu_{m,\infty}}\right) + \frac{\mathrm{d}\mu_m}{\mathrm{d}\eta_m} x^{-1} \left(\frac{1}{2}\frac{w_{x,\infty} x}{\nu_{m,\infty}}\right) w_{xm,\infty} \frac{\mathrm{d}W_{xm}(\eta_m)}{\mathrm{d}\eta_m} x^{-1}.$$

The above equation divided by $\frac{1}{2} w_{xv,\infty}^2 \cdot x^{-1} \frac{\mu_m}{\nu_{m,\infty}}$, and simplified to

$$\frac{\nu_{m,\infty}}{\nu_m} \left[W_{xm}(\eta_m) \left(-\eta_m \frac{\mathrm{d}W_{xm}(\eta_m)}{\mathrm{d}\eta_m} \right) + 2W_{ym}(\eta_m) \frac{\mathrm{d}W_{xm}(\eta_m)}{\mathrm{d}\eta_m} \right]$$

$$= \frac{\mathrm{d}^2 W_{xm}(\eta_m)}{\mathrm{d}\eta_m^2} + \frac{1}{\mu_m} \frac{\mathrm{d}\mu_m}{\mathrm{d}\eta_m} \frac{\mathrm{d}W_{xm}(\eta_m)}{\mathrm{d}\eta_m},$$

i.e.

$$\frac{\nu_{m,\infty}}{\nu_m} [-\eta_m \cdot W_{xm}(\eta_m) + 2W_{ym}(\eta_m)] \frac{\mathrm{d}W_{xm}(\eta_m)}{\mathrm{d}\eta_m} = \frac{\mathrm{d}^2 W_{xm}(\eta_m)}{\mathrm{d}\eta_m^2} + \frac{1}{\mu_m} \frac{\mathrm{d}\mu_m}{\mathrm{d}\eta_m} \frac{\mathrm{d}W_{xm}(\eta_m)}{\mathrm{d}\eta_m}. \tag{12.32}$$

12.4.2.3 Transformation of (12.6)

Equation (12.6) is changed as

$$\rho_m \left[w_{xm} c_{p_m} \frac{\partial t_m}{\partial x} + w_{xm} t_m \frac{\partial c_{p_m}}{\partial x} + w_{ym} c_{p_m} \frac{\partial t_m}{\partial y} + w_{ym} t_m \frac{\partial c_{p_m}}{\partial y} \right]$$

$$= \lambda_m \frac{\partial^2 t_m}{\partial y^2} + \frac{\partial \lambda_m}{\partial y} \cdot \frac{\partial t_m}{\partial y} + D_v (c_{p_v} - c_{p_g}) \frac{\partial}{\partial y} \left[\rho_m \frac{\partial C_{mv}}{\partial y} t_m \right],$$

or

$$\rho_m \left[w_{xm} c_{p_m} \frac{\partial t_m}{\partial x} + w_{xm} t_m \frac{\partial c_{p_m}}{\partial x} + w_{ym} c_{p_m} \frac{\partial t_m}{\partial y} + w_{ym} t_m \frac{\partial c_{p_m}}{\partial y} \right]$$

$$= \lambda_m \frac{\partial^2 t_m}{\partial y^2} + \frac{\partial \lambda_m}{\partial y} \cdot \frac{\partial t_m}{\partial y} \tag{12.6a}$$

$$+ D_v (c_{p_v} - c_{p_g}) \left[\rho_m \frac{\partial C_{mv}}{\partial y} \frac{\partial t_m}{\partial y} + t_m \rho_m \frac{\partial^2 C_{mv}}{\partial y^2} + t_m \frac{\partial C_{mv}}{\partial y} \frac{\partial \rho_m}{\partial y} \right].$$

With (12.25), (12.26), (kk) to (pp), and (q) to (s), (12.6a) is changed to

$$\rho_m \left[\begin{array}{l} -\frac{1}{2} \eta_m x^{-1} (t_{s,\mathrm{int}} - t_\infty) \frac{\mathrm{d}\theta_m(\eta_m)}{\mathrm{d}\eta_m} w_{xm,\infty} W_{xm}(\eta_m) \cdot c_{p_m} - \frac{1}{2} \frac{\mathrm{d}c_{p_m}}{\mathrm{d}\eta_m} \eta_m x^{-1} w_{xm,\infty} W_{xm}(\eta_m) \cdot t_m \\ + w_{xm,\infty} \left(\frac{1}{2} \mathrm{Re}_{xm,\infty} \right)^{-1/2} W_{ym}(\eta_m) c_{p_m} x^{-1} \left(\frac{1}{2} \mathrm{Re}_{xm,\infty} \right)^{1/2} (t_{s,\mathrm{int}} - t_\infty) \frac{\mathrm{d}\theta_m(\eta_m)}{\mathrm{d}\eta_m} \\ + w_{xm,\infty} \left(\frac{1}{2} \mathrm{Re}_{xm,\infty} \right)^{-1/2} W_{ym}(\eta_m) \cdot t_m \frac{\mathrm{d}c_{p_m}}{\mathrm{d}\eta_m} x^{-1} \left(\frac{1}{2} \mathrm{Re}_{xm,\infty} \right)^{1/2} \end{array} \right].$$

$$= \lambda_m x^{-2} \left(\frac{1}{2} \mathrm{Re}_{xm,\infty} \right) (t_{s,\mathrm{int}} - t_\infty) \frac{\mathrm{d}^2 \theta_m(\eta_m)}{\mathrm{d}^2 \eta_m}$$

$$+ \frac{\mathrm{d}\lambda_m}{\mathrm{d}\eta_m} x^{-1} \left(\frac{1}{2} \mathrm{Re}_{xm,\infty} \right)^{1/2} \cdot x^{-1} \left(\frac{1}{2} \mathrm{Re}_{xm,\infty} \right)^{1/2} (t_{s,\mathrm{int}} - t_\infty) \frac{\mathrm{d}\theta_m(\eta_m)}{\mathrm{d}\eta_m}$$

$$+ D_v (c_{p_v} - c_{p_g}) \left[\begin{array}{l} \rho_m (C_{mv,s} - C_{mv,\infty}) \frac{\mathrm{d}\Gamma_{mv}(\eta_m)}{\mathrm{d}\eta_v} x^{-1} \left(\frac{1}{2} \mathrm{Re}_{xm,\infty} \right)^{1/2} x^{-1} \left(\frac{1}{2} \mathrm{Re}_{xm,\infty} \right)^{1/2} (t_{s,\mathrm{int}} - t_\infty) \frac{\mathrm{d}\theta_m(\eta_m)}{\mathrm{d}\eta_m} \\ + T_m \rho_m (C_{mv,s} - C_{mv,\infty}) \frac{\mathrm{d}^2 \Gamma_{mv}(\eta_m)}{\mathrm{d}\eta_v^2} x^{-2} \left(\frac{1}{2} \mathrm{Re}_{xm,\infty} \right) \\ + T_m (C_{mv,s} - C_{mv,\infty}) \frac{\mathrm{d}\Gamma_{mv}(\eta_m)}{\mathrm{d}\eta_m} x^{-1} \left(\frac{1}{2} \mathrm{Re}_{xm,\infty} \right)^{1/2} \frac{\mathrm{d}\rho_m}{\mathrm{d}\eta_m} x^{-1} \left(\frac{1}{2} \mathrm{Re}_{xm,\infty} \right)^{1/2} \end{array} \right].$$

With definition of $\mathrm{Re}_{xm,\infty}$, the above equation is simplified to

$$\rho_m \left[\begin{array}{l} -\frac{1}{2} \eta_m x^{-1} (t_{s,\mathrm{int}} - t_\infty) \frac{\mathrm{d}\theta_m(\eta_m)}{\mathrm{d}\eta_m} w_{xm,\infty} W_{xm}(\eta_m) \cdot c_{p_m} - \frac{1}{2} \frac{\mathrm{d}c_{p_m}}{\mathrm{d}\eta_m} \eta_m x^{-1} w_{xm,\infty} W_{xm}(\eta_m) \cdot t_m \\ + w_{xm,\infty} W_{ym}(\eta_m) c_{p_m} x^{-1} (t_{s,\mathrm{int}} - t_\infty) \frac{\mathrm{d}\theta_m(\eta_m)}{\mathrm{d}\eta_m} + w_{xm,\infty} W_{ym}(\eta_m) \cdot t_m \frac{\mathrm{d}c_{p_m}}{\mathrm{d}\eta_m} x^{-1} \end{array} \right].$$

$$= \lambda_m x^{-2} \left(\frac{1}{2} \frac{w_{xm,\infty} x}{\nu_{m,\infty}} \right) (t_{s,\mathrm{int}} - t_\infty) \frac{\mathrm{d}^2 \theta_m(\eta_m)}{\mathrm{d}^2 \eta_m} + \frac{\mathrm{d}\lambda_m}{\mathrm{d}\eta_m} x^{-1} \left(\frac{1}{2} \frac{w_{xm,\infty} x}{\nu_{m,\infty}} \right) \cdot x^{-1} (t_{s,\mathrm{int}} - t_\infty) \frac{\mathrm{d}\theta_m(\eta_m)}{\mathrm{d}\eta_m}$$

$$+D_v(c_{p_v}-c_{p_g})\left[\begin{array}{l}\rho_m(C_{mv,s}-C_{mv,\infty})\dfrac{\mathrm{d}\Gamma_{mv}(\eta_m)}{\mathrm{d}\eta_m}x^{-1}\left(\dfrac{1}{2}\dfrac{w_{xm,\infty}x}{\nu_{m,\infty}}\right)x^{-1}(t_{s,\mathrm{int}}-t_\infty)\dfrac{\mathrm{d}\theta_m(\eta_m)}{\mathrm{d}\eta_m}\\ +t_m\rho_m(C_{mv,s}-C_{mv,\infty})\dfrac{\mathrm{d}^2\Gamma_{mv}(\eta_m)}{\mathrm{d}\eta_m^2}x^{-2}\left(\dfrac{1}{2}\dfrac{w_{xm,\infty}x}{\nu_{m,\infty}}\right)\\ +T_m(C_{mv,s}-C_{mv,\infty})\dfrac{\mathrm{d}\Gamma_{mv}(\eta_m)}{\mathrm{d}\eta_m}x^{-1}\left(\dfrac{1}{2}\dfrac{w_{xm,\infty}x}{\nu_{m,\infty}}\right)\dfrac{\mathrm{d}\rho_m}{\mathrm{d}\eta_m}x^{-1}\end{array}\right].$$

The above equation is divided by $\frac{1}{2}w_{xv,\infty}(t_{s,\mathrm{int}}-t_\infty)x^{-1}$, and simplified to

$$\rho_m\left[\begin{array}{l}-\eta_m\dfrac{d\theta_m(\eta_m)}{d\eta_m}W_{xm}(\eta_m)\cdot c_{p_m}-\dfrac{dc_{p_m}}{d\eta_m}\eta_m W_{xm}(\eta_m)\cdot\dfrac{t_m}{t_{s,\mathrm{int}}-t_\infty}\\ +2W_{ym}(\eta_m)c_{p_m}\dfrac{\mathrm{d}\theta_m(\eta_m)}{\mathrm{d}\eta_m}+2W_{ym}(\eta_m)\cdot\dfrac{t_m}{t_{s,\mathrm{int}}-t_\infty}\dfrac{\mathrm{d}c_{p_m}}{\mathrm{d}\eta_m}\end{array}\right]$$

$$=\lambda_m\left(\frac{1}{\nu_{m,\infty}}\right)\frac{d^2\theta_m(\eta_m)}{d^2\eta_m}+\frac{d\lambda_m}{d\eta_m}\left(\frac{1}{\nu_{m,\infty}}\right)\cdot\frac{d\theta_m(\eta_m)}{d\eta_m}$$

$$+D_v(c_{p_v}-c_{p_g})\left[\begin{array}{l}\rho_m(C_{mv,s}-C_{mv,\infty})\dfrac{\mathrm{d}\Gamma_{mv}(\eta_m)}{\mathrm{d}\eta_m}\left(\dfrac{1}{\nu_{m,\infty}}\right)\dfrac{\mathrm{d}\theta_m(\eta_m)}{\mathrm{d}\eta_m}\\ +\dfrac{t_m}{t_{s,\mathrm{int}}-t_\infty}\rho_m(C_{mv,s}-C_{mv,\infty})\dfrac{\mathrm{d}^2\Gamma_{mv}(\eta_m)}{\mathrm{d}\eta_m^2}\left(\dfrac{1}{\nu_{m,\infty}}\right)\\ +\dfrac{t_m}{t_{s,\mathrm{int}}-t_\infty}(C_{mv,s}-C_{mv,\infty})\dfrac{\mathrm{d}\Gamma_{mv}(\eta_m)}{\mathrm{d}\eta_m}\left(\dfrac{1}{\nu_{m,\infty}}\right)\dfrac{\mathrm{d}\rho_m}{\mathrm{d}\eta_m}\end{array}\right].$$

The above equation is divided by $\frac{\lambda_m}{\nu_{m,\infty}}$, and further simplified to

$$\frac{\nu_{m,\infty}\rho_m}{\lambda_m}\left[\begin{array}{l}-\eta_m\dfrac{\mathrm{d}\theta_m(\eta_m)}{\mathrm{d}\eta_m}W_{xm}(\eta_m)\cdot c_{p_m}-\dfrac{\mathrm{d}c_{p_m}}{\mathrm{d}\eta_m}\eta_m W_{xm}(\eta_m)\cdot\dfrac{t_m}{t_{s,\mathrm{int}}-t_{\infty\infty}}\\ +2W_{ym}(\eta_m)c_{p_m}\dfrac{\mathrm{d}\theta_m(\eta_m)}{\mathrm{d}\eta_m}+2W_{ym}(\eta_m)\cdot\dfrac{t_m}{t_{s,\mathrm{int}}-t_\infty}\dfrac{\mathrm{d}c_{p_m}}{\mathrm{d}\eta_m}\end{array}\right]$$

$$=\frac{\mathrm{d}^2\theta_m(\eta_m)}{\mathrm{d}^2\eta_m}+\frac{1}{\lambda_m}\frac{\mathrm{d}\lambda_m}{\mathrm{d}\eta_m}\cdot\frac{\mathrm{d}\theta_m(\eta_m)}{\mathrm{d}\eta_m}$$

$$+\frac{\nu_{m,\infty}}{\lambda_m}D_v(c_{p_v}-c_{p_g})\left[\begin{array}{l}\rho_m(C_{mv,s}-C_{mv,\infty})\dfrac{\mathrm{d}\Gamma_{mv}(\eta_m)}{\mathrm{d}\eta_m}(\dfrac{1}{\nu_{m,\infty}})\dfrac{\mathrm{d}\theta_m(\eta_m)}{\mathrm{d}\eta_m}\\ +\dfrac{t_m}{t_{s,\mathrm{int}}-t_\infty}\rho_m(C_{mv,s}-C_{mv,\infty})\dfrac{\mathrm{d}^2\Gamma_{mv}(\eta_m)}{\mathrm{d}\eta_m^2}(\dfrac{1}{\nu_{m,\infty}})\\ +\dfrac{t_m}{t_{s,\mathrm{int}}-t_\infty}(C_{mv,s}-C_{mv\infty})\dfrac{\mathrm{d}\Gamma_{mv}(\eta_m)}{\mathrm{d}\eta_m}(\dfrac{1}{\nu_{m,\infty}})\dfrac{\mathrm{d}\rho_m}{\mathrm{d}\eta_m}\end{array}\right].$$

With $\frac{t_m}{t_{s,\text{int}} - t_\infty} = \theta_m + \frac{t_\infty}{t_{s,\text{int}} - t_\infty}$, the above equation is further changed to

$$\frac{\nu_{m,\infty}\rho_m c_{p_m}}{\lambda_m}\begin{bmatrix} -\eta_m \frac{\mathrm{d}\theta_m(\eta_m)}{\mathrm{d}\eta_m} W_{xm}(\eta_m) - \frac{1}{c_{p_m}}\frac{\mathrm{d}c_{p_m}}{\mathrm{d}\eta_m}\eta_m W_{xm}(\eta_m)\cdot\left(\theta_m + \frac{t_\infty}{t_{s,\text{int}} - t_\infty}\right) \\ +2W_{ym}(\eta_m)\frac{\mathrm{d}\theta_m(\eta_m)}{\mathrm{d}\eta m} + 2W_{ym}(\eta_m)\cdot\left(\theta_m + \frac{t_\infty}{t_{s,\text{int}} - t_\infty}\right)\frac{1}{c_{p_m}}\frac{\mathrm{d}c_{p_m}}{\mathrm{d}\eta_m}\end{bmatrix}$$

$$= \frac{\mathrm{d}^2\theta_m(\eta_m)}{\mathrm{d}^2\eta_m} + \frac{1}{\lambda_m}\frac{\mathrm{d}\lambda_m}{\mathrm{d}\eta_m}\cdot\frac{\mathrm{d}\theta_m(\eta_m)}{\mathrm{d}\eta_m}$$

$$+\frac{1}{\lambda_m}D_v(c_{p_v} - c_{p_g})\begin{bmatrix} \rho_m(C_{mv,s} - C_{mv,\infty})\frac{\mathrm{d}\Gamma_{mv}(\eta_m)}{\mathrm{d}\eta_m}\frac{\mathrm{d}\theta_m(\eta_m)}{\mathrm{d}\eta_m} \\ +\left(\theta_m + \frac{t_\infty}{t_{s,\text{int}} - t_\infty}\right)\rho_m(C_{mv,s} - C_{mv,\infty})\frac{\mathrm{d}^2\Gamma_{mv}(\eta_m)}{\mathrm{d}\eta_m^2} \\ +\left(\theta_m + \frac{t_\infty}{t_{s,\text{int}} - t_\infty}\right)(C_{mv,s} - C_{mv,\infty})\frac{d\Gamma_{mv}(\eta_m)}{d\eta_m}\frac{d\rho_m}{d\eta_m}\end{bmatrix},$$

i.e.

$$\frac{\nu_{m,\infty}}{\nu_m}\begin{bmatrix} -\eta_m \frac{\mathrm{d}\theta_m(\eta_m)}{\mathrm{d}\eta_m} W_{xm}(\eta_m) - \frac{1}{c_{p_m}}\frac{\mathrm{d}c_{p_m}}{\mathrm{d}\eta_m}\eta_m W_{xm}(\eta_m)\cdot(\theta_m + \frac{t_\infty}{t_{s,\text{int}} - t_\infty}) \\ +2W_{ym}(\eta_m)\frac{\mathrm{d}\theta_m(\eta_m)}{\mathrm{d}\eta_m} + 2W_{ym}(\eta_m)\cdot\left(\theta_m + \frac{t_\infty}{t_{s,\text{int}} - t_\infty}\right)\frac{1}{c_{p_m}}\frac{\mathrm{d}c_{p_m}}{\mathrm{d}\eta_m}\end{bmatrix}$$

$$= \frac{1}{\mathrm{Pr}_m}\left[\frac{\mathrm{d}^2\theta_m(\eta_m)}{\mathrm{d}^2\eta_m} + \frac{1}{\lambda_m}\frac{\mathrm{d}\lambda_m}{\mathrm{d}\eta_m}\cdot\frac{\mathrm{d}\theta_m(\eta_m)}{\mathrm{d}\eta_m}\right]$$

$$+\frac{1}{\mathrm{Pr}_m\lambda_m}D_v(c_{p_v} - c_{p_g})\begin{bmatrix} \rho_m(C_{mv,s} - C_{mv,\infty})\frac{\mathrm{d}\Gamma_{mv}(\eta_m)}{\mathrm{d}\eta_m}\frac{\mathrm{d}\theta_m(\eta_m)}{\mathrm{d}\eta_m} \\ +\left(\theta_m + \frac{t_\infty}{t_{s,\text{int}} - t_\infty}\right)\rho_m(C_{mv,s} - C_{mv,\infty})\frac{\mathrm{d}^2\Gamma_{mv}(\eta_m)}{\mathrm{d}\eta_m^2} \\ +\left(\theta_m + \frac{t_\infty}{t_{s,\text{int}} - t_\infty}\right)(C_{mv,s} - C_{mv,\infty})\frac{\mathrm{d}\Gamma_{mv}(\eta_m)}{\mathrm{d}\eta_m}\frac{\mathrm{d}\rho_m}{\mathrm{d}\eta_m}\end{bmatrix}.$$

The above equation is simplified to

$$\frac{\nu_{m,\infty}}{\nu_m}\left[\begin{array}{l} -\eta_m\dfrac{\mathrm{d}\theta_m(\eta_m)}{\mathrm{d}\eta_m}W_{xm}(\eta_m)-\dfrac{1}{c_{p_m}}\dfrac{\mathrm{d}c_{p_m}}{\mathrm{d}\eta_m}\eta_m W_{xm}(\eta_m)\cdot\left(\theta_m+\dfrac{t_\infty}{t_{s,\mathrm{int}}-t_\infty}\right)\\ +2W_{ym}(\eta_m)\dfrac{\mathrm{d}\theta_m(\eta_m)}{\mathrm{d}\eta_m}+2W_{ym}(\eta_m)\cdot\left(\theta_m+\dfrac{t_\infty}{t_{s,\mathrm{int}}-t_\infty}\right)\dfrac{1}{c_{p_m}}\dfrac{\mathrm{d}c_{p_m}}{\mathrm{d}\eta_m} \end{array}\right]$$

$$=\frac{1}{\mathrm{Pr}_m}\left[\frac{\mathrm{d}^2\theta_m(\eta_m)}{\mathrm{d}^2\eta_m}+\frac{1}{\lambda_m}\frac{\mathrm{d}\lambda_m}{\mathrm{d}\eta_m}\cdot\frac{\mathrm{d}\theta_m(\eta_m)}{\mathrm{d}\eta_m}\right]$$

$$+\frac{\rho_m}{\mu_m}D_v\frac{c_{p_v}-c_{p_g}}{c_{p_m}}\left[\begin{array}{l} (C_{mv,s}-C_{mv,\infty})\dfrac{\mathrm{d}\Gamma_{mv}(\eta_m)}{\mathrm{d}\eta_m}\dfrac{\mathrm{d}\theta_m(\eta_m)}{\mathrm{d}\eta_m}\\ +(\theta_m+\dfrac{t_\infty}{t_{s,\mathrm{int}}-t_\infty})(C_{mv,s}-C_{mv,\infty})\dfrac{\mathrm{d}^2\Gamma_{mv}(\eta_m)}{\mathrm{d}\eta_m^2}\\ +(\theta_m+\dfrac{t_\infty}{t_{s,\mathrm{int}}-t_\infty})(C_{mv,s}-C_{mv,\infty})\dfrac{\mathrm{d}\Gamma_{mv}(\eta_m)}{\mathrm{d}\eta_m}\dfrac{1}{\rho_m}\dfrac{\mathrm{d}\rho_m}{\mathrm{d}\eta_m} \end{array}\right],$$

i.e.

$$\frac{\nu_{m,\infty}}{\nu_m}\left\{\begin{array}{l} [-\eta_m W_{xm}(\eta_m)+2W_{ym}(\eta_m)]\dfrac{\mathrm{d}\theta_m(\eta_m)}{\mathrm{d}\eta_m}\\ +[-\eta_m W_{xm}(\eta_m)+2W_{ym}(\eta_m)]\dfrac{1}{c_{p_m}}\dfrac{\mathrm{d}c_{p_m}}{\mathrm{d}\eta_m}\left(\theta_m+\dfrac{t_\infty}{t_{s,\mathrm{int}}-t_\infty}\right) \end{array}\right\}$$

$$=\frac{1}{\mathrm{Pr}_m}\left[\frac{\mathrm{d}^2\theta_m(\eta_m)}{\mathrm{d}^2\eta_m}+\frac{1}{\lambda_m}\frac{\mathrm{d}\lambda_m}{\mathrm{d}\eta_m}\cdot\frac{\mathrm{d}\theta_m(\eta_m)}{\mathrm{d}\eta_m}\right]$$

$$+\frac{1}{Sc_{m,\infty}}\frac{\nu_{m,\infty}}{\nu_m}\frac{c_{p_v}-c_{p_g}}{c_{p_m}}(C_{mv,s}-C_{mv,\infty})\left[\begin{array}{l} \dfrac{\mathrm{d}\Gamma_{mv}(\eta_m)}{\mathrm{d}\eta_m}\dfrac{\mathrm{d}\theta_m(\eta_m)}{\mathrm{d}\eta_m}\\ +(\theta_m+\dfrac{t_\infty}{t_{s,\mathrm{int}}-t_\infty})\dfrac{\mathrm{d}^2\Gamma_{mv}(\eta_m)}{\mathrm{d}\eta_v^2}\\ +(\theta_m+\dfrac{t_\infty}{t_{s,\mathrm{int}}-t_\infty})\dfrac{1}{\rho_m}\dfrac{\mathrm{d}\rho_m}{\mathrm{d}\eta_m}\dfrac{\mathrm{d}\Gamma_{mv}(\eta_m)}{\mathrm{d}\eta_m} \end{array}\right], \tag{12.33}$$

where $Sc_{m,\infty}=\dfrac{\nu_{m,\infty}}{D_v}$ is defined as local Schmitt number of the vapour-gas mixture bulk.

12.4.2.4 Transformation of (12.7)

Equation (12.7) is changed to

$$w_{xm}\left[\frac{\partial(\rho_m C_{mv})}{\partial x}\right]+\rho_m C_{mv}\frac{\partial w_{xm}}{\partial x}+w_{yv}\left[\frac{\partial(\rho_m C_{mv})}{\partial y}\right]+\rho_m C_{mv}\frac{\partial w_{y_m}}{\partial y}$$

$$=D_v\left(\rho_m\frac{\partial^2 C_{mv}}{\partial y^2}+\frac{\partial\rho_m}{\partial y}\frac{\partial C_{mv}}{\partial y}\right)$$

i.e.

$$w_{xm}\left[\rho_m\frac{\partial(C_{mv})}{\partial x}+C_{mv}\frac{\partial\rho_m}{\partial x}\right]+\rho_m C_{mv}\frac{\partial w_{xm}}{\partial x}$$

$$+w_{ym}\left[\rho_m\frac{\partial(C_{mv})}{\partial y}+C_{mv}\frac{\partial\rho_{mv}}{\partial y}\right]+\rho_m C_{mv}\frac{\partial w_{ym}}{\partial y},$$

$$=D_v\left(\rho_m\frac{\partial^2 C_{mv}}{\partial y^2}+\frac{\partial\rho_m}{\partial y}\frac{\partial C_{mv}}{\partial y}\right)$$

or

$$w_{xv}\rho_m\frac{\partial(C_{mv})}{\partial x}+w_{yv}\rho_m\frac{\partial(C_{mv})}{\partial y}$$

$$+C_{mv}\left(w_{xm}\frac{\partial\rho_m}{\partial x}+\rho_m\frac{\partial w_{xv}}{\partial x}+w_{ym}\frac{\partial\rho_m}{\partial y}+\rho_m\frac{\partial w_{yv}}{\partial y}\right).$$

$$=D_v\left(\rho_m\frac{\partial^2 C_{mv}}{\partial y^2}+\frac{\partial\rho_m}{\partial y}\frac{\partial C_{mv}}{\partial y}\right)$$

With (12.4a), the above equation is changed to

$$w_{xm}\rho_m\frac{\partial(C_{mv})}{\partial x}+w_{ym}\rho_m\frac{\partial(C_{mv})}{\partial y}=D_v\left[\rho_m\frac{\partial^2 C_{mv}}{\partial y^2}+\frac{\partial\rho_m}{\partial y}\frac{\partial C_{mv}}{\partial y}\right]$$

i.e.

$$w_{xm}\frac{\partial(C_{mv})}{\partial x}+w_{ym}\frac{\partial(C_{mv})}{\partial y}=D_v\left[\frac{\partial^2 C_{mv}}{\partial y^2}+\frac{1}{\rho_m}\frac{\partial\rho_m}{\partial y}\frac{\partial C_{mv}}{\partial y}\right]. \tag{12.7a}$$

With (12.25), (12.26), (0) to (q), and (ff), (12.7a) is changed to

$$-\frac{1}{2}\eta_m x^{-1}(C_{mv,s}-C_{mv,\infty})\frac{\mathrm{d}\Gamma_{mv}(\eta_m)}{\mathrm{d}\eta_m}w_{xm,\infty}W_{xm}(\eta_m)$$

$$+ w_{xm,\infty}\left(\frac{1}{2}\mathrm{Re}_{xm,\infty}\right)^{-1/2} W_{ym}(\eta_m)(C_{mv,s}-C_{mv,\infty})\frac{\mathrm{d}\Gamma_{mv}(\eta_m)}{\mathrm{d}\eta_m}x^{-1}\left(\frac{1}{2}\mathrm{Re}_{xm,\infty}\right)^{1/2}$$

$$= D_v\left[\begin{array}{l}(C_{mv,s}-C_{mv,\infty})\dfrac{\mathrm{d}^2\Gamma_{mv}(\eta_m)}{\mathrm{d}\eta_m^2}x^{-2}\left(\dfrac{1}{2}\mathrm{Re}_{xm,\infty}\right)\\ +\dfrac{1}{\rho_m}\dfrac{\mathrm{d}\rho_m}{\mathrm{d}\eta_m}x^{-1}\left(\dfrac{1}{2}\mathrm{Re}_{xm,\infty}\right)^{1/2}(C_{mv,s}-C_{mv,\infty})\dfrac{\mathrm{d}\Gamma_{mv}(\eta_m)}{\mathrm{d}\eta_m}x^{-1}\left(\dfrac{1}{2}\mathrm{Re}_{xm,\infty}\right)^{1/2}\end{array}\right]$$

With definition of $\mathrm{Re}_{xm,\infty}$, the above equation is further simplified to $\dfrac{w_{xm,\infty}x}{\nu_{m,\infty}}$

$$-\frac{1}{2}\eta_m x^{-1}(C_{mv,s}-C_{mv,\infty})\frac{\mathrm{d}\Gamma_{mv}(\eta_m)}{\mathrm{d}\eta_m}w_{xm,\infty}W_{xm}(\eta_m)$$

$$+ w_{xm,\infty}W_{ym}(\eta_m)(C_{mv,s}-C_{mv,\infty})\frac{\mathrm{d}\Gamma_{mv}(\eta_m)}{\mathrm{d}\eta_m}x^{-1}$$

$$= D_v\left[\begin{array}{l}(C_{mv,s}-C_{mv,\infty})\dfrac{\mathrm{d}^2\Gamma_{mv}(\eta_m)}{\mathrm{d}\eta_m^2}x^{-2}\left(\dfrac{1}{2}\dfrac{w_{xm,\infty}x}{\nu_{m,\infty}}\right)\\ +\dfrac{1}{\rho_m}\dfrac{\mathrm{d}\rho_m}{\mathrm{d}\eta_m}x^{-1}\left(\dfrac{1}{2}\dfrac{w_{xm,\infty}x}{\nu_{m,\infty}}\right)(C_{mv,s}-C_{mv,\infty})\dfrac{\mathrm{d}\Gamma_{mv}(\eta_m)}{\mathrm{d}\eta_m}x^{-1}\end{array}\right].$$

The above equation is divided by $\frac{1}{2}w_{xv,\infty}\cdot x^{-1}(C_{mv,s}-C_{mv,\infty})$, and becomes

$$[-\eta_m W_{xm}(\eta_m)+2W_{ym}(\eta_m)]\frac{d\Gamma_{mv}(\eta_m)}{d\eta_m}$$

$$= D_v\left[\frac{d^2\Gamma_{mv}(\eta_m)}{d\eta_m^2}\left(\frac{1}{\nu_{m,\infty}}\right)+\frac{1}{\rho_m}\frac{d\rho_m}{d\eta_m}\left(\frac{1}{\nu_{m,\infty}}\right)\frac{d\Gamma_{mv}(\eta_m)}{d\eta_m}\right],$$

i.e.

$$\left[-\eta_m W_{xm}(\eta_m)+2W_{ym}(\eta_m)\right]\frac{\mathrm{d}\Gamma_{mv}(\eta_m)}{\mathrm{d}\eta_m}$$
$$=\frac{1}{Sc_{m,\infty}}\left[\frac{\mathrm{d}^2\Gamma_{mv}(\eta_m)}{\mathrm{d}\eta_m^2}+\frac{1}{\rho_m}\frac{\mathrm{d}\rho_m}{\mathrm{d}\eta_m}\frac{\mathrm{d}\Gamma_{mv}(\eta_m)}{\mathrm{d}\eta_m}\right]. \tag{12.34}$$

where $Sc_{m,\infty}(=\nu_{m,\infty}/D_v)$ is defined as local Schmidt number.

12.4.3 For Boundary Conditions

12.4.3.1 Transformation of (12.8) and (12.9)

With (12.24), (12.25), and (12.26), (12.8) can be easily transformed to

$$\eta_l = 0: \ W_{xl} = 0, \ W_{yl} = 0, \ \theta_l = 1. \tag{12.35}$$

With (11.20) and (12.25), (12.9) can be easily transformed to

$$W_{xm,\infty}(\eta_m) = W_{xl,s}(\eta_l). \tag{12.36}$$

12.4.3.2 Transformation of (12.10)

With (11.17), we have

$$\delta_l = \eta_{l\delta}\left(\frac{1}{2}\mathrm{Re}_{xl,s}\right)^{-1/2} x. \tag{t}$$

At the interface between liquid and vapour–gas mixture films

Then, at the interface between liquid and vapour–gas mixture films, we have

$$\frac{\partial \delta_l}{\partial x} = \frac{1}{2}\eta_{l\delta}\left(\frac{1}{2}\mathrm{Re}_{xl,s}\right)^{-1/2}. \tag{u}$$

With similar derivation, we have

$$\delta_m = \eta_{m0}\left(\frac{1}{2}\mathrm{Re}_{xm,\infty}\right)^{-1/2} x,$$

and

$$\left(\frac{\partial \delta_m}{\partial x}\right)_s = \frac{1}{2}\eta_{m0}\left(\frac{1}{2}\mathrm{Re}_{xm,\infty}\right)^{-1/2}, \tag{v}$$

at the interface between liquid and vapour–gas mixture films.

With (12.20), (12.21), (12.25), (12.26), (u) and (v), (12.10) is changed to

$$\rho_{l,s}\left[w_{xm,\infty} W_{xl}(\eta_l)\frac{1}{2}\eta_{l\delta}\left(\frac{1}{2}\mathrm{Re}_{xl,s}\right)^{-1/2} - w_{xm,\infty}\left(\frac{1}{2}\mathrm{Re}_{xl,s}\right)^{-1/2} W_{yl}(\eta_l)\right]_{l,s}$$
$$= \rho_{m,s}C_{m,vs}\left[w_{xm,\infty} W_{xm}(\eta_m)\frac{1}{2}\eta_{m0}\left(\frac{1}{2}\mathrm{Re}_{xm,\infty}\right)^{-1/2} - w_{xm,\infty}\left(\frac{1}{2}\mathrm{Re}_{xm,\infty}\right)^{-1/2} W_{ym}(\eta_m)\right]_{v,s}.$$

Since $\eta_{m0} = 0$, the above equation become

$$\rho_{l,s}\left(\frac{1}{2}\mathrm{Re}_{xl,s}\right)^{-1/2}\left[w_{xm,\infty}W_{xl,s}(\eta_l)\frac{1}{2}\eta_{l\delta} - w_{xm,\infty}W_{yl,s}(\eta_l)\right]$$

$$= -\rho_{m,s}C_{mv,s}w_{xm,\infty}\left(\frac{1}{2}\mathrm{Re}_{xm,\infty}\right)^{-1/2} W_{ym,s}(\eta_m).$$

Then

$$-W_{ym,s}(\eta_m) = \frac{\rho_{l,s}}{\rho_{m,s}C_{mv,s}}\frac{\left(\frac{1}{2}\mathrm{Re}_{xm,\infty}\right)^{1/2}}{\left(\frac{1}{2}\mathrm{Re}_{xl,s}\right)^{1/2}}[W_{xl,s}(\eta_l)\frac{1}{2}\eta_{l\delta} - W_{yl,s}(\eta_l)].$$

With definitions of $\mathrm{Re}_{xl,s}$ and $\mathrm{Re}_{xm,\infty}$, the above equation is changed to

$$-W_{ym,s}(\eta_m) = \frac{1}{C_{mv,s}}\frac{\rho_{l,s}}{\rho_{m,s}}\frac{\left(\frac{1}{2}\frac{w_{xm,\infty}x}{\nu_{m,\infty}}\right)^{1/2}}{\left(\frac{1}{2}\frac{w_{xm,\infty}x}{\nu_{l,s}}\right)^{1/2}}[W_{xl,s}(\eta_l)\frac{1}{2}\eta_{l\delta} - W_{yl,s}(\eta_l)].$$

The above equation is simplified to

$$W_{ym,s}(\eta_m) = \frac{1}{2C_{mv,s}}\frac{\rho_{l,s}}{\rho_{m,s}}\left(\frac{\nu_{l,s}}{\nu_{m,\infty}}\right)^{1/2}[-\eta_{l\delta}\cdot W_{xl,s}(\eta_l) + 2W_{yl,s}(\eta_l)]. \quad (12.37)$$

12.4.3.3 Transformation of (12.11)

With (12.g) and (12.gg), (12.11) is changed to

$$\mu_{l,s}w_{xm,\infty}\left(\frac{\mathrm{d}W_x(\eta_l)}{\mathrm{d}\eta_l}\right)_s x^{-1}\left(\frac{1}{2}\mathrm{Re}_{xl,s}\right)^{1/2} = \mu_{m,s}w_{xm,\infty}\left(\frac{\mathrm{d}W_{xm}(\eta_m)}{\mathrm{d}\eta_m}\right)_s x^{-1}\left(\frac{1}{2}\mathrm{Re}_{xm,\infty}\right)^{1/2},$$

or

$$\mu_{l,s}\left(\frac{dW_x(\eta_l)}{\mathrm{d}\eta_l}\right)_s\left(\frac{1}{2}\mathrm{Re}_{xl,s}\right)^{1/2} = \mu_{m,s}\left(\frac{\mathrm{d}W_{xm}(\eta_m)}{\mathrm{d}\eta_m}\right)_s\left(\frac{1}{2}\mathrm{Re}_{xm,\infty}\right)^{1/2}.$$

With definitions of $\mathrm{Re}_{xl,s}$ and $\mathrm{Re}_{xm,\infty}$, the above equation is changed to

$$\mu_{l,s}\left(\frac{\mathrm{d}W_x(\eta_l)}{\mathrm{d}\eta_l}\right)_s\left(\frac{1}{2}\frac{w_{xm,\infty}x}{\nu_{l,s}}\right)^{1/2} = \mu_{m,s}\left(\frac{\mathrm{d}W_{xm}(\eta_m)}{\mathrm{d}\eta_m}\right)_s\left(\frac{1}{2}\frac{w_{xm,\infty}x}{\nu_{m,\infty}}\right)^{1/2}.$$

The above equation is further simplified to

$$\mu_{l,s}\left(\frac{\mathrm{d}W_x(\eta_l)}{\mathrm{d}\eta_l}\right)_s\left(\frac{1}{\nu_{l,s}}\right)^{1/2}=\mu_{m,s}\left(\frac{\mathrm{d}W_{xm}(\eta_m)}{\mathrm{d}\eta_m}\right)_s\left(\frac{1}{\nu_{m,\infty}}\right)^{1/2}.$$

Then

$$\left(\frac{\mathrm{d}W_{xm}(\eta_m)}{\mathrm{d}\eta_m}\right)_s=\frac{\mu_{l,s}}{\mu_{m,s}}\left(\frac{\nu_{m,\infty}}{\nu_{l,s}}\right)^{1/2}\left(\frac{\mathrm{d}W_x(\eta_l)}{\mathrm{d}\eta_l}\right)_s. \tag{12.38}$$

12.4.3.4 Transformation of (12.12)

With (12.25), (12.26), (l), (ll), and (v), (12.12) is changed to

$$\lambda_{l,s}x^{-1}\left(\frac{1}{2}\mathrm{Re}_{xl,s}\right)^{1/2}(t_w-t_{s,\mathrm{int}})\left(\frac{\mathrm{d}\theta_l(\eta_l)}{\mathrm{d}\eta_l}\right)_s$$

$$=\lambda_{m,s}x^{-1}\left(\frac{1}{2}\mathrm{Re}_{xm,\infty}\right)^{1/2}(t_{s,\mathrm{int}}-t_\infty)\left(\frac{\mathrm{d}\theta_m(\eta_m)}{\mathrm{d}\eta_m}\right)_s$$

$$+h_{fg}\rho_{m,s}C_{mv,s}\begin{bmatrix}w_{xm,\infty}W_{xm}(\eta_m)\frac{1}{2}\eta_{m0}\left(\frac{1}{2}\mathrm{Re}_{xm,\infty}\right)^{-1/2}\\-w_{xm,\infty}\left(\frac{1}{2}\mathrm{Re}_{xm,\infty}\right)^{-1/2}W_{ym}(\eta_m)\end{bmatrix}_s.$$

With $\eta_{m0}=0$, the above equation becomes

$$\lambda_{l,s}x^{-1}\left(\frac{1}{2}\mathrm{Re}_{xl,s}\right)^{1/2}(t_w-t_{s,\mathrm{int}})\left(\frac{\mathrm{d}\theta_l(\eta_l)}{\mathrm{d}\eta_l}\right)_s$$

$$=\lambda_{m,s}x^{-1}\left(\frac{1}{2}\mathrm{Re}_{xm,\infty}\right)^{1/2}(t_{s,\mathrm{int}}-t_\infty)\left(\frac{\mathrm{d}\theta_m(\eta_m)}{\mathrm{d}\eta_m}\right)_s$$

$$-h_{fg}\rho_{m,s}C_{mv,s}\left[w_{xm,\infty}\left(\frac{1}{2}\mathrm{Re}_{xm,\infty}\right)^{-1/2}W_{ym}(\eta_m)\right]_s.$$

With definitions of $\mathrm{Re}_{xl,s}$ and $\mathrm{Re}_{xm,\infty}$, the above equation becomes

$$\lambda_{l,s}x^{-1}\left(\frac{1}{2}\frac{w_{xm,\infty}x}{\nu_{l,s}}\right)^{1/2}(t_w-t_{s,\mathrm{int}})\left(\frac{\mathrm{d}\theta_l(\eta_l)}{\mathrm{d}\eta_l}\right)_s$$

$$= \lambda_{m,s} x^{-1} \left(\frac{1}{2} \frac{w_{xm,\infty} x}{\nu_{m,\infty}} \right)^{1/2} (t_{s,\mathrm{int}} - t_\infty) \left(\frac{\mathrm{d}\theta_m(\eta_m)}{\mathrm{d}\eta_m} \right)_s$$

$$- h_{fg} \rho_{m,s} C_{mv,s} \left[w_{xm,\infty} \left(\frac{1}{2} \frac{w_{xm,\infty} x}{\nu_{m,\infty}} \right)^{-1/2} W_{ym}(\eta_m) \right]_s .$$

The above equation is further simplified to

$$\lambda_{l,s} \left(\frac{1}{2} \frac{1}{\nu_{l,s}} \right)^{1/2} (t_w - t_{s,\mathrm{int}}) \left(\frac{\mathrm{d}\theta_l(\eta_l)}{\mathrm{d}\eta_l} \right)_s = \lambda_{m,s} \left(\frac{1}{2} \frac{1}{\nu_{m,\infty}} \right)^{1/2} (t_{s,\mathrm{int}} - t_\infty) \left(\frac{\mathrm{d}\theta_m(\eta_m)}{\mathrm{d}\eta_m} \right)_s$$

$$- h_{fg} \rho_{m,s} C_{mv,s} \left[\left(\frac{1}{2} \frac{1}{\nu_{m,\infty}} \right)^{-1/2} W_{ym}(\eta_m) \right]_s ,$$

i.e.

$$\left(\frac{\mathrm{d}\theta_m(\eta_m)}{\mathrm{d}\eta_m} \right)_s$$

$$= \frac{\lambda_{l,s} \left(\frac{1}{2} \frac{1}{\nu_{l,s}} \right)^{1/2} (t_w - t_{s,\mathrm{int}}) \left(\frac{\mathrm{d}\theta_l(\eta_l)}{\mathrm{d}\eta_l} \right)_s + h_{fg} \rho_{m,s} C_{mv,s} \left[\left(\frac{1}{2} \frac{1}{\nu_{m,\infty}} \right)^{-1/2} W_{ym}(\eta_m) \right]_s}{\lambda_{m,s} \left(\frac{1}{2} \frac{1}{\nu_{m,\infty}} \right)^{1/2} (t_{s,\mathrm{int}} - t_\infty)} .$$

Then,

$$\left(\frac{\mathrm{d}\theta_m(\eta_m)}{\mathrm{d}\eta_m} \right)_s$$

$$= \frac{\lambda_{l,s} \left(\frac{\nu_{m,\infty}}{\nu_{l,s}} \right)^{1/2} (t_w - t_{s,\mathrm{int}}) \left(\frac{\mathrm{d}\theta_l(\eta_l)}{\mathrm{d}\eta_l} \right)_s + 2 h_{fg} \rho_{m,s} \nu_{m,\infty} C_{mv,s} W_{ym,s}(\eta_m)}{\lambda_{m,s} (t_{s,\mathrm{int}} - t_\infty)} . \quad (12.39)$$

12.4.3.5 Transformation of (12.13) and (12.14)

With (12.19) and (12.24), and (12.27) (12.13) and (12.14) can be easily transformed to following equations, respectively:

$\eta_m = 0$

$$\theta_{l,s}(\eta_{l\delta}) = 0, \quad \theta_{m,s}(\eta_{m0}) = 1 \quad (12.40)$$

$$\Gamma_{mv,s}(\eta_{m0}) = 1, \; C_{mv} = C_{mv,s} \quad (12.41)$$

12.4.3.6 Transformation of (12.15)

$$\rho_{m,s} C_{mv,s} \left[w_{xm} \frac{1}{2} \eta_{m0} \left(\frac{1}{2} \mathrm{Re}_{xm,\infty} \right)^{-1/2} - w_{ym} \right]_s$$

$$= D_v \rho_{m,s} (C_{mv,s} - C_{mv,\infty}) \left(\frac{\mathrm{d}\Gamma_{mv}(\eta_m)}{\mathrm{d}\eta_m} \right)_s x^{-1} \left(\frac{1}{2} \mathrm{Re}_{xm,\infty} \right)^{1/2}.$$

With $\eta_{m0} = 0$, the above equation is changed to

$$\rho_{m,s} C_{mv,s} [-w_{ym}]_s = D_v \rho_{m,s} (C_{mv,s} - C_{mv,\infty}) \left(\frac{\mathrm{d}\Gamma_{mv}(\eta_m)}{\mathrm{d}\eta_m} \right)_s x^{-1} \left(\frac{1}{2} \mathrm{Re}_{xm,\infty} \right)^{1/2}.$$

With (12.23) and (12.26), the above equation is changed to

$$- \rho_{m,s} C_{mv,s} w_{xm,\infty} \left(\frac{1}{2} \mathrm{Re}_{xm,\infty} \right)^{-1/2} W_{ym,s}(\eta_{m0})$$

$$= D_v \rho_{m,s} (C_{mv,s} - C_{mv,\infty}) \left(\frac{\mathrm{d}\Gamma_{mv}(\eta_m)}{\mathrm{d}\eta_m} \right)_s x^{-1} \left(\frac{1}{2} \mathrm{Re}_{xm,\infty} \right)^{1/2}.$$

With definition of $\mathrm{Re}_{xm,\infty}$, the above equation becomes

$$- \rho_{m,s} C_{mv,s} w_{xm,\infty} \left(\frac{1}{2} \frac{w_{xm,\infty} x}{\nu_{m,\infty}} \right)^{-1/2} W_{ym,s}(\eta_{m0})$$

$$= D_v \rho_{m,s} (C_{mv,s} - C_{mv,\infty}) \left(\frac{\mathrm{d}\Gamma_{mv}(\eta_m)}{\mathrm{d}\eta_m} \right)_s x^{-1} \left(\frac{1}{2} \frac{w_{xm,\infty} x}{\nu_{m,\infty}} \right)^{1/2},$$

i.e.

$$\left(\frac{d\Gamma_{mv}(\eta_m)}{d\eta_m} \right)_s = \frac{-C_{mv,s} \left(\frac{1}{2} \frac{1}{\nu_{m,\infty}} \right)^{-1/2} W_{ym,s}(\eta_{m0})}{D_v (C_{mv,s} - C_{mv,\infty}) \left(\frac{1}{2} \frac{1}{\nu_{m,\infty}} \right)^{1/2}}$$

The above equation is further simplified to

$$\left(\frac{\mathrm{d}\Gamma_{mv}(\eta_m)}{\mathrm{d}\eta_m} \right)_s = - \frac{2 \nu_{m,\infty} C_{mv,s} W_{ym,s}(\eta_{m0})}{D_v (C_{mv,s} - C_{mv,\infty})}.$$

With local Schmidt number $Sc_{m,\infty} = \dfrac{\nu_{m,\infty}}{D_v}$, the above equation becomes

$$\left(\frac{\mathrm{d}\Gamma_{mv}(\eta_m)}{\mathrm{d}\eta_m}\right)_s = -2Sc_{m,\infty}\frac{C_{mv,s}}{C_{mv,s} - C_{mv,\infty}} W_{ym,s}(\eta_{m0}). \quad (12.42)$$

12.4.3.7 Transformation of (12.16)

With (12.22), (12.24), (12.25), and (12.27), (12.16) becomes

$$\eta_m \to \infty: \quad W_{xm,\infty}(\eta_{m,\infty}) = 1, \quad \theta_{m,\infty} = 0, \quad \Gamma_{mv,\infty} = 0 \quad (12.43)$$

Now, the governing similarity mathematical model is summarized as follows:

For liquid film

$$-\eta_l \frac{\mathrm{d}W_{xl}(\eta_l)}{\mathrm{d}\eta_l} + 2\frac{\mathrm{d}W_{yl}(\eta_l)}{\mathrm{d}\eta_l} + \frac{1}{\rho_l}\frac{\mathrm{d}\rho_l}{\mathrm{d}\eta_l}[-\eta_l W_{xl}(\eta_l) + 2W_{yl}(\eta_l)] = 0, \quad (12.28)$$

$$\frac{\nu_{l,s}}{\nu_l}[-\eta_l W_{xl}(\eta_l) + 2(W_{yl}(\eta_l)]\frac{\mathrm{d}W_{xl}(\eta_l)}{\mathrm{d}\eta_l} = \frac{\mathrm{d}^2 W_{xl}(\eta_l)}{\mathrm{d}\eta_l^2} + \frac{1}{\mu_l}\frac{\mathrm{d}\mu_l}{\mathrm{d}\eta_l}\frac{\mathrm{d}W_{xl}(\eta_l)}{\mathrm{d}\eta_l}, \quad (12.29)$$

$$\mathrm{Pr}_l \frac{\nu_{l,s}}{\nu_l}\left\{[-\eta_l W_{xl}(\eta_l) + 2W_{yl}(\eta_l)]\frac{\mathrm{d}\theta_l(\eta_l)}{\mathrm{d}\eta_l}\right.$$

$$\left. + [-\eta_l W_{xl}(\eta_l) + 2W_{yl}(\eta_l)]\left(\theta_l(\eta_l) + \frac{t_{s,\mathrm{int}}}{t_w - t_{s,\mathrm{int}}}\right)\frac{1}{c_{p_l}}\frac{\mathrm{d}c_{p_l}}{\mathrm{d}\eta_l}\right\}. \quad (12.30)$$

$$= \frac{\mathrm{d}^2\theta_l(\eta_l)}{\mathrm{d}\eta_l^2} + \frac{1}{\lambda_l}\frac{\partial\lambda_l}{\partial\eta_l}\frac{\mathrm{d}\theta_l(\eta_l)}{\mathrm{d}\eta_l}$$

For vapour film

$$-\eta_m \frac{\mathrm{d}W_{xm}(\eta_m)}{\mathrm{d}\eta_m} + 2\frac{\mathrm{d}W_{ym}(\eta_m)}{\mathrm{d}\eta_m} + \frac{1}{\rho_m}\frac{\mathrm{d}\rho_m}{\mathrm{d}\eta_m}[-\eta_m \cdot W_{xm}(\eta_m) + 2W_{ym}(\eta_m)] = 0, \quad (12.31)$$

$$\frac{\nu_{m,\infty}}{\nu_m}[-\eta_m \cdot W_{xm}(\eta_m) + 2W_{ym}(\eta_m)]\frac{\mathrm{d}W_{xm}(\eta_m)}{\mathrm{d}\eta_m} = \frac{\mathrm{d}^2 W_{xm}(\eta_m)}{\mathrm{d}\eta_m^2} + \frac{1}{\mu_m}\frac{\mathrm{d}\mu_m}{\mathrm{d}\eta_m}\frac{\mathrm{d}W_{xm}(\eta_m)}{\mathrm{d}\eta_m}, \quad (12.32)$$

$$\frac{\nu_{m,\infty}}{\nu_m}\left\{\begin{array}{l}[-\eta_m W_{xm}(\eta_m)+2W_{ym}(\eta_m)]\dfrac{\mathrm{d}\theta_m(\eta_m)}{\mathrm{d}\eta_m}\\ +[-\eta_m W_{xm}(\eta_m)+2W_{ym}(\eta_m)]\dfrac{1}{c_{p_m}}\dfrac{\mathrm{d}c_{p_m}}{\mathrm{d}\eta_m}\left(\theta_m+\dfrac{t_\infty}{t_{s,\mathrm{int}}-t_\infty}\right)\end{array}\right\}$$

$$=\frac{1}{\mathrm{Pr}_m}\left[\frac{\mathrm{d}^2\theta_m(\eta_m)}{\mathrm{d}^2\eta_m}+\frac{1}{\lambda_m}\frac{\mathrm{d}\lambda_m}{\mathrm{d}\eta_m}\cdot\frac{\mathrm{d}\theta_m(\eta_m)}{\mathrm{d}\eta_m}\right]$$

$$+\frac{1}{Sc_{m,\infty}}\frac{\nu_{m,\infty}}{\nu_m}\frac{c_{p_v}-c_{p_g}}{c_{p_m}}(C_{mv,s}-C_{mv,\infty})\left[\begin{array}{l}\dfrac{\mathrm{d}\Gamma_{mv}(\eta_m)}{\mathrm{d}\eta_m}\dfrac{\mathrm{d}\theta_m(\eta_m)}{\mathrm{d}\eta_m}\\ +\left(\theta_m+\dfrac{t_\infty}{t_{s,\mathrm{int}}-t_\infty}\right)\dfrac{\mathrm{d}^2\Gamma_{mv}(\eta_m)}{\mathrm{d}\eta_m^2}\\ +\left(\theta_m+\dfrac{t_\infty}{t_{s,\mathrm{int}}-t_\infty}\right)\dfrac{1}{\rho_m}\dfrac{\mathrm{d}\rho_m}{\mathrm{d}\eta_m}\dfrac{\mathrm{d}\Gamma_{mv}(\eta_m)}{\mathrm{d}\eta_m}\end{array}\right] \tag{12.33}$$

$$[-\eta_m W_{xm}(\eta_m)+2W_{ym}(\eta_m)]\frac{d\Gamma_{mv}(\eta_m)}{d\eta_m}=\frac{1}{Sc_{m,\infty}}\left[\frac{d^2\Gamma_{mv}(\eta_m)}{d\eta_m^2}+\frac{1}{\rho_m}\frac{d\rho_m}{d\eta_m}\frac{d\Gamma_{mv}(\eta_m)}{d\eta_m}\right] \tag{12.34}$$

For boundary conditions

$$\eta_l=0,\qquad W_{xl}=0,\ W_{yl}=0,\ \theta_l=1 \tag{12.35}$$

$$\eta_l=\eta_{l\delta},\qquad W_{xm,\infty}(\eta_m)=W_{xl,s}(\eta_l) \tag{12.36}$$

$$W_{ym,s}(\eta_m)=\frac{1}{2C_{mv,s}}\frac{\rho_{l,s}}{\rho_{m,s}}\left(\frac{\nu_{l,s}}{\nu_{m,\infty}}\right)^{1/2}[-\eta_{l\delta}\cdot W_{xl,s}(\eta_l)+2W_{yl,s}(\eta_l)] \tag{12.37}$$

$$\left(\frac{\mathrm{d}W_{xm}(\eta_m)}{\mathrm{d}\eta_m}\right)_s=\frac{\mu_{l,s}}{\mu_{m,s}}\left(\frac{\nu_{m,\infty}}{\nu_{l,s}}\right)^{1/2}\left(\frac{\mathrm{d}W_x(\eta_l)}{\mathrm{d}\eta_l}\right)_s \tag{12.38}$$

$$\left(\frac{\mathrm{d}\theta_m(\eta_m)}{\mathrm{d}\eta_m}\right)_s$$

$$=\frac{\lambda_{l,s}\left(\dfrac{\nu_{m,\infty}}{\nu_{l,s}}\right)^{1/2}(t_w-t_{s,\mathrm{int}})\left(\dfrac{\mathrm{d}\theta_l(\eta_l)}{\mathrm{d}\eta_l}\right)_s+2h_{fg}\rho_{m,s}\nu_{m,\infty}C_{mv,s}W_{ym,s}(\eta_m)}{\lambda_{m,s}(t_{s,\mathrm{int}}-t_\infty)} \tag{12.39}$$

$$\theta_{l,s}(\eta_{l\delta})=0,\quad \theta_{m,s}(\eta_{m0})=1 \tag{12.40}$$

$$\Gamma_{mv,s}(\eta_{m0}) = 1, \quad C_{mv} = C_{mv,s} \tag{12.41}$$

$$\left(\frac{d\Gamma_{mv}(\eta_m)}{d\eta_m}\right)_s = -2Sc_{m,\infty}\frac{C_{mv,s}}{C_{mv,s} - C_{mv,\infty}}W_{ym,s}(\eta_{m0}) \tag{12.42}$$

$$\eta_m \to \infty : \quad W_{xm,\infty}(\eta_{m,\infty}) = 1, \quad \theta_{m,\infty} = 0, \quad \Gamma_{mv,\infty} = 0 \tag{12.43}$$

12.5 Remarks

With the new similarity analysis method, the complete similarity mathematical model is developed for laminar forced film condensation of vapour–gas mixture. While, the similarity models for the liquid and vapour film flows are connected by the seven interfacial matching conditions respectively for two-dimensional velocity component balances, shear force balance, mass flow rate balance, temperature balance, heat transfer balance, concentration condition, as well as the balance between the condensate mass flow and vapour mass diffusion. The work of this chapter is the successful application of the new similarity analysis method in the laminar forced film condensation of vapour–gas mixture, and the developed similarity mathematical model is completely dimensionless.

In the complete similarity mathematical model both for liquid and vapour–gas mixture films, the variable physical properties such as temperature-dependent density, thermal conductivity, viscosity, and specific heat are considered. In addition, the concentration-dependent physical properties are taken into account for vapour–gas film flow. These variable physical properties exist in the transformed governing similarity mathematical model as the related dimensionless forms.

Such developed new similarity dimensionless mathematical model demonstrates again that the present new similarity method is a better alternative to the traditional Falkner-Skan type transformation for extensive investigation of laminar film condensation.

Exercises

1. Compare the governing mathematical models of present chapter and Chapter 9, and point out the essential differences of the governing partial differential equations for laminar forced film condensation of vapour–gas mixture from those for laminar forced film condensation of vapour.
2. Compare the governing mathematical models of present chapter and Chapter 9, and point out the essential differences of the similarity variables for laminar forced film condensation of vapour–gas mixture from those for laminar forced film condensation of vapour.
3. Compare the governing mathematical models of present chapter and Chapter 9, and point out the essential differences of the governing similarity ordinary

differential equations for laminar forced film condensation of vapour–gas mixture from those for laminar forced film condensation of vapour.

4. Compare the governing similarity mathematical models of present chapter based on the present new similarity analysis method for laminar forced film condensation of vapour–gas mixture with the related existing governing similarity mathematical models and the similarity variables based on the Falkner–Skan-type transformation method, and point out their essential differences.
5. Please indicate the difference and relation between the interfacial vapour saturation temperature for laminar forced film condensation of vapour–mass mixture and the vapour saturation temperature for laminar forced film condensation of vapour.

References

1. D.G. Kroger, W.M. Rohsenow, Condensation heat transfer in presence of non-condensable gas. Int. J. Heat Mass Transfer **11**(1), 15–26 (1968)
2. L. Slegers, R.A. Seban, Laminar film condensation of steam containing small concentration of air. Int. J. Heat Mass Transfer **13**, 1941–1947 (1970)
3. V.E. Denny, A.F. Mills, V.J. Jusionis, Laminar film condensation from a steam-air mixture undergoing forced flow down a vertical surface. J. Heat Transfer **93**, 297–304 (1971)
4. H.K. Al-Diwany, J.W. Rose, Free convection film condensation of steam in the presence of non-condensing gases. Int. J. Heat Mass Transfer **16**, 1359–1369 (1972)
5. R.K. Oran, C.J. Chen, Effect of lighter noncond-Nsa3l -Gas on laminar film condensation over a vertical plate. Int. J. Heat Mass Transfer **18**, 993–996 (1975)
6. K. Lucks, Combined body forced and forced-convection in laminar-film condensation of mixed vapour-Integral and finite-difference treatment. Int. J. Heat Mass Transfer **19**(11), 1273–1280 (1976)
7. V.M. Borishanskiy, O.P. Volkov, Effect of uncondensable gases content on heat transfer in steam condensation in a vertical tube. Heat Transf.-Soviet Res **9**, 35–41 (1977)
8. J.W. Rose, Approximate equations for forced-convection in the presence of a non- condensing gas on a flat-plate and horizontal tube. Int. J. Heat Mass Transfer **23**, 539–545 (1980)
9. C.D. Morgan, C.G. Rush, Experimental measurements of condensation heat transfer with noncondensable gases present, in: 21st National Heat Transfer Conference, vol. 27, Seattle, (Washington, DC, ASME HTD, 1983) July 24–28
10. S. Kotake, Effect of a small amount of noncondensable gas on film condensation of multicomponent mixtures. Int. J. Heat Mass Transfer **28**(2), 407–414 (1985)
11. P.J. Vernier, P. Solignac, **A** test of some condensation models in the presence of a noncondensable gas against the ecotra experiment. Nucl. Technol. **77**(1), 82–91 (1987)
12. A.C. Bannwart,. A. Bontemps, Condensation of a vapour with incondensables: an improved gas phase film model accounting for the effect of mass transfer on film thickness. Int. J. Heat Mass Transfer **33**, 1465–1473 (1990)
13. S.M. Chiaasaan, B.K. Kamboj, S.I. Abdel-Khalik, Two-fluid modeling of condensation in the presence of noncondensables in two-phase channel flows. Nucl. Sci. Eng. **119**, 1–17 (1995)
14. Y.S. Chin, S.J. Ormiston, H.M. Soliman, A two-phase boundary-layer model for laminar mixed-convection condensation with a noncondensable gas on inclined plates. Heat Mass Transfer 34(4) 271–277 (1998)
15. V. Srzic, H.M. Soliman, S.J. Ormiston, Analysis of laminar mixedconvection condensation on isothermal plates using the full boundarylayer equations: Mixtures of a vapor and a lighter gas. Int. J. Heat Mass Transfer **42**(4) 685–695 (1999).

16. E.C. Siow, S.J. Ormiston, H.M. Soliman, A two-phase model for laminar film condensation from steam–air mixtures in vertical parallel-plate channels. Heat Mass Transfer **40**(5) 365–375 (2004)
17. S.T.Revankar, D. Pollock, Laminar film condensation in a vertical tube in the presence of noncondensable gas. Appl. Math. Model. **29**(4), 341–359 (2005)
18. J.M. Martin-Valdepenas, M.A. Jimenez, F. Martin-Fuertes, J.A. Fernandez. Benitez, Comparison of film condensation models in presence of non-condensable gases implemented in a CFD Code. Heat Mass Transfer **41**(11), 961–976 (2005)
19. B.J. Chung, S. Kim, M.C. Kim, Film condensations of flowing mixtures of steam and air on an inclined flat plate. Int. Commun. Heat Mass **32**(1–2), 233–239 (2005)
20. S. Oh, S.T. Revankar, Experimental and theoretical investigation of film condensation with noncondensable gas. Int. J. Heat Mass Transfer **49**(15–16, July), 2523–2534 (2006)
21. E.P. Volchkov, V.V. Terekhov, V.I. Terekhov, Heat and mass transfer in a boundary layer during vapor condensation in the presence of noncondensing gas. Heat Transfer Res. **28**, 296–304 (1997)
22. P.K. Sarma, M.A. Reddy, A.E. Bergles, S. Kakac, Condensation of vapours on a fin in the presence of non-condensable gas. Int. J. Heat Mass Transfer, **44**, 3233–3240 (2001)
23. E.C. Siow, S.J. Ormiston, H.M. Soliman, A two-phase model for laminar film condensation from steam-air mixtures in vertical parallel-plate channels. Heat Mass Transfer **40**(5), 365–375 (2004)
24. I. Tamasawa, *Advances in Condensation Heat Transfer, Advances in Heat Transfer* (Academic, New York, 1991) pp. 55–139
25. K. Stephan, *Heat Transfer in Condensation and Boiling* (Springer, New York, NY, 1992)
26. T. Fujii, *Theory of Laminar Film Condensation* (Springer, New York, NY, 1991)
27. G.F. Hewitt, G.L. Shires, T.R. Bott, *Process Heat Transfer* (CRC Press, Inc., Boca Raton, FL, 1994)
28. D.R. Webb, *Condensation of vapour mixtures. Heat Exchanger Design Handbook Update* (Begell House Publishers, New York, NY, 1995)
29. D.Y. Shang, L.C.Zhong, Extensive study on laminar free film condensation from vapor–gas mixture. Int. J. Heat and Mass Transfer **51**, 4300–4314 (2008)

Chapter 13
Velocity, Temperature, and Concentration Fields on Laminar Forced Film Condensation of Vapour–Gas Mixture

Abstract A series of treatments on concentration- and temperature-dependent physical properties are successfully treated for laminar forced film condensation of vapour–gas mixture. The physical property factors of liquid film are temperature-dependent. The physical property factors of vapour–gas mixture film are concentration-dependent in the governing similarity mathematical model; however, after the analysis, it is found that they cover the temperature-dependent physical property factors of vapour and gas. Obviously, the concentration-dependent physical properties of vapour–gas mixture are closely dependent on the temperature-dependent physical properties of vapour and gas. Finally, and a set of physical property factors of governing similarity mathematical models becomes functions of the dimensionless temperature and concentration for convenient numerical calculation. Seven interfacial matching conditions, such as those for two-dimensional velocity component balances, shear force balance, mass flow rate balance, temperature balance, heat transfer balance, concentration condition, as well as the balance between the condensate mass flow and vapour mass diffusion, are considered and rigorously satisfied in the numerical calculation. Take the laminar forced film condensation of water vapour–air mixture as an example of the condensation, a formulation for rigorous determination of the water vapour condensate saturated temperature with variation of the vapour partial pressure is applied for rigorous determination of interfacial vapour saturation temperature. By means of the provided iterative calculation procedure, interfacial vapour saturation temperature is evaluated finally. All these lead to rigorous and reliable numerical results on the laminar forced film condensation of vapour–gas mixture. A system of rigorous numerical results is obtained for laminar forced film condensation of water vapour–air mixture, including velocity, temperature and concentration fields of the two-phase film flows. It is found that the noncondensable gas strongly reduced the wall subcooled grade, because it strongly reduces the interfacial vapour saturation temperature. Increasing the wall temperatures t_w causes decreasing the condensate liquid film thickness obviously, increasing the wall temperature gradient obviously, and increasing the vapour–gas mixture film thickness at accelerative pace. Furthermore, increasing the bulk water vapour mass fraction $C_{mv,\infty}$ causes increasing the condensate liquid film velocity and thickness obviously, decreasing the wall temperature gradient obviously, and increasing the

D.Y. Shang, *Theory of Heat Transfer with Forced Convection Film Flows*,
Heat and Mass Transfer, DOI 10.1007/978-3-642-12581-2_13,

vapour–gas mixture film thickness obviously. However, with increasing the wall temperature, the effect of the bulk vapour mass fraction on the two-phase film flows will decrease. The study results of this work for laminar forced film condensation from vapour–gas mixture is a basis on a successive investigation for its heat and mass transfer.

13.1 Introduction

In Chap. 12, I derived out the complete similarity mathematical model for laminar forced film condensation of vapour–gas mixture. Such similarity mathematical model consists of two parts respectively for liquid film flow and vapour–gas mixture film flow with consideration of variable physical properties. Such two parts of model are combined by the set of the interfacial matching conditions, i.e. two-dimensional velocity component balances, shear force balance, mass flow rate balance, temperature balance, heat transfer balance, concentration condition, as well as the balance between the condensate mass flows and the vapour mass diffusion. Meanwhile, the variable physical properties of gas and liquid exist in the form of the related dimensionless physical properties factors in the governing similarity mathematical model, so that the provided governing mathematical model becomes completely dimensionless for convenient solution.

In this Chapter, I will further investigate the available solution approach and numerical calculation for the velocity and temperature fields of the two-phase film flows. For this purpose, the system of the treatment of the variable physical properties of liquid and vapour–gas mixture will be done, and then a series of the temperature- and concentration-dependent physical property factors becomes the functions of the dimensionless temperatures and concentration. On these bases, the governing similarity mathematical model on the laminar forced film condensation of vapour–gas mixture will be solved smoothly.

The laminar forced film condensation of water vapour–air mixture on a horizontal flat plate will be taken as an example for numerical calculation. The system of the calculation results on velocity, temperature, and concentration fields of the two-phase film flows will be obtained, and the important characteristics of the laminar forced film condensation of vapour–gas mixture will be further clarified.

13.2 Treatment of Variable Physical Properties

For the present work on the two-phase flow problem, the variable physical properties exist respectively in liquid and vapour–gas mixture film flows. Take the laminar forced film condensation of water–air mixture, the treatment approaches related to such variable physical properties are presented as follows, respectively.

13.2.1 Treatment of Temperature-Dependent Physical Properties of Liquid Film

Obviously, the treatment approach of the variable physical properties of liquid film flow on laminar forced film condensation of vapour–gas mixture is same as that on laminar forced film condensation of pure vapour reported in Chap. 10. The related temperature-dependent physical properties of water, as well as the corresponding physical property factors, are respectively expressed by (10.1), (10.2), (10.3), (10.4), (10.5), (10.6), (10.7), and (10.8).

13.2.2 Treatment of Concentration-Dependent Densities of Vapour–Gas Mixture

In ref. [1] for extensive study on laminar-free film condensation of vapour–mass mixture, we briefly reported the treatment of concentration-dependent density of vapour–gas mixture. In this chapter, I will describe in detail the treatment approach of the concentration-dependent densities of vapour–gas mixture for laminar forced film condensation of vapour–gas mixture.

13.2.2.1 Density Equations of Vapour–Gas Mixture

Here, the work focuses on derivation of expression for vapour–gas mixture density. For this purpose, we take a unit volume of vapour–gas mixture, and the following equations can be obtained:

First, the vapour–gas mixture density ρ_m should be sum of local vapour density ρ_{mv} and local gas density ρ_{mg}, i.e.

$$\rho_m = \rho_{mv} + \rho_{mg}. \tag{13.1}$$

The local vapour density ρ_{mv} should equal the following

$$\rho_{mv} = C_{mv}\rho_m. \tag{13.2}$$

With the volume fraction of vapour, $\frac{\rho_{mv}}{\rho_v}$, and the volume fraction of gas, $\frac{\rho_{mg}}{\rho_g}$, we have the following relation:

$$\frac{\rho_{mv}}{\rho_v} + \frac{\rho_{mg}}{\rho_g} = 1, \tag{13.3}$$

where ρ_v and ρ_g are densities of vapour and gas, respectively.

For (13.1), (13.2), and (13.3), C_{mv}, ρ_v and ρ_g can be given variables, and the solutions $\rho_{m,v}$, $\rho_{m,g}$ and ρ_m are found out as follows.

$$\rho_{mv} = \frac{C_{mv}\rho_v\rho_g}{(1-C_{mv})\rho_v + C_{mv}\rho_g}, \tag{13.4}$$

$$\rho_{mg} = \frac{(1-C_{mv})\rho_v\rho_g}{(1-C_{mv})\rho_v + C_{mv}\rho_g}, \tag{13.5}$$

$$\rho_m = \frac{\rho_v\rho_g}{(1-C_{mv})\rho_v + C_{mv}\rho_g}. \tag{13.6}$$

With (13.4), (13.5), and (13.6), the variables $\rho_{m,v}$, $\rho_{m,g}$, and ρ_m should be evaluated by the given conditions C_{mv}, ρ_v, and ρ_g.

13.2.2.2 Expression for $\frac{\mathrm{d}\rho_m}{\mathrm{d}\eta_m}$

With (13.16), we have

$$\begin{aligned}
\frac{\mathrm{d}\rho_m}{\mathrm{d}\eta_m} &= \frac{\mathrm{d}}{\mathrm{d}\eta_m}\left(\frac{\rho_v\rho_g}{(1-C_{mv})\rho_v + C_{m,v}\rho_g}\right) \\
&= \frac{\rho_v\dfrac{\mathrm{d}\rho_g}{\mathrm{d}\eta_m} + \rho_g\dfrac{\mathrm{d}\rho_v}{\mathrm{d}\eta_m}}{(1-C_{mv})\,\rho_v + C_{mv}\rho_g} - \frac{\rho_v\rho_g}{\left((1-C_{mv})\,\rho_v + C_{mv}\rho_g\right)^2} \\
&\quad \times \left[(1-C_{mv})\frac{\mathrm{d}\rho_v}{\mathrm{d}\eta_m} - \rho_v\frac{\mathrm{d}C_{mv}(\eta_m)}{\mathrm{d}\eta_m} + C_{mv}\frac{\mathrm{d}\rho_g}{\mathrm{d}\eta_m} + \rho_g\frac{\mathrm{d}C_{mv}(\eta_m)}{\mathrm{d}\eta_m}\right].
\end{aligned}$$

Then,

$$\begin{aligned}
\frac{1}{\rho_m}\frac{\mathrm{d}\rho_m}{\mathrm{d}\eta_m} &= \frac{\rho_v\dfrac{\mathrm{d}\rho_g}{\mathrm{d}\eta_m} + \rho_g\dfrac{\mathrm{d}\rho_v}{\mathrm{d}\eta_m}}{\rho_v\rho_g} - \frac{1}{(1-C_{mv})\,\rho_v + C_{mv}\rho_g} \\
&\quad \times \left[(1-C_{mv})\,\frac{\mathrm{d}\rho_v}{\mathrm{d}\eta_m} - \rho_v\frac{\mathrm{d}C_{mv}}{\mathrm{d}\eta_m} + C_{mv}\frac{\mathrm{d}\rho_g}{\mathrm{d}\eta_m} + \rho_g\frac{\mathrm{d}C_{mv}\,(\eta_m)}{\mathrm{d}\eta_m}\right].
\end{aligned}$$

The above equation is further simplified to

$$\begin{aligned}
\frac{1}{\rho_m}\frac{\mathrm{d}\rho_m}{\mathrm{d}\eta_m} &= \frac{1}{\rho_g}\frac{\mathrm{d}\rho_g}{\mathrm{d}\eta_m} + \frac{1}{\rho_v}\frac{\mathrm{d}\rho_v}{\mathrm{d}\eta_m} - \frac{C_{mv}\left(\rho_v - \rho_g\right)}{(1-C_{mv})\,\rho_v + C_{mv}\rho_g} \\
&\times \left[\frac{(1-C_{mv})}{C_{mv}}\frac{\rho_v}{\left(\rho_v-\rho_g\right)}\frac{1}{\rho_v}\frac{\mathrm{d}\rho_v}{\mathrm{d}\eta_m} + \frac{\rho_g}{\left(\rho_v-\rho_g\right)}\frac{1}{\rho_g}\frac{\mathrm{d}\rho_g}{\mathrm{d}\eta_m} - \frac{1}{C_{mv}}\frac{\mathrm{d}C_{mv}\,(\eta_m)}{\mathrm{d}\eta_m}\right].
\end{aligned}$$

From (12.27) we have

$$C_{mv} = \left(C_{mv,s} - C_{mv,\infty}\right)\Gamma_{mv}(\eta_m) + C_{mv,\infty}, \tag{a}$$

$$\frac{\mathrm{d}C_{mv}}{\mathrm{d}\eta_m} = \left(C_{mv,s} - C_{mv,\infty}\right)\frac{\mathrm{d}\Gamma_{mv}(\eta_m)}{\mathrm{d}\eta_m}. \tag{b}$$

Then

$$\begin{aligned}\frac{1}{\rho_m}\frac{\mathrm{d}\rho_m}{\mathrm{d}\eta_m} &= \frac{1}{\rho_g}\frac{\mathrm{d}\rho_g}{\mathrm{d}\eta_m} + \frac{1}{\rho_v}\frac{\mathrm{d}\rho_v}{\mathrm{d}\eta_m} - \frac{C_{mv}\left(\rho_v - \rho_g\right)}{\left(1 - C_{mv}\right)\rho_v + C_{mv}\rho_g}\\ &\times\left[\frac{(1 - C_{mv})}{C_{mv}}\frac{\rho_v}{\left(\rho_v - \rho_g\right)}\frac{1}{\rho_v}\frac{\mathrm{d}\rho_v}{\mathrm{d}\eta_m} + \frac{\rho_g}{\left(\rho_v - \rho_g\right)}\frac{1}{\rho_g}\frac{\mathrm{d}\rho_g}{\mathrm{d}\eta_m}\right.\\ &\left.- \frac{C_{mv,s} - C_{mv,\infty}}{C_{mv}}\frac{\mathrm{d}\Gamma_{mv}(\eta_m)}{\mathrm{d}\eta_m}\right].\end{aligned} \tag{13.7}$$

Equation (13.7) demonstrates that the concentration-dependent density factor $\frac{1}{\rho_m}\frac{\mathrm{d}\rho_m}{\mathrm{d}\eta_v}$ of the vapour–gas mixture covers the temperature-dependent gas density factor $\frac{1}{\rho_g}\frac{\mathrm{d}\rho_g}{\mathrm{d}\eta_m}$ and temperature-dependent vapor density factor $\frac{1}{\rho_v}\frac{\mathrm{d}\rho_v}{\mathrm{d}\eta_m}$.

13.2.3 Treatment of Other Concentration-Dependent Physical Properties of Vapour–Gas Mixture

The treatment of other concentration-dependent physical properties, such as absolute viscosity, thermal conductivity, specific heat, and Prandtl number, is a difficult issue at present, due to the absence of the related experimental data. Here, I take the following weighted sum [1] to approximately describe them:

$$\mu_m = C_{mv}\mu_v + (1 - C_{mv})\mu_g, \tag{13.8}$$

$$\lambda_m = C_{mv}\lambda_v + (1 - C_{mv})\lambda_g, \tag{13.9}$$

$$c_{p_m} = C_{mv}c_{p_v} + (1 - C_{mv})c_{p_g}, \tag{13.10}$$

$$\mathrm{Pr}_m = \mu_m C_{pm}/\lambda_m. \tag{13.11}$$

Then

$$\begin{aligned}\frac{\mathrm{d}\mu_m}{\mathrm{d}\eta_m} &= \frac{\mathrm{d}}{\mathrm{d}\eta_v}\left[C_{mv}\mu_v + (1 - C_{mv})\mu_g\right]\\ &= C_{mv}\frac{\mathrm{d}\mu_w}{\mathrm{d}\eta_m} + \mu_v\frac{\mathrm{d}C_{mv}}{\mathrm{d}\eta_m} + (1 - C_{mv})\frac{\mathrm{d}\mu_g}{\mathrm{d}\eta_m} - \mu_g\frac{\mathrm{d}C_{mv}}{\mathrm{d}\eta_m}.\end{aligned}$$

With (b) the above equation is further changed to

$$\frac{\mathrm{d}\mu_m}{\mathrm{d}\eta_m} = C_{mv}\frac{\mathrm{d}\mu_v}{\mathrm{d}\eta_m} + (1 - C_{mv})\frac{\mathrm{d}\mu_g}{\mathrm{d}\eta_m} + (\mu_v - \mu_g)(C_{mv,s} - C_{mv,\infty})\frac{\mathrm{d}\Gamma_{mv}(\eta_m)}{\mathrm{d}\eta_m},$$

i.e.

$$\frac{1}{\mu_m}\frac{\mathrm{d}\mu_m}{\mathrm{d}\eta_m} = C_{mv}\frac{\mu_v}{\mu_m}\frac{1}{\mu_v}\frac{\mathrm{d}\mu_v}{\mathrm{d}\eta_m} + (1 - C_{mv})\frac{\mu_g}{\mu_m}\frac{1}{\mu_g}\frac{d\mu_g}{\mathrm{d}\eta_m} + \frac{\mu_v - \mu_g}{\mu_m}(C_{mv,s} - C_{mv,\infty})\frac{\mathrm{d}\Gamma_{mv}}{\mathrm{d}\eta_m}, \tag{13.12}$$

Equation (13.12) shows that the concentration-dependent absolute viscosity factor $\frac{1}{\mu_m}\frac{\mathrm{d}\mu_m}{\mathrm{d}\eta_m}$ for the vapour–gas mixture covers the temperature-dependent gas absolute viscosity factor $\frac{1}{\mu_g}\frac{\mathrm{d}\mu_g}{\mathrm{d}\eta_m}$ and temperature-dependent vapour absolute viscosity factor $\frac{1}{\mu_v}\frac{\mathrm{d}\mu_v}{\mathrm{d}\eta_m}$.

Similar to derivation of (13.12), the thermal conductivity factor of the vapour–gas mixture, $\frac{\mathrm{d}\lambda_m}{\mathrm{d}\eta_m}$, can be expressed as

$$\frac{1}{\lambda_m}\frac{\mathrm{d}\lambda_m}{\mathrm{d}\eta_m} = C_{mv}\frac{\lambda_v}{\lambda_m}\frac{1}{\lambda_v}\frac{\mathrm{d}\lambda_v}{\mathrm{d}\eta_m} + \left(1 - C_{m,v}\right)\frac{\lambda_g}{\lambda_m}\frac{1}{\lambda_g}\frac{\mathrm{d}\lambda_g}{\mathrm{d}\eta_m} + \frac{\lambda_v - \lambda_g}{\lambda_m}\left(C_{mv,s} - C_{mv,\infty}\right)\frac{\mathrm{d}\Gamma_{mv}}{\mathrm{d}\eta_m}, \tag{13.13}$$

Equation (13.13) shows that the concentration-dependent thermal conductivity factor $\frac{1}{\lambda_m}\frac{\mathrm{d}\lambda_m}{\mathrm{d}\eta_m}$ for the vapour–gas mixture covers the temperature-dependent gas thermal conductivity factor $\frac{1}{\lambda_g}\frac{\mathrm{d}\lambda_g}{\mathrm{d}\eta_m}$ and temperature-dependent vapour thermal conductivity factor $\frac{1}{\lambda_v}\frac{\mathrm{d}\lambda_v}{\mathrm{d}\eta_m}$.

13.2.4 Treatment of Temperature-Dependent Physical Properties of Vapour–Gas Mixture

In above equations for treatment of concentration-dependent physical properties of the medium of vapour–gas mixture film, the temperature-dependent physical properties of vapour and gas exist. For treatment of such related temperature-dependent physical properties, the temperature parameter method will be applied. If the bulk temperature T_∞ is taken as the reference temperature, the related temperature-dependent physical properties, such as density, absolute viscosity, and thermal conductivity are expressed as follows, respectively, for vapour and gas.

For vapour and gas densities we have

$$\frac{\rho_v}{\rho_{v,\infty}} = \left(\frac{T_v}{T_\infty}\right)^{-1} \tag{13.14}$$

$$\frac{\rho_g}{\rho_{g,\infty}} = \left(\frac{T_v}{T_\infty}\right)^{-1} \tag{13.15}$$

For vapour and gas absolute viscosities we have

$$\frac{\mu_v}{\mu_{v,\infty}} = \left(\frac{T}{T_\infty}\right)^{n_{\mu,v}} \quad (13.16)$$

$$\frac{\mu_g}{\mu_{g,\infty}} = \left(\frac{T}{T_\infty}\right)^{n_{\mu,g}} \quad (13.17)$$

For vapour and gas thermal conductivities we have

$$\frac{\lambda_v}{\lambda_{v,\infty}} = \left(\frac{T}{T_\infty}\right)^{n_{\lambda,v}} \quad (13.18)$$

$$\frac{\lambda_g}{\lambda_{g,\infty}} = \left(\frac{T}{T_\infty}\right)^{n_{\lambda,g}} \quad (13.19)$$

where $n_{\mu,v}$ and $n_{\mu,g}$ denote viscosity parameters of vapour and gas, respectively, and $\lambda_{\mu,v}$ and $\lambda_{\mu,g}$ denote thermal conductivity parameters of vapour and gas, respectively.

With (12.22) and (12.24), the vapour and gas physical property factors are derived as follows based on the above equations:

For density factors

$$\frac{1}{\rho_v}\frac{\mathrm{d}\rho_v}{\mathrm{d}\eta_m} = -\frac{(T_{s,\mathrm{int}}/T_\infty - 1)\mathrm{d}\theta_m/\mathrm{d}\eta_m}{(T_{s,\mathrm{int}}/T_\infty - 1)\theta_m + 1} \quad (13.20)$$

$$\frac{1}{\rho_g}\frac{\mathrm{d}\rho_g}{\mathrm{d}\eta_m} = -\frac{(T_{s,\mathrm{int}}/T_\infty - 1)\mathrm{d}\theta_m/\mathrm{d}\eta_m}{(T_{s,\mathrm{int}}/T_\infty - 1)\theta_m + 1} \quad (13.21)$$

For absolute viscosity factors

$$\frac{1}{\mu_v}\frac{\mathrm{d}\mu_v}{\mathrm{d}\eta_m} = \frac{n_{\mu,v}(T_{s,\mathrm{int}}/T_\infty - 1)\mathrm{d}\theta_m/\mathrm{d}\eta_m}{(T_{s,\mathrm{int}}/T_\infty - 1)\theta_m + 1} \quad (13.22)$$

$$\frac{1}{\mu_g}\frac{\mathrm{d}\mu_g}{\mathrm{d}\eta_m} = \frac{n_{\mu,g}(T_{s,\mathrm{int}}/T_\infty - 1)\mathrm{d}\theta_m/\mathrm{d}\eta_m}{(T_{s,\mathrm{int}}/T_\infty - 1)\theta_m + 1} \quad (13.23)$$

For thermal conductivity factors

$$\frac{1}{\lambda_v}\frac{\mathrm{d}\lambda_v}{\mathrm{d}\eta_m} = \frac{n_{\lambda,v}(T_{s,\mathrm{int}}/T_\infty - 1)\mathrm{d}\theta_m/\mathrm{d}\eta_m}{(T_{s,\mathrm{int}}/T_\infty - 1)\theta_m + 1} \quad (13.24)$$

$$\frac{1}{\lambda_g}\frac{\mathrm{d}\lambda_g}{\mathrm{d}\eta_m} = \frac{n_{\lambda,g}(T_{s,\mathrm{int}}/T_\infty - 1)\mathrm{d}\theta_m/\mathrm{d}\eta_m}{(T_{s,\mathrm{int}}/T_\infty - 1)\theta_m + 1} \quad (13.25)$$

So far, the set of equations of the physical property factors involed in the similarity mathematical model of laminar forced film condensation of vapour–gas mixture are completely described on temperature-dependent physical properties of the liquid film medium and the temperature and concentration-dependent physical properties of the vapour–gas mixture film media. All these are basis for rigorous and reliable numerical solution of the governing mathematical model of laminar forced film condensation of vapour–gas mixture.

13.3 Numerical Calculation Procedure

Although a solution of the governing equations for laminar film condensation of vapour is a difficult issue, the solution for the film condensation of vapour–gas mixture is even a challenge. The biggest challenge of the solution is to satisfy the whole set of the matching conditions between the liquid–vapour interface (for short, interfacial matching conditions). To this end, some of our previous literatures have dealt with this issue for laminar-free film condensation of pure vapour [2–4] as well as for laminar-free film condensation of vapour–gas mixture [1]. Furthermore, in Chap. 10, I obtained the solution for laminar forced film condensation of pure vapour. In this chapter, I will further treat the challenging issue on solution of laminar forced film condensation of vapour–gas mixture. The whole solution is divided into the following steps:

Step 1: Calculation for liquid film
In the first step, the governing equations of liquid film are calculated. The solutions of (12.28), (12.39), and (12.30) are found out by using boundary condition equations (12.35), as well as the supposed initial values of condensate liquid film thickness $\eta_{l\delta}$ and liquid interfacial velocity component $W_{xl,s}$ as the boundary conditions. Then, the second step of calculation begins.

Step 2: Calculation for vapour film
In this step of calculation, first (12.36) and (12.37) are used for evaluation of $W_{xv,s}$ and $W_{xv,s}$ respectively by using the solutions of $W_{xl,s}$ and $W_{xy,s}$ obtained from the step 1. Then, with equations (12.40), (12.41), and (12.43), as well as the evaluated values of $W_{xv,s}$ and $W_{xv,s}$ as the boundary conditions, (12.31), (12.32), (12.33), and (12.34) are solved.

Step 3: Judgment on the convergence of the numerical calculation
In this step of calculation, first (12.38), (12.39), and (12.42) are applied as judgment equations for verification of the solution convergence. By means of these judgment equations, the calculation is iterated with appropriate change of the values $\eta_{l\delta}$ and $W_{xl,s}$. In each iteration, calculation on step 1 for liquid film equations (12.28), (12.29), and (12.30) and calculation on step 2 for vapour film equations (12.31), (12.32), (12.33), and (12.34) are done successively. In each step of calculations, the governing equations are solved by a shooting method with seventh-order Runge-Kutta integration. For solving such very strong nonlinear problem, a special variable

mesh approach is applied to the numerical calculation programs, in order to obtain accurate numerical solutions.

It should be emphasized that satisfying all interfacial physical matching conditions is important for obtaining reliable solutions for the laminar forced film condensation with two-phase flows. Even one of the interfacial physical matching conditions is ignored, the solutions will be multiple, and it is never possible to obtain the sole set of correct solutions.

13.4 Numerical Solutions

13.4.1 Interfacial Vapour Saturation Temperature

The interfacial saturation temperature $T_{s,\text{int}}$ of the vapour–gas mixture depends on the interfacial vapour mass fraction $C_{mv,s}$. In this present work, the laminar forced film condensation of water vapour in the presence of air is taken as example for the numerical calculation. The vapour–gas mixture temperature in the bulk is set as $t_\infty = 100\,°\text{C}$. According to the experimental data, the following equation [1] is taken for determination of the interfacial water vapour condensate saturated temperature $T_{s,\text{int}}$ dependent on its interfacial vapour mass fraction $C_{mv,s}$ at atmospheric pressure as the system pressure:

$$T_{s,\text{int}} = T_\text{s} \cdot C_{mv,s}{}^{0.063}, \tag{13.26}$$

where $T_\text{s} = 373\,\text{K}$ is the vapour saturation temperature (with $C_{mv,s} = 1$) at the atmospheric pressure as the system pressure.

It should be indicated that according to the experimental data, for any system pressure the relationship of the interfacial saturation temperature $T_{s,int}$ to its interfacial vapour mass fraction is similar to that shown in (13.26). Here in this book, the pressure in the system is taken as atmospheric pressure, and (13.26) keep valid in this case.

Equation (13.26) shows that the interfacial vapour mass fraction $C_{mv,s}$ dominates the interfacial vapour saturation temperature. However, $C_{mv,s}$ is solution of the governing equations and found out in iterative process for solving the governing equations.

13.4.2 Effect of the Interfacial Vapour Saturation Temperature on Wall Subcooled Temperature

For laminar forced film condensation of water vapour in the presence of air on a horizontal flat plate, a system of numerical calculations is carried out for different wall temperatures t_w and water vapour bulk mass fraction, $C_{mv,\infty}$. According to Chap. 7, the absolute viscosity parameters $n_{\mu,v}$ and $n_{\mu,g}$ are taken as $n_{\mu,v} = 1.04$

and $n_{\lambda,v} = 1.185$, respectively, for water vapour and air, and thermal conductivity parameters $n_{\lambda,v}$ and $n_{\lambda,g}$ are taken as $n_{\mu,g} = 0.68$ and $n_{\lambda,g} = 0.81$, respectively, for water vapour and air, in the numerical calculation. While the water vapour mass diffusion coefficient in air is taken as 0.256×10^{-4}, m^2/s, and the latent heat of water vapour condensation is taken as $h_{fg} = 2257.3\,\mathrm{kJ/kg}$.

With the numerical solutions on interfacial vapour saturation temperature $t_{s,\mathrm{int}}$ for laminar forced film condensation of water vapour–gas mixture with the bulk temperature $t_\infty = 100°\mathrm{C}$ and at atmosphere as the system pressure, the wall subcooled temperature $t_{s,\mathrm{int}} - t_w$ is found out, and shown in Fig. 13.1 with variations of wall temperature and interfacial vapour mass fraction $C_{mv,s}$. Obviously, with a special wall temperature, the wall subcooled temperature $t_{s,\mathrm{int}} - t_w$ depends on the interfacial vapour saturation temperature $t_{s,\mathrm{int}}$. It is seen from Fig. 13.1 that with decreasing the interfacial vapour mass fraction $C_{mv,s}$, the wall subcooled temperature $t_{s,\mathrm{int}} - t_w$ will decrease at an accelerative pace, and such decreasing will be even obvious with decreasing the wall temperature t_w. Since the wall subcooled temperature is the force of the condensate heat band mass transfer, such effect of the interfacial vapour saturation temperature on the wall subcooled temperature will dominate the condensate heat and mass transfer.

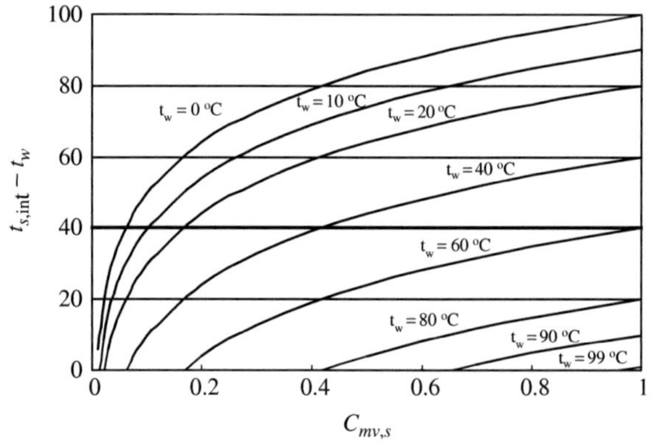

Fig. 13.1 Evaluated wall subcooled temperature $t_{s,\mathrm{int}} - t_w$ with different wall temperature t_w and interfacial vapour mass faction $C_{mv,s}$ for laminar forced film condensation of water vapour–gas mixture with the bulk temperature $t_\infty = 100°\mathrm{C}$ and at atmosphere as the system pressure

13.4.3 Velocity, Concentration, and Temperature Fields of the Two-Phase Film Flows

The laminar forced film condensation of water vapour in the presence of air on a horizontal flat plate is taken as an example for the practical solution. The given conditions for solution are wall temperatures t_w and the bulk vapour mass fraction $C_{mv,\infty}$ with the bulk temperature $t_\infty = 100°\mathrm{C}$ and at atmosphere as the system pressure.

With the available numerical procedure, the governing ordinary equations (12.28), (12.29), (12.30), (12.31), (12.32), (12.33), and (12.34), the boundary conditions, (12.35), (12.36), (12.37), (12.38), (12.39), (12.40), (12.41), (12.42), and

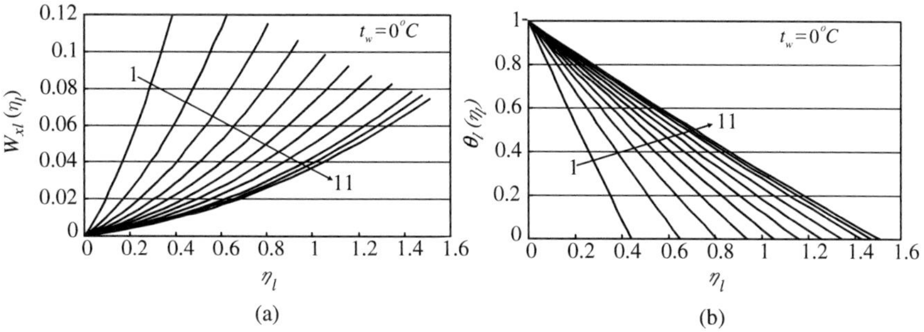

Fig. 13.2 Numerical results of (**a**) velocity $W_{xl}(\eta_l)$ and (**b**) temperature $\theta_l(\eta_l)$ profiles of liquid film for laminar forced film condensation of water–air mixture on a horizontal flat plate with $t_w =$ 0°C (lines 1 to 11 for $C_{mv,\infty} = 0.1, 0.2, 0.3, 0.4, 0.5, 0.6, 0.7, 0.8, 0.9, 0.95$, and 0.99. respectively)

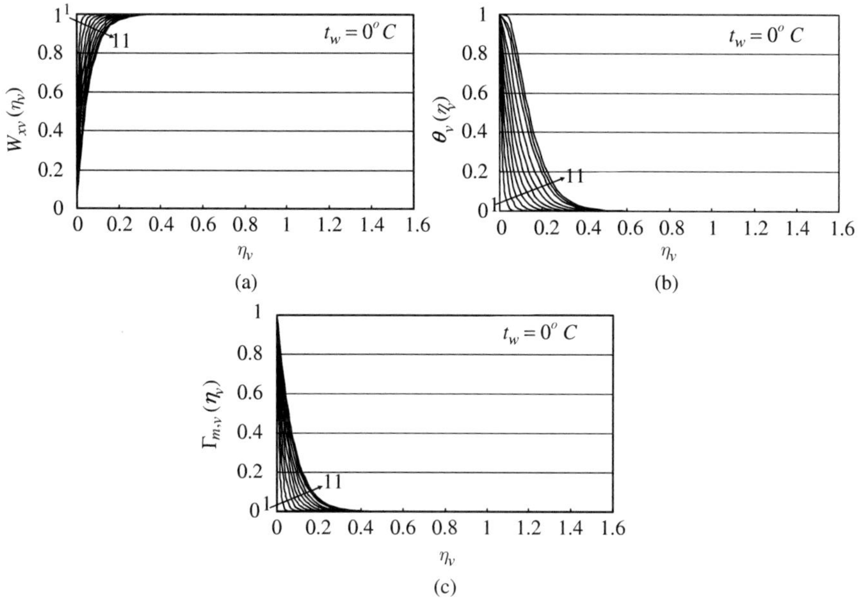

Fig. 13.3 Numerical results on (**a**) velocity component $W_{xv}(\eta_v)$, (**b**) temperature $\theta_v(\eta_v)$, and (**c**) relative concentration $\Gamma_{m,v}(\eta_v)$ profiles of vapour–gas mixture film for laminar forced film condensation of water-air mixture on a horizontal flat plate with $t_w = 0$°C (lines 1 to 11 for $C_{mv,\infty} = 0.1$, 0.2, 0.3, 0.4, 0.5, 0.6, 0.7, 0.8, 0.9, 0.95, and 0.99. respectively)

(12.43), and the physical property equations (13.7), (13.8), (13.9), (13.10), (13.11), (13.12), and (13.13), and (13.20), (13.21), (13.22), (13.23), (13.24), and (13.25) are rigorously calculated, and a system of solutions for velocity, concentration, and temperature fields is obtained, and plotted in Figs. 13.2, 13.3, 13.4, 13.5, 13.6, 13.7, 13.8, and 13.9 with variation of the wall temperatures t_w and bulk vapour mass fraction $C_{mv,\infty}$, respectively. From Figs. 13.1, 13.2, 13.3, 13.4, 13.5, 13.6, 13.7, 13.8,

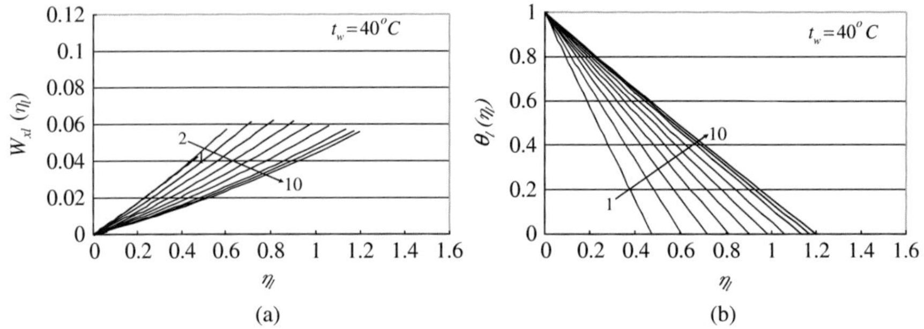

Fig. 13.4 Numerical results of (**a**) velocity $W_{xl}(\eta_l)$ and (**b**) temperature $\theta_l(\eta_l)$ profiles of liquid film for laminar forced film condensation of water–air mixture on a horizontal flat plate with $t_w =$ 40°C (lines 1 to 10 for $C_{mv,\infty} = 0.2$, 0.3, 0.4, 0.5, 0.6, 0.7, 0.8, 0.9, 0.95, and 0.99. respectively)

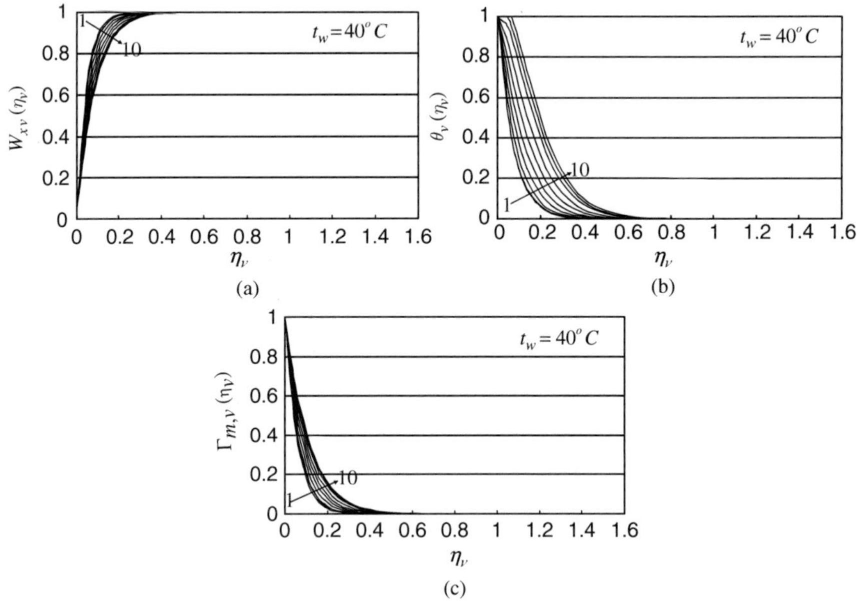

Fig. 13.5 Numerical results of (**a**) velocity component $W_{xv}(\eta_v)$, (**b**) temperature $\theta_v(\eta_v)$, and (**c**) relative concentration $\Gamma_{m,v}(\eta_v)$ profiles of vapour–gas mixture film for laminar forced film condensation of water-air mixture on a horizontal flat plate with $t_w = 40°C$ (lines 1 to 10 for $C_{mv,\infty} = 0.2$, 0.3, 0.4, 0.5, 0.6, 0.7, 0.8, 0.9, 0.95, and 0.99. respectively)

and 13.9 it is seen that the following effect of the given conditions t_w and $C_{mv,\infty}$ on velocity and temperature fields of the two-phase forced film condensation.

First, it is found that the bulk vapour mass fraction $C_{mv,\infty}$ has a decisive effect on the laminar forced film condensation of vapour–gas mixture, and decreasing the bulk vapour mass fraction $C_{mv,\infty}$ (i.e. increasing the bulk gas mass fraction) causes

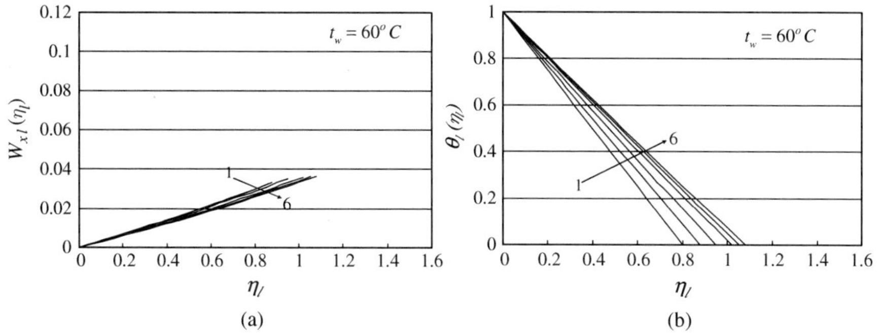

Fig. 13.6 Numerical results of (**a**) velocity $W_{xl}(\eta_l)$ and (**b**) temperature $\theta_l(\eta_l)$ profiles of liquid film for laminar forced film condensation of water–air mixture on a horizontal flat plate with $t_w =$ 60°C (lines 1 to 6 for $C_{mv,\infty} = 0.6$, 0.7, 0.8, 0.9, 0.95, and 0.99. respectively)

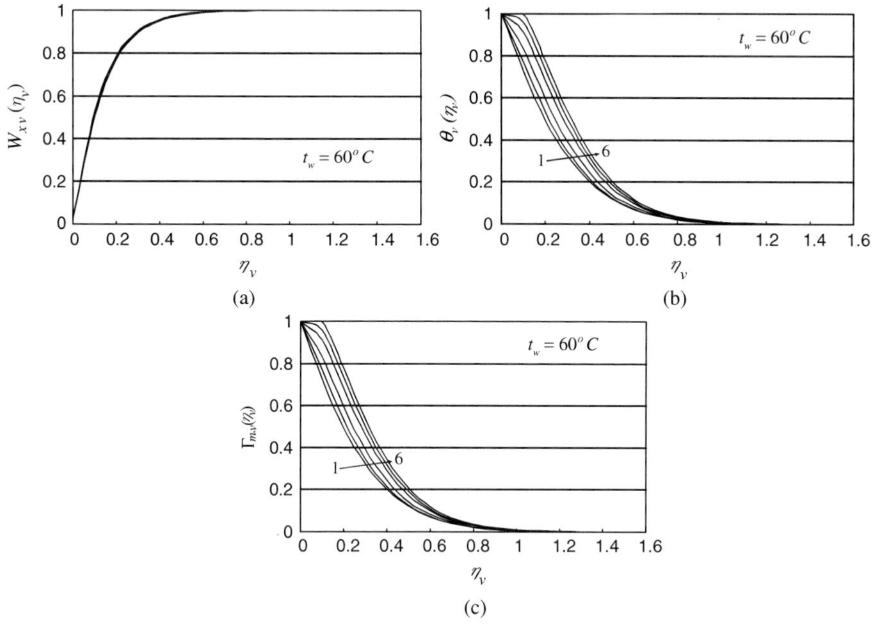

Fig. 13.7 Numerical results of (**a**) velocity component $W_{xv}(\eta_v)$, (**b**) temperature $\theta_v(\eta_v)$, and (**c**) relative concentration $\Gamma_{m,v}(\eta_v)$ profiles of vapour–gas mixture film for laminar forced film condensation of water–air mixture on a horizontal flat plate with $t_w = 60$°C (lines 1 to 6 for $C_{mv,\infty} = 0.6$, 0.7, 0.8, 0.9, 0.95, and 0.99. respectively)

(1) decreasing the condensate liquid film thickness;
(2) increasing the condensate liquid film temperature gradient;
(3) decreasing the wall subcooled temperature $t_{s,\text{int}} - t_w$ at an accelerative pace, which will lead to decreasing the condensate heat transfer at an accelerative pace;

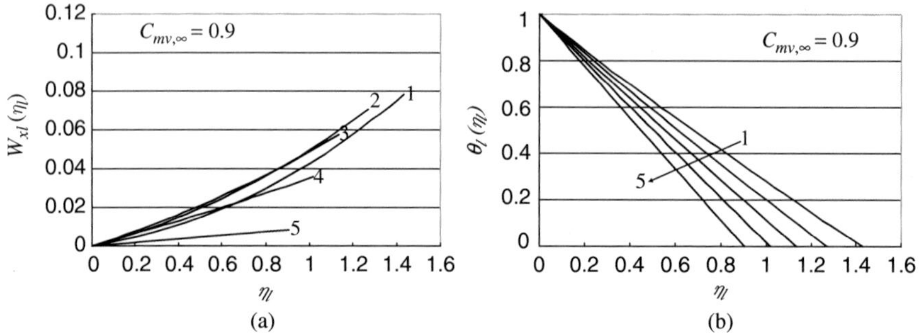

Fig. 13.8 Numerical results of (**a**) velocity $W_{xl}(\eta_l)$ and (**b**) temperature $\theta_l(\eta_l)$ profiles of liquid film for laminar forced film condensation of water–air mixture on a horizontal flat plate with $C_{mv,\infty} = 0.9$ (lines 1 to 5 for $t_w = 0$, 20, 40, 60 and 80°C, respectively)

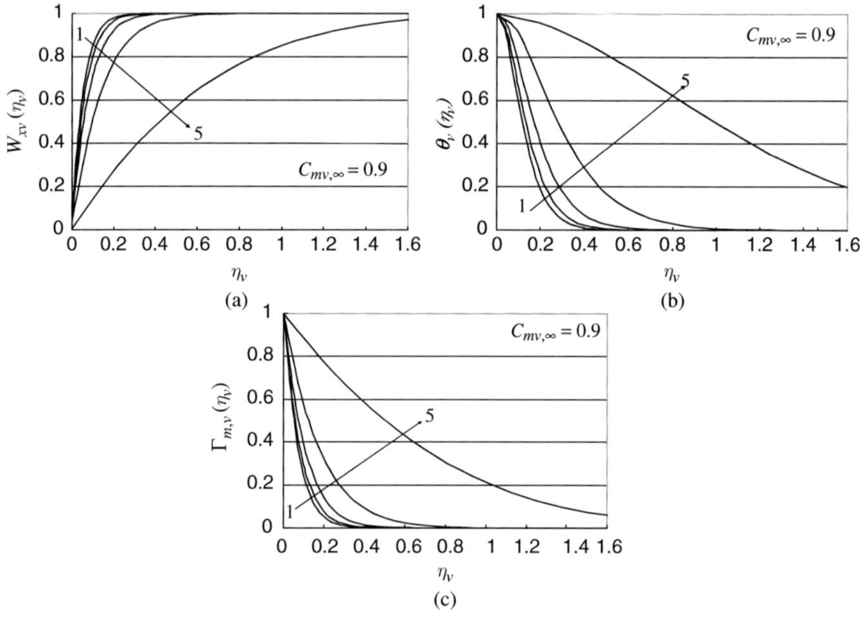

Fig. 13.9 Numerical results of (**c**) velocity component $W_{xv}(\eta_v)$, (**d**) temperature $\theta_v(\eta_v)$, and (**e**) relative concentration $\Gamma_{m,v}(\eta_v)$ profiles of vapour–gas mixture film for laminar forced film condensation of water-air mixture on a horizontal flat plate with $C_{mv,\infty} = 0.9$ (lines 1 to 5 for $t_w = 0$, 20, 40, 60 and 80°C, respectively)

(4) increasing the vapour–gas mixture film flow velocity level;
(5) increasing the temperature gradient of the vapour–gas mixture film;
(6) decreasing the vapour–gas mixture velocity, concentration, and temperature boundary layer thicknesses.

While, increasing the wall temperature t_w causes

(1) decreasing the condensate liquid film thickness obviously;
(2) decreasing the condensate liquid film velocity level at an accelerative pace;
(3) increasing the wall temperature gradient obviously;
(4) decreasing the wall subcooled temperature $t_{s,\text{int}} - t_w$ obviously, which will lead to decreasing the condensate heat transfer obviously;
(5) decreasing the vapour–gas mixture film flow velocity level at an accelerative pace;
(6) decreasing the dimensionless temperature gradient of vapour–gas mixture film flow at an accelerative pace;
(7) increasing the vapour–gas mixture film flow velocity, concentration, and temperature boundary layer thicknesses at an accelerative pace.

13.5 Remarks

In this chapter, a series of treatments of concentration and temperature-dependent physical properties of the governing similarity mathematical model is done, before the numerical calculation. The physical property factors of liquid film are temperature-dependent. The physical property factors of vapour–gas mixture film are concentration-dependent in the governing similarity mathematical model; however, after the analysis, it is found that they cover the temperature-dependent physical property factors of vapour and gas. Obviously, the concentration-dependent physical properties of vapour–gas mixture are closely dependent on the temperature-dependent physical properties of vapour and gas. Finally, a set of physical property factors covered by the governing similarity mathematical model is transformed to the functions of the dimensionless temperature and dimensionless concentration for convenient numerical calculation. Seven matching conditions at the liquid–vapor interface, including two-dimensional velocity component balances, shear force balance, mass flow rate balance, temperature balance, heat transfer balance, concentration condition, as well as the balance between the condensate mass flow and vapour mass diffusion are considered and rigorously satisfied in the numerical calculation. In fact, it is necessary to satisfy all these interfacial physical matching conditions for obtaining reliable solutions of the two-point boundary value problem for the laminar forced film condensation of vapour–gas mixture. Otherwise, it is never possible to obtain any reliable solution of the governing similarity mathematical model.

Furthermore, a useful approach is provided to rigorously determine the vapour condensate saturated temperature $t_{s,\text{int}}$ at the liquid–vapour interface. While, the vapour condensate saturated temperature $t_{s,\text{int}}$ is dependent on interfacial vapour partial pressure or vapour mass fraction. On this basis, the vapour condensate saturated temperature $t_{s,\text{int}}$ is evaluated iteratively by using the provided iterative procedure. All these approaches lead to rigorous and reliable numerical results on the laminar forced film condensation of vapour–gas mixture.

The laminar forced film condensation of water vapour in the presence of air is taken as an example for the numerical calculation. A system of rigorous numerical results is obtained, including velocity and temperature fields of the condensate liquid film, as well as the velocity, temperature and concentration fields of the vapour–gas mixture film. From these numerical results, it is found that the bulk vapour mass fraction has a decisive effect on the laminar forced film condensation of vapour in the presence of noncondensable gas, since the noncondensable gas strongly reduced the wall subcooled temperature.

Furthermore, the effects of the given conditions, the bulk vapour mass fraction, and wall temperature on the two-phase film flow condensation of vapour–gas mixture are clarified in detail.

The bulk vapour mass fraction $C_{mv,\infty}$ has a decisive effect on the laminar forced film condensation of vapour–gas mixture. Decreasing the bulk vapour mass fraction $C_{mv,\infty}$ (i.e. increasing the bulk gas mass fraction) causes

(1) decreasing the condensate liquid film thickness;
(2) increasing the condensate liquid film temperature gradient;
(3) decreasing the wall subcooled temperature $t_{s,\text{int}} - t_w$ at an accelerative pace, which will lead to decreasing the condensate heat transfer at an accelerative pace;
(4) increasing the vapour–gas mixture film flow velocity level;
(5) increasing the temperature gradient of the vapour–gas mixture film;
(6) decreasing the vapour–gas mixture velocity, concentration, and temperature boundary layer thicknesses.

While, increasing the wall temperature t_w causes

(1) decreasing the condensate liquid film thickness obviously;
(2) decreasing the condensate liquid film velocity level at an accelerative pace;
(3) increasing the wall temperature gradient obviously;
(4) decreasing the wall subcooled temperature $t_{s,\text{int}} - t_w$ obviously, which will lead to decreasing the condensate heat transfer obviously;
(5) decreasing the vapour–gas mixture film flow velocity level at an accelerative pace;
(6) decreasing the temperature gradient of vapour–gas mixture film flow at an accelerative pace;
(7) increasing the vapour–gas mixture film flow velocity, concentration, and temperature boundary layer thicknesses at an accelerative pace.

The study results of this work for laminar forced film condensation from vapour–gas mixture can be taken as a basis on a successive investigation for its heat and mass transfer.

Exercises

1. Which parts of contents does the complete similarity governing mathematical model cover for laminar forced film condensation of vapour–gas mixture?
2. Please provide the related physical analysis to explain the close relationship between the concentration-dependent density of vapour–gas mixture and the temperature-dependent densities of vapour and gas.

References

1. D.Y. Shang, L.C. Zhong, Extensive study on laminar free film condensation from vapor–gas mixture. Int. J. Heat Mass Transfer **51**, 4300–4314 (2008)
2. D.Y. Shang, T. Adamek, Study on laminar film condensation of saturated steam on a vertical flat plate for consideration of various physical factors including variable thermophysical properties. Wärme- und Stoffübertragung **30**, 89–100 (1994)
3. D.Y. Shang, B.X. Wang, An extended study on steady-state laminar film condensation of a superheated vapor on an isothermal vertical plate. Int. J. Heat Mass Transfer **40**(4), 931–941 (1997)
4. D.Y. Shang, *Free convection film flows and heat transfer* (Springer-Verlag, Berlin, Heidelberg and New York, 2006)

Chapter 14
Heat and Mass Transfer on Laminar Forced Film Condensation of Vapour–Gas Mixture

Abstract Through the heat and mass transfer analyses, the equations of condensate heat and mass transfer rates are provided, where only dimensionless wall temperature gradient, interfacial vapour saturation temperature and defined condensate mass flow rate parameter are non-given conditions. The defined mass flow rate parameter depends on condensate liquid film thickness as well as the interfacial condensate liquid velocity components. Then, the laminar forced film condensation of water vapour in the presence of air on a horizontal flat plate is taken as an example, and a system of numerical results of the wall dimensionless temperature gradient and mass flow rate parameter are obtained. It is found that although decreasing the bulk vapour mass fraction (i.e. increasing the bulk gas mass fraction) causes increase of the wall dimensionless temperature gradient, it decreases the condensate mass flow rate parameter. However increasing the reference wall subcooled grade causes decrease of the wall temperature gradient and increase of the condensate mass flow rate parameter. These phenomena are closely related to the effect of the non-condensable gas on the condensation. The system of the rigorous key solutions of the wall dimensionless temperature gradient and condensate mass flow rate paramater is formulated using the simple and reliable equations for the laminar film condensation of water vapour–air mixture on a horizontal flat plate. In combination with the provided theoretical heat transfer equation, the formulated equation of the wall dimensionless temperature gradient and condensate mass flow rate parameter can be used for simple and reliable evaluation of the condensate heat transfer rate. The interfacial vapour saturation temperature, necessary for correct prediction of laminar film condensation of vapour–gas mixture, is deeply investigated here. By taking the laminar forced film condensation of water vapour–air mixture as an example, three methods are reported for evaluation of the interfacial vapour saturation temperature: (1) the numerical calculation method, (2) prediction with the formulation equation and (3) prediction by solving the condensate mass–energy transformation equation. The calculated results of the interfacial vapour saturation temperature related to different methods are well coincident. It proves that the similarity analysis method, the similarity mathematical model, the numerical calculation and treatment method of variable physical properties reported in this chapter are valid for extensive investigation of heat and mass transfer of laminar forced film condensation of vapour–gas

D.Y. Shang, *Theory of Heat Transfer with Forced Convection Film Flows*,
Heat and Mass Transfer, DOI 10.1007/978-3-642-12581-2_14,

mixture. The author believes that the analysis and calculation methods reported in this work can be conveniently extended to investigate other different types of laminar forced film condensation from vapour–gas mixture.

14.1 Introduction

In Chaps. 12 and 13, I presented the complete similarity mathematical model and the numerical results of velocity, temperature and concentration fields on laminar forced film condensation from vapour–gas mixture, respectively. In this chapter, further investigation will be done for heat and mass transfer. For this purpose, the appropriate heat and mass transfer analyses will be done for providing the related appropriate theoretical equations. In these theoretical equations, the key solutions, i.e. the wall dimensionless temperature gradient and the induced condensate mass flow rate parameter become the unknown variables on evaluation of condensate heat and mass transfer respectively. Furthermore, the key solutions to the wall dimensionless temperature gradient and the condensate mass flow rate parameter of the governing similarity mathematical model will be formulated based on the rigorous numerical results so that the theoretical equations of condensate heat and mass transfer can be conveniently used for simple and reliable prediction.

14.2 Heat Transfer Analysis

Consulting the related derivation in Chap. 11 for heat and mass transfer analyses of laminar forced film condensation of pure vapour, the heat and transfer analyses of the laminar forced film condensation of vapour–gas mixture will be done briefly as below.

The local heat transfer rate $q_{x,\mathrm{w}}$ with consideration of variable physical properties for the film condensation at position x per unit area on the plate can be calculated by applying Fourier's law as

$$q_{x,\mathrm{w}} = -\lambda_{l,\mathrm{w}} \left(\frac{\partial t}{\partial y} \right)_{y=0}$$

With (12.17), (12.18) and (12.19), $q_{x,\mathrm{w}}$ is described as

$$q_{x,\mathrm{w}} = \lambda_{l,\mathrm{w}} \left(t_\mathrm{w} - t_{\mathrm{s,int}} \right) \left(\frac{1}{2} \mathrm{Re}_{xl,\mathrm{s}} \right)^{1/2} x^{-1} \left(-\frac{\mathrm{d}\theta_l(\eta_l)}{\mathrm{d}\eta_l} \right)_{\eta_l=0} \tag{14.1}$$

where the subscript w notes the local value at the wall with consideration of variable physical properties and $t_{\mathrm{s,int}}$ denotes the interfacial vapour saturation temperature dependent on the interfacial vapour mass fraction $C_{\mathrm{mv,s}}$.

Then, the local heat transfer coefficient, defined as $q_{x,\mathrm{w}} = \alpha_x (t_\mathrm{w} - t_{\mathrm{s,int}})$, will be

$$\alpha_{x,\mathrm{w}} = \lambda_{l,\mathrm{w}} \left(\frac{1}{2}\mathrm{Re}_{xl,\mathrm{s}}\right)^{1/2} x^{-1} \left(-\frac{\mathrm{d}\theta_l(\eta_l)}{\mathrm{d}\eta_l}\right)_{\eta_l=0} \tag{14.2}$$

The local Nusselt number, defined as $\mathrm{Nu}_{x,\mathrm{w}} = \alpha_{x,\mathrm{w}} x/\lambda_{l,\mathrm{w}}$, is expressed by

$$\mathrm{Nu}_{x,\mathrm{w}} = \left(\frac{1}{2}\mathrm{Re}_{xl,\mathrm{s}}\right)^{1/2} \left(-\frac{\mathrm{d}\theta_l(\eta_l)}{\mathrm{d}\eta_l}\right)_{\eta_l=0} \tag{14.3}$$

The total heat transfer rate from position $x = 0$ to x with width of b on the plate is an integration

$$Q_{x,\mathrm{w}} = \iint_A q_{x,\mathrm{w}}\, \mathrm{d}A = \int_0^x q_{x,\mathrm{w}} b\, \mathrm{d}x$$

where the plate area related to condensate heat transfer is described as $A = b \times x$, where b is the plate width related to heat transfer.

With (14.1), the above equation is integrated as follows:

$$Q_{x,\mathrm{w}} = \int_0^x b\lambda_{l,\mathrm{w}} \left(t_\mathrm{w} - t_{\mathrm{s,int}}\right) \left(\frac{1}{2}\mathrm{Re}_{xl,\mathrm{s}}\right)^{1/2} x^{-1} \left(-\frac{\mathrm{d}\theta_l\,(\eta_l)}{\mathrm{d}\eta_l}\right)_{\eta_l=0} \mathrm{d}x$$

i.e.

$$Q_{x,\mathrm{w}} = \sqrt{2} b\lambda_{l,\mathrm{w}} \left(t_\mathrm{w} - t_{\mathrm{s,int}}\right) \left(\mathrm{Re}_{xl,\mathrm{s}}\right)^{1/2} \left(-\frac{\mathrm{d}\theta_l(\eta_l)}{\mathrm{d}\eta_l}\right)_{\eta_l=0} \tag{14.4}$$

The average heat transfer coefficient $\overline{\alpha}_{x,\mathrm{w}}$ defined as $Q_{x,\mathrm{w}} = \overline{\alpha}_{x,\mathrm{w}}(t_\mathrm{w} - t_{\mathrm{s,int}})A$ is expressed as

$$\overline{\alpha}_{x,\mathrm{w}} = \sqrt{2}\lambda_{l,\mathrm{w}} \left(\mathrm{Re}_{xl,\mathrm{s}}\right)^{1/2} x^{-1} \left(-\frac{\mathrm{d}\theta_l(\eta_l)}{\mathrm{d}\eta_l}\right)_{\eta_l=0} \tag{14.5}$$

The average Nusselt number defined as $\overline{\mathrm{Nu}}_{xl,\mathrm{w}} = (\overline{\alpha}_{x,\mathrm{w}} \cdot x)/\lambda_{l,\mathrm{w}}$ will be

$$\overline{\mathrm{Nu}}_{xl,\mathrm{w}} = \sqrt{2} \left(\mathrm{Re}_{xl,\mathrm{s}}\right)^{1/2} \left(-\frac{\mathrm{d}\theta_l\,(\eta_l)}{\mathrm{d}\eta_l}\right)_{\eta_l=0} \tag{14.6}$$

It is indicated that the local Reynolds number in (14.1), (14.2), (14.3), (14.4), (14.5) and (14.6) is $\mathrm{Re}_{xl,\mathrm{s}} = w_{x\mathrm{m},\infty} x/\nu_{l,\mathrm{s}}$ with the vapour–gas mixture velocity $w_{x\mathrm{m},\infty}$ for the laminar forced film condensation of vapour–gas mixture, which is different from the local Reynolds number $\mathrm{Re}_{xl,\mathrm{s}} = w_{x\mathrm{v},\infty} x/\nu_{l,\mathrm{s}}$ with vapour bulk velocity $w_{x\mathrm{v},\infty}$ for the laminar forced film condensation of pure vapour. While the wall subcooled temperature related to condensate heat transfer of laminar forced film condensation of vapour–gas mixture is $t_{s,int} - t_w$, which strongly reduces the condensate heat transfer due to the presence of the non-condensable gas.

From these theoretical equations of heat transfer, it is seen that there are two no-given variables for prediction of condensate heat transfer of the laminar forced film condensation of vapour–gas mixture. They are wall dimensionless temperature gradient

$$\left(-\frac{d\theta_l(\eta_l)}{d\eta_l}\right)_{\eta_l=0}$$

and the interfacial vapour saturation temperature $t_{s,int}$. In the next sections, the evaluation approach of the two variables will be further investigated for prediction of heat transfer of the laminar forced film condensation of vapour–gas mixture.

14.3 Wall Dimensionless Temperature Gradient

Equations (14.1), (14.2), (14.3), (14.4), (14.5) and (14.6) demonstrate that wall dimensionless temperature gradient

$$\left(-\frac{d\theta_l(\eta_l)}{d\eta_l}\right)_{\eta_l=0}$$

is a key solution of the governing similarity mathematical model for evaluation of heat transfer of the film condensation of vapour–gas mixture.

The laminar forced film condensation of water vapour–air mixture on a horizontal flat plate is taken as an example for solution. A system of key numerical solutions of the wall dimensionless temperature gradient

$$\left(-\frac{d\theta_l(\eta_l)}{d\eta_l}\right)_{\eta_l=0}$$

are plotted in Fig. 14.1, and some of the typical solutions are listed in Table 14.1 with variation of reference wall subcooled grade

$$\frac{(\Delta t_w)_s}{t_s}\left(=\frac{t_s - t_w}{t_s}\right)$$

and bulk vapour mass fraction $C_{mv,\infty}$. Here, t_s is the reference vapour saturation temperature (corresponding to $C_{mv,s} = 1$) and $(\Delta t_w)_s = t_s - t_w$ is defined as reference wall subcooled temperature (corresponding to $C_{mv,s} = 1$). The blank space in Table 14.1 which is located above the lines of Fig. 14.1 is the region with $t_{s,int} - t_w < 0$.

It is seen that decreasing the bulk vapour mass fraction causes increase of the wall dimensionless temperature gradient at an accelerative pace. On the other hand,

Table 14.1 Numerical results $\left(-\frac{d\theta_l(\eta_l)}{d\eta_l}\right)_{\eta_l=0}$ of laminar forced film condensation of water vapour–air mixture on a horizontal flat plate at bulk temperature $t_\infty = 100°C$ (The blank space in the table corresponds to the non-condensable region with the wall subcooled temperature $\Delta t_w = t_{s,int} - t_w < 0$)

			$C_{m,v\infty}$							
			0.4	0.5	0.6	0.7	0.8	0.9	0.99	0.999
$(\Delta t_w)_s/t_s$	t_w (°C)	Type of data	$-\left(\frac{d\theta_l}{d\eta_l}\right)_{\eta_l=0}$							
0.05	95	(1)							1.634255	1.599003
		(2)	Region for $\Delta t_w = t_{s,int} - t_w < 0$						1.608437	1.568207
		(3)							−0.0158	−0.01926
0.08	92	(1)						1.612457	1.342983	1.324167
		(2)	Region for $\Delta t_w = t_{s,int} - t_w < 0$					1.640572	1.366957	1.343479
		(3)						0.017436	0.017851	0.014584
0.13	87	(1)					1.495319	1.267659	1.138875	1.12789
		(2)	Region for $\Delta t_w = t_{s,int} - t_w < 0$				1.526569	1.310914	1.158915	1.145433
		(3)					0.020898	0.034122	0.017597	0.015554
0.2	80	(1)			1.675939	1.379882	1.215886	1.1084	1.0366	1.0274
		(2)	Region for $\Delta t_w = t_{s,int} - t_w < 0$		1.620295	1.404829	1.241487	1.113247	1.019238	1.010736
		(3)			−0.0332	0.018079	0.021055	0.004373	−0.01675	−0.01622
0.4	60	(1)	1.5979	1.3999	1.2604	1.1539	1.069	0.9982	0.9436	0.9384
		(2)	1.653211	1.441955	1.289549	1.173329	1.081155	1.005882	0.948825	0.943578
		(3)	0.034615	0.030042	0.023126	0.016837	0.011371	0.007696	0.005537	0.005518
0.6	40	(1)	1.4307	1.2659	1.145606	1.0545	0.9815	0.9181	0.8698	0.8653
		(2)	1.422804	1.264364	1.148098	1.058178	0.985999	0.92643	0.880872	0.876664
		(3)	−0.00552	−0.00121	0.002175	0.003488	0.004583	0.009073	0.012729	0.013133
0.8	20	(1)	1.29	1.1488	1.0456	0.9653	0.9002	0.8454	0.8026	0.7986
		(2)	1.323317	1.174005	1.064604	0.9801	0.91234	0.856471	0.813776	0.809835
		(3)	0.025827	0.021941	0.018176	0.015332	0.013486	0.013096	0.013925	0.014068
1	0	(1)	1.1691	1.0473	0.9575	0.8872	0.8298	0.7814	0.7434	0.7398
		(2)	1.170609	1.047796	0.957078	0.886533	0.829641	0.782497	0.746316	0.742969
		(3)	0.00129	0.000473	−0.00044	−0.00075	−0.00019	0.001404	0.003923	0.004284

Note: (1) numerical solutions, (2) results evaluated by (14.7), and (3) deviation evaluated by (14.7)

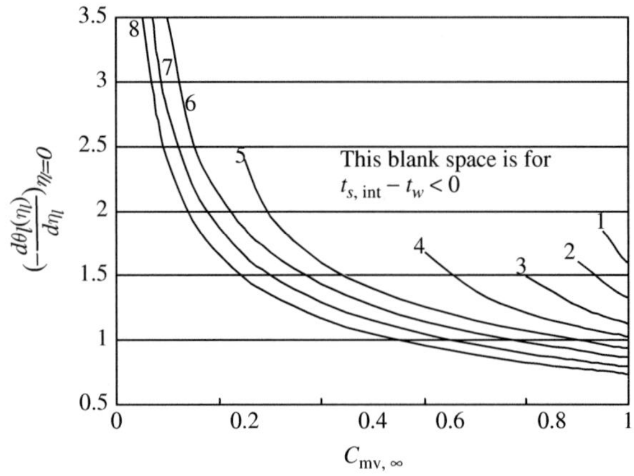

Fig. 14.1 Numerical results $\left(-\frac{d\theta_l(\eta_l)}{d\eta_l}\right)_{\eta_l=0}$ of laminar forced film condensation of water vapour-air mixture on a horizontal flat plate at bulk temperature $t_\infty = 100°C$ (lines 1 → 8 denote the reference wall subcooled grade $\frac{(\Delta t_w)_s}{t_s}\left(=\frac{t_s - t_w}{t_s}\right) = 0.05$, 0.08, 0.13, 0.2, 0.4, 0.6, 0.8 and 1, respectively)

increasing the reference wall subcooled grade will cause decrease of the wall dimensionless temperature gradient. These phenomena will dominate condensate heat transfer of the laminar forced film condensation of vapour–gas mixture.

Furthermore, the rigorous numerical solutions of the wall dimensionless temperature gradient $(d\theta_l/d\eta_l)_{\eta_l=0}$ of laminar forced film condensation of water–air mixture on a horizontal flat plate are formulated by using a curve-fitting approach, and the formulated equation of

$$\left(-\frac{d\theta_l\,(\eta_l)}{d\eta_l}\right)_{\eta_l=0}$$

is shown below with the bulk vapour mass fraction and the reference wall subcooled grade:

$$-\left(\frac{d\theta_l}{d\eta_l}\right)_{\eta_l=0} = AC_{\mathrm{mv},\infty}{}^{B} \qquad \left(0.4 \le C_{\mathrm{mv},\infty} \le 0.999\right) \tag{14.7}$$

where

$$A = 0.5868\left[\frac{(\Delta t_w)_s}{t_s}\right]^{-0.3272} \qquad \left(0.05 \le \frac{(\Delta t_w)_s}{t_s} \le 0.2\right)$$

$$A = -0.334\frac{(\Delta t_w)_s}{t_s} + 1.0766 \qquad \left(0.2 \le \frac{(\Delta t_w)_s}{t_s} \le 1\right)$$

$$B = -0.2486\left[\frac{(\Delta t_w)_s}{t_s}\right]^{-0.8082} \qquad \left(0.05 \le \frac{(\Delta t_w)_s}{t_s} \le 0.2\right)$$

$$B = 2.8813\left[\frac{(\Delta t_w)_s}{t_s}\right]^3 - 6.3245\left[\frac{(\Delta t_w)_s}{t_s}\right]^2 + 4.5528\frac{(\Delta t_w)_s}{t_s} - 1.6063 \qquad \left(0.2 \le \frac{(\Delta t_w)_s}{t_s} \le 1\right)$$

The results of the wall dimensionless temperature gradient

$$\left(-\frac{\mathrm{d}\theta_l\,(\eta_l)}{\mathrm{d}\eta_l}\right)_{\eta_l=0}$$

predicted by using (14.7) are listed in Table 14.1. It is seen that the predicted value of

$$\left(-\frac{\mathrm{d}\theta_l\,(\eta_l)}{\mathrm{d}\eta_l}\right)_{\eta_l=0}$$

is very well coincident to the related numerical solutions. Therefore, (14.7) is reliable for laminar forced film condensation of water–air mixture on a horizontal flat plate.

14.4 Determination of Interfacial Vapour Saturation Temperature

The prerequisite of correct determination of the interfacial vapour saturation temperature $t_{\mathrm{s,int}}$ is the correct prediction of the corresponding interfacial vapour mass fraction $C_{\mathrm{mv,s}}$. With the numerical calculation in Chap. 13, a system of values of the interfacial vapour mass fraction $C_{\mathrm{mv,s}}$ have been obtained for the laminar film condensation of water vapour–air mixture on a horizontal flat plate. Now, these calculated results of the interfacial vapour mass fraction $C_{\mathrm{mv,s}}$ are plotted in Fig. 14.2 with the related reference wall subcooled grade and bulk vapour mass fraction, and

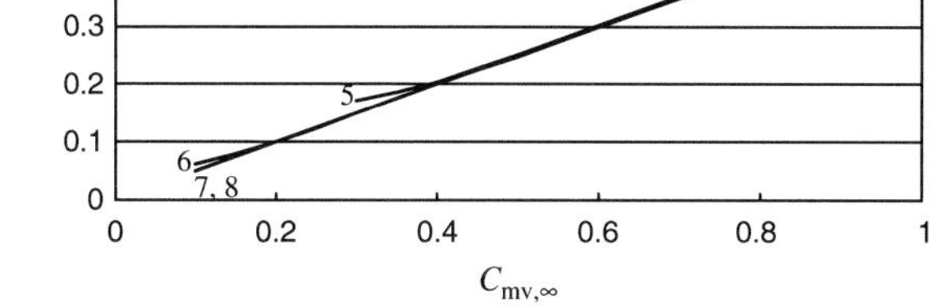

Fig. 14.2 Numerical results of interfacial vapour mass fraction $C_{\mathrm{mv,s}}$ of the laminar film condensation of water vapour–air mixture on a horizontal flat plate at bulk temperature $t_\infty = 100°\mathrm{C}$ (lines $1 \to 8$ denote the reference wall subcooled grade $\frac{(\Delta t_{\mathrm{w}})_{\mathrm{s}}}{t_{\mathrm{s}}} = 0.05, 0.08, 0.13, 0.2, 0.4, 0.6, 0.8$ and 1, respectively)

some of their typical values are listed in Table 14.2. In fact, the blank space in Table 14.2 is the region with $t_{s,\text{int}} - t_w < 0$, which is located above the lines of Fig. 14.2.

It is seen that increasing the bulk vapour mass fraction $C_{m,v\infty}$ will cause increasing the interfacial vapour mass fraction $C_{mv,s}$. In addition, increasing the reference wall subcooled temperature $(\Delta t_w)_s/t_s$ will cause decrease of the interfacial vapour mass fraction $C_{mv,s}$.

With the above calculated results of the interfacial vapour mass fraction $C_{mv,s}$ and by using (13.26), the system of the interfacial vapour saturation temperature $t_{s,\text{int}}$ is evaluated for the laminar forced film condensation of water vapour–air mixture on a horizontal flat plate. These evaluated values of the interfacial vapour saturation temperature $t_{s,\text{int}}$ are plotted in Fig. 14.3, and some of their typical values are listed in Table 14.3. In fact, the blank space in Table 14.3 is the non-condensable region with $t_{s,\text{int}} - t_w < 0$.

Obviously, the interfacial vapour saturation temperature $t_{s,\text{int}}$ also depends on the bulk vapour mass fraction $C_{mv,\infty}$ and reference wall subcooled grade $(\Delta t_w)_s/t_s$. Increasing the bulk vapour mass fraction $C_{m,v\infty}$ will cause increase of the interfacial vapour saturation temperature $t_{s,\text{int}}$. Increasing the reference wall subcooled temperature $(\Delta t_w)_s/t_s$ will cause decrease of the interfacial vapour saturation temperature $t_{s,\text{int}}$. Furthermore, with increasing reference wall subcooled grade $(\Delta t_w)_s/t_s$, the effect of reference wall subcooled grade $(\Delta t_w)_s/t_s$ on the interfacial vapour saturation temperature $t_{s,\text{int}}$ will be weaker and weaker.

The system numerical results of the interfacial vapour saturation temperature $t_{s,\text{int}}$ are further formulated by using a curve-fitting approach, and the related formulated equations are shown below.

For $0.4 \le C_{mv,\infty} \le 0.999$

$$t_{s,\text{int}} = A C_{mv,\infty}{}^{B} \qquad \left(0.05 \le \frac{(\Delta t_w)_s}{t_s} \le 0.4\right) \tag{14.8}$$

$$A = 177.5\left[\frac{(\Delta t_w)_s}{t_s}\right]^2 - 112.12\frac{(\Delta t_w)_s}{t_s} + 100.59$$

$$B = -4.9374\left[\frac{(\Delta t_w)_s}{t_s}\right]^3 + 3.6734\left[\frac{(\Delta t_w)_s}{t_s}\right]^2 - 0.0445\frac{(\Delta t_w)_s}{t_s} + 0.02437$$

and

$$t_{s,\text{int}} = 21.79\ln\left(C_{mv,\infty}\right) + 84.076 \qquad \left(0.4 \le \frac{(\Delta t_w)_s}{t_s} \le 1\right) \tag{14.9}$$

Obviously, (14.8) and (14.9) are suitable for laminar forced film condensation of water vapour–air mixture on a horizontal flat plate at bulk temperature $t_\infty = 100°\text{C}$.

Table 14.2 Numerical results of interfacial vapour mass fraction $C_{mv,s}$ of the laminar forced film condensation of water vapour–air mixture on a horizontal flat plate at bulk temperature $t_\infty = 100°C$ (The blank space in the table corresponds to the non-condensable region with the wall subcooled temperature $\Delta t_w = t_{s,int} - t_w < 0$)

			$C_{m,v\infty}$							
			0.4	0.5	0.6	0.7	0.8	0.9	0.99	0.999
$(\Delta t_w)_s$	t_w (°C)	$\lambda_{l,w}$ (W/(m°C))	$C_{mv,s}$							
0.05	95	0.6750			$\Delta t_w = t_{s,int} - t_w < 0$				0.81964	0.82057
0.08	92	0.6737		$\Delta t_w = t_{s,int} - t_w < 0$				0.72180	0.73334	0.7346
0.13	87	0.6712		$\Delta t_w = t_{s,int} - t_w < 0$			0.586295	0.601520	0.62019	0.622433
0.2	80	0.6669	$\Delta t_w = t_{s,int} - t_w < 0$		0.427822	0.4419	0.46432	0.496232	0.53165	0.5354
0.4	60	0.6506	0.2042	0.2524	0.30230	0.3524	0.40261	0.453	0.4984	0.5030
0.6	40	0.6278	0.20022	0.25032	0.30118	0.35064	0.40086	0.4511	0.49635	0.5009
0.8	20	0.5986	0.2001	0.025015	0.30025	0.350365	0.40051	0.45067	0.49587	0.50037
1	0	0.563	0.200065	0.250125	0.3002	0.35027	0.40038	0.45052	0.49567	0.50017

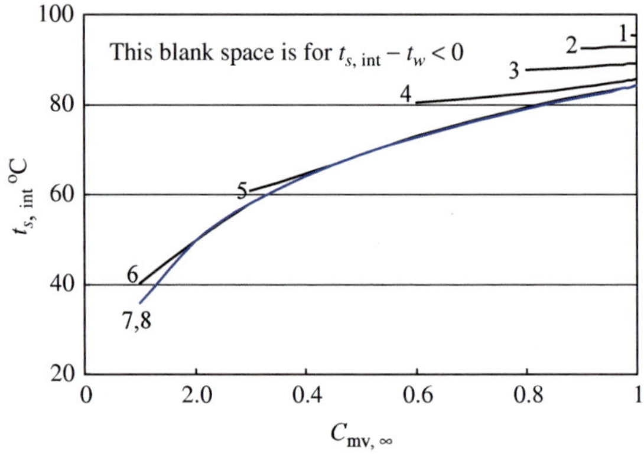

Fig. 14.3 Numerical results of interfacial vapour saturation temperature $t_{s,int}$ of the laminar film condensation of water vapour–air mixture on a horizontal flat plate at bulk temperature $t_\infty = 100°C$ (lines 1 → 8 denote wall subcooled grade $(\Delta t_w)_s/t_s = 0.05, 0.08, 0.13, 0.2, 0.4, 0.6, 0.8$ and 1, respectively)

The values of the interfacial vapour saturation temperature $t_{s,int}$ predicted by using (14.8) and (14.9), and the relative predicted deviation from the numerical results are listed in Table 14.3. It is seen that the predicted values are in very good agreement with the related numerical results.

14.5 Simple and Reliable Prediction Equations of Heat Transfer

With the formulated equation (14.7), the theoretical equations (14.1), (14.2), (14.3), (14.4), (14.5) and (14.6) become available for prediction of heat transfer of laminar forced film condensation of water vapour–air mixture on a horizontal flat plate. For example (14.4) and (14.6) become:

$$Q_x = \sqrt{2}b\lambda_{l,w}\left(t_w - t_{s,int}\right)\left(\mathrm{Re}_{xl,s}\right)^{1/2} AC_{mv,\infty}{}^B \quad (0.4 \le C_{mv,\infty} \le 0.999) \tag{14.10}$$

$$\overline{\mathrm{Nu}}_{xl,w} = \sqrt{2}\left(\mathrm{Re}_{xl,s}\right)^{1/2} AC_{mv,\infty}{}^B \quad (0.4 \le C_{mv,\infty} \le 0.999) \tag{14.11}$$

where

$$A = 0.5868\left[\frac{(\Delta t_w)_s}{t_s}\right]^{-0.3272} \quad \left(0.05 \le \frac{(\Delta t_w)_s}{t_s} \le 0.2\right)$$

$$A = -0.334\frac{(\Delta t_w)_s}{t_s} + 1.0766 \quad \left(0.2 \le \frac{(\Delta t_w)_s}{t_s} \le 1\right)$$

$$B = -0.2486\left[\frac{(\Delta t_w)_s}{t_s}\right]^{-0.8082} \quad \left(0.05 \le \frac{(\Delta t_w)_s}{t_s} \le 0.2\right)$$

Table 14.3 Numerical results of interfacial vapour saturation temperature $t_{s,\mathrm{int}}$ of the laminar film condensation of water vapour–air mixture on a horizontal flat plate at bulk temperature $t_\infty = 100°\mathrm{C}$ (the blank space in the table corresponds to the non-condensable region with the wall subcooled temperature $\Delta t_\mathrm{w} = t_{\mathrm{s,int}} - t_\mathrm{w} < 0$)

			$C_{\mathrm{mv},\infty}$							
			0.4	0.5	0.6	0.7	0.8	0.9	0.99	0.999
$(\Delta t_\mathrm{w})_\mathrm{s}$	t_w (°C)	Type of data	$t_{\mathrm{s,int}}$ (°C)							
0.05	95	(1)	$\Delta t_\mathrm{w} = t_{\mathrm{s,int}} - t_\mathrm{w} < 0$						95.3554	95.3818
		(2)							95.3984	95.42485
		(3)							0.00045	0.00045
0.08	92	(1)	$\Delta t_\mathrm{w} = t_{\mathrm{s,int}} - t_\mathrm{w} < 0$					92.4173	92.7826	92.8222
		(2)						92.3496	92.7175	92.7525
		(3)						−0.00073	−0.00070	−0.00075
0.13	87	(1)	$\Delta t_\mathrm{w} = t_{\mathrm{s,int}} - t_\mathrm{w} < 0$				87.6618	88.2448	88.9411	89.0234
		(2)					87.6395	88.3624	88.9518	89.0079
		(3)					−0.00025	0.0013	0.00012	−0.00017
0.2	80	(1)	$\Delta t_\mathrm{w} = t_{\mathrm{s,int}} - t_\mathrm{w} < 0$		80.5724	81.2944	82.4007	83.8921	85.4456	85.6044
		(2)			80.0801	81.6109	82.9606	84.1696	85.1608	85.2555
		(3)			−0.0061	0.0039	0.0068	0.0033	−0.00333	−0.0041
0.4	60	(1)	64.4756	69.0113	72.92	76.2787	79.2221	81.8485	83.9902	84.1968
		(2)	65.2057	69.38276	72.9934	76.1923	79.0765	81.7110	83.9070	84.1186
		(3)	0.0113	0.0054	0.0010	−0.0011	−0.00185	−0.00174	−0.00099	−0.00093
0.6	40	(1)	64.0573	68.833	72.8397	76.1685	79.1254	81.7546	83.8975	84.1027
		(2)	64.1100	68.9723	72.9451	76.30417	79.2137	81.7802	83.8570	84.0542
		(3)	0.00082	0.00202	0.00145	0.00178	0.00112	0.00031	−0.00048	−0.00058
0.8	20	(1)	64.0446	68.8201	72.7723	76.1513	79.106	81.7333	83.8757	84.0789
		(2)	64.1100	68.9723	72.9451	76.30417	79.2137	81.7802	83.8570	84.0542
		(3)	0.0010	0.0022	0.00237	0.0020	0.00136	0.000574	−0.00022	−0.00029
1	0	(1)	64.0409	68.8163	72.7687	76.1453	79.0988	81.7258	83.8666	84.0699
		(2)	64.1100	68.9723	72.9451	76.30417	79.2137	81.7802	83.8570	84.0542
		(3)	0.00108	0.00227	0.00242	0.00208	0.00145	0.000666	−0.000114	−0.000187

Note: (1) numerical solution, (2) evaluated results by using (14.8) and (14.9) and (3) deviation evaluated by (14.8) and (14.9)

$$B = 2.8813\left[\frac{(\Delta t_\mathrm{w})_\mathrm{s}}{t_\mathrm{s}}\right]^3 - 6.3245\left[\frac{(\Delta t_\mathrm{w})_\mathrm{s}}{t_\mathrm{s}}\right]^2 + 4.5528\frac{(\Delta t_\mathrm{w})_\mathrm{s}}{t_\mathrm{s}} - 1.6063 \qquad \left(0.2 \le \frac{(\Delta t_\mathrm{w})_\mathrm{s}}{t_\mathrm{s}} \le 1\right)$$

14.6 Condensate Mass Transfer Analysis

Consulting the mass transfer analysis of laminar forced film condensation of pure vapour in Chap. 11, the related mass transfer analysis of laminar forced condensation of vapour–gas mixture is given below.

The local mass flow rate g_x entering the liquid film through the interface of liquid and vapour–gas mixture at position x corresponding to unit area of the plate is expressed as

$$g_x = \rho_{l,\mathrm{s}}\left(w_{xl,\mathrm{s}}\frac{\mathrm{d}\delta_l}{\mathrm{d}x} - w_{yl,\mathrm{s}}\right)_\mathrm{s}$$

With (12.17), (12.18), (12.20) and (12.21) for the assumed dimensionless coordinate variable $\eta_{l\delta}$, local Reynolds number $\mathrm{Re}_{xl,\mathrm{s}}$ and dimensionless velocity components $W_{xl,\mathrm{s}}$ and $W_{yl,\mathrm{s}}$ of the liquid film, the above equation is changed to

$$g_x = \rho_{l,\mathrm{s}}\left[w_{x\mathrm{m},\infty}W_{xl,\mathrm{s}}\frac{\mathrm{d}\delta_l}{\mathrm{d}x} - w_{x\mathrm{m},\infty}\left(\frac{1}{2}\mathrm{Re}_{xl,\mathrm{s}}\right)^{-1/2}W_{yl,\mathrm{s}}\right]_\mathrm{s} \tag{14.12}$$

where

$$\delta_l = \eta_{l\delta}\left(\frac{1}{2}\frac{w_{x\mathrm{m},\infty}x}{\nu_{l,\mathrm{s}}}\right)^{-1/2}x$$

$$\frac{\mathrm{d}\delta_l}{\mathrm{d}x} = \frac{1}{2}\eta_{l\delta}\left(\frac{1}{2}\mathrm{Re}_{xl,\mathrm{s}}\right)^{-1/2} \tag{14.13}$$

Then,

$$g_x = \rho_{l,\mathrm{s}}\left[w_{x\mathrm{m},\infty}W_{xl,\mathrm{s}}\frac{1}{2}\eta_{l\delta}\left(\frac{1}{2}\mathrm{Re}_{xl,\mathrm{s}}\right)^{-1/2} - w_{x\mathrm{v},\infty}\left(\frac{1}{2}\mathrm{Re}_{xl,\mathrm{s}}\right)^{-1/2}W_{yl,\mathrm{s}}\right]$$

i.e.

$$g_x = \frac{1}{2}\rho_{l,\mathrm{s}}w_{x\mathrm{m},\infty}\left(\frac{1}{2}\mathrm{Re}_{xl,\mathrm{s}}\right)^{-1/2}\left[\eta_{l\delta}W_{xl,\mathrm{s}} - 2W_{yl,\mathrm{s}}\right]$$

The above equation is rewritten as

$$g_x = \frac{1}{2}\rho_{l,s} w_{xm,\infty} (\frac{1}{2}\mathrm{Re}_{xl,s})^{-1/2} \Phi_s \tag{14.14}$$

where Φ_s is defined as mass flow rate parameter of laminar forced film condensation with the following expression:

$$\Phi_s = (\eta_{l\delta} W_{xl,s} - 2W_{yl,s}) \tag{14.15}$$

The total mass flow rate G_x entering the liquid film through the liquid–vapour interface at position $x = 0$ to x with width b of the plate should be the following integration:

$$G_x = \iint_A g_x \mathrm{d}A = b\int_0^x g_x \mathrm{d}x$$

where $A = b{\cdot}x$ is the related area of the plate with b as the plate width corresponding to the condensation.

With (14.14), the above equation is changed to

$$G_x = \frac{1}{2}b\int_0^x \rho_{l,s} w_{xm,\infty} \left(\frac{1}{2}\mathrm{Re}_{xl,s}\right)^{-1/2} \Phi_s \,\mathrm{d}x$$

Then,

$$G_x = 2\cdot\frac{1}{2}b\rho_{l,s} w_{xm,\infty} x^{1/2} \left(\frac{1}{2}\frac{w_{xm,\infty}}{\nu_{l,s}}\right)^{-1/2} \Phi_s$$

i.e.

$$G_x = 2\cdot\frac{1}{2}b\rho_{l,s} w_{xm,\infty} x^{1/2} \left(2\frac{\nu_{l,s}}{w_{xm,\infty}}\right)^{1/2} \Phi_s$$

$$G_x = 2b\rho_{l,s} \left(\frac{1}{2}\frac{w_{xm,\infty}\cdot \nu_{l,s} x^{1/2}}{1}\right)^{1/2} \Phi_s$$

$$G_x = 2b\rho_{l,s}\nu_{l,s} \left(\frac{1}{2}\frac{w_{xm,\infty}\cdot x^{1/2}}{\nu_{l,s}}\right)^{1/2} \Phi_s$$

or

$$G_x = \sqrt{2}b\mu_{l,s}\mathrm{Re}_{xl,s}{}^{1/2}\Phi_s \tag{14.16}$$

It is seen that (14.16) for the laminar forced film condensation of vapour-gas mixture is the same as (11.17) for the laminar forced film condensation of pure vapour. However, such two equations have different expression of local Reynolds number in definition. For (14.16), the definition of the local Reynolds number is

$\mathrm{Re}_{xl,\mathrm{s}} = w_{x\mathrm{m},\infty}x/\nu_{l,\mathrm{s}}$ with the vapour–gas mixture bulk velocity $w_{x\mathrm{m},\infty}$, while for (11.17), the definition of the local Reynolds number is $\mathrm{Re}_{xl,s} = w_{x\mathrm{v},\infty}x/\nu_{l,\mathrm{s}}$ with the related vapour bulk velocity $w_{x\mathrm{v},\infty}$. In addition, although both the kinetic viscosity of Reynolds number on laminar forced film condensation from vapour-gas mixture and that from vapour are defined at the vapour interfacial saturation temperature, the former is corresponding to any existing interfacial vapour saturation mass fraction, and the latter is only corresponding to the unit interfacial vapour saturation mass fraction.

14.7 Mass Flow Rate Parameter

Equations (14.14) and (14.16) show that only the condensate mass flow rate parameter Φ_s, the key solution to the similarity governing mathematical model, is the unknown variable for evaluation of the condensate mass flow rate of the laminar forced film condensation of vapour–gas mixture.

Equation (11.15) demonstrates that the condensate mass flow rate parameter Φ_s is the function of the key solutions $\eta_{l\delta}$, $W_{xl,\mathrm{s}}$ and $W_{yl,\mathrm{s}}$. The rigorous numerical solutions $\eta_{l\delta}$, $W_{xl,\mathrm{s}}$ and $W_{yl,\mathrm{s}}$ obtained in Chap. 13 for laminar forced film condensation of water vapour–air mixture on a horizontal flat plate are plotted in Figs. 14.4, 14.5 and 14.6, respectively, and some of the typical values are listed in Tables 14.4, 14.5 and 14.6, respectively, with variation of the reference wall subcooled grade $(\Delta t_\mathrm{w})_\mathrm{s}/t_\mathrm{s}$ and bulk vapour mass fraction $C_{\mathrm{mv},\infty}$.

With the numerical solutions of $\eta_{l\delta}$, $W_{xl,\mathrm{s}}(\eta_{l\delta})$ and $W_{yl,\mathrm{s}}(\eta_{l\delta})$, a system of numerical solutions of the condensate mass flow rate parameter Φ_s are found out with (14.15), and some of the typical results are plotted in Fig. 14.7 and listed in Table 14.7. From these numerical results, the following points are addressed:

Decreasing the bulk vapour mass fraction causes decrease of the condensate mass flow rate parameter Φ_s, and such decrease of the condensate mass flow rate parameter Φ_s will be even obvious with increasing reference wall subcooled grade.

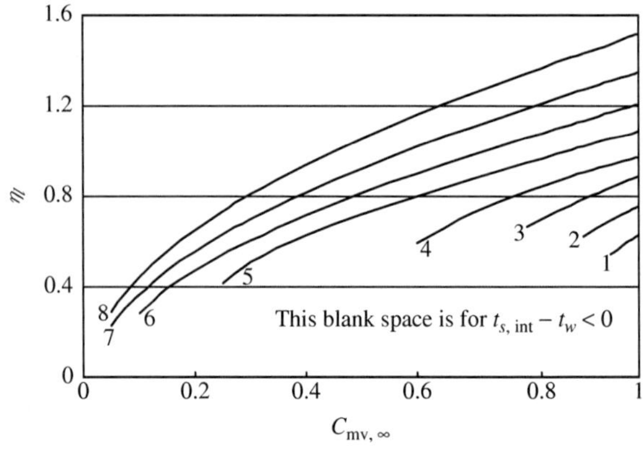

Fig. 14.4 Numerical results $\eta_{l\delta}$ of laminar forced film condensation of water vapour–air mixture on a horizontal flat plate at bulk temperature $t_\infty = 100°\mathrm{C}$ (lines 1 → 8 denote the reference wall subcooled grade $(\Delta t_\mathrm{w})_\mathrm{s}/t_\mathrm{s} = 0.05$, 0.08, 0.13, 0.2, 0.4, 0.6, 0.8 and 1 respectively)

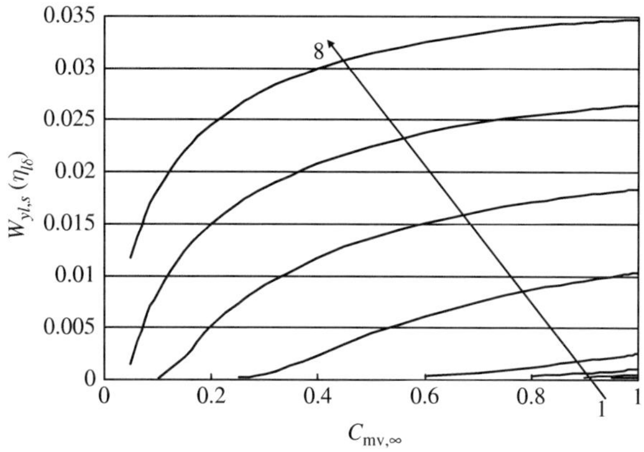

Fig. 14.5 Numerical results $W_{yl,s}(\eta_{l\delta})$ of laminar forced film condensation of water vapour–air mixture on a horizontal flat plate at bulk temperature $t_\infty = 100°C$ (lines 1 → 5 denote the reference wall subcooled grade $(\Delta t_w)_s/t_s = 0.2, 0.4, 0.6, 0.8$ and 1 respectively)

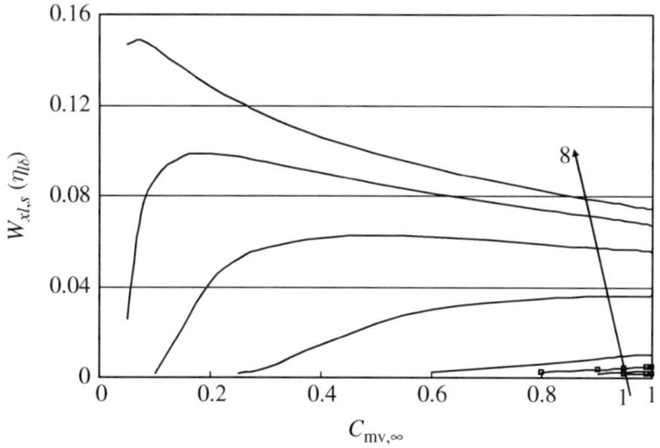

Fig. 14.6 Numerical results $W_{xl,s}(\eta_{l\delta})$ of laminar forced film condensation of water vapour–air mixture on a horizontal flat at bulk temperature $t_\infty = 100°C$ (lines 1 → 8 denote the reference wall subcooled grade $(\Delta t_w)_s/t_s = 0.2, 0.4, 0.6, 0.8$ and 1 respectively)

On the other hand, the condensate mass flow rate parameter Φ_s will increase with increasing reference wall subcooled grade.

The numerical solutions of the condensate mass flow rate parameter Φ_s are formulated with (14.17) by means of a curve-fitting approach. The predicted results by (14.17) are also listed in Table 14.7, which are obviously coincident with the related numerical solutions.

$$\Phi_s = A \ln\left(\frac{(\Delta t_w)_s}{t_s}\right) + B \qquad \left(0.4 \le \frac{(\Delta t_w)_s}{t_s} \le 1\right) \tag{14.17}$$

where

$$A = 0.0272\left(C_{mv,\infty}\right)^{-0.3865} \qquad (0.4 \le C_{mv,\infty} \le 0.999)$$

$$B = -0.0112\left(C_{mv,\infty}\right)^2 + 0.0215\left(C_{mv,\infty}\right) + 0.0334 \qquad (0.4 \le C_{mv,\infty} \le 0.999)$$

Table 14.4 Numerical results $\eta_{l\delta}$ of laminar forced film condensation of water vapour–air mixture on a horizontal flat plate at bulk temperature $t_{\infty} = 100°C$ (the blank space in the table corresponds to the region with the wall subcooled temperature $\Delta t_w = t_{s,int} - t_w < 0$)

		$C_{mv,\infty}$							
		0.4	0.5	0.6	0.7	0.8	0.9	0.99	0.999
$(\Delta t_w)_s/t_s$	t_w (°C)	$\eta_{l\delta}$							
0.05	95	$\Delta t_w = t_{s,int} - t_w < 0$						0.612	0.6255
0.08	92	$\Delta t_w = t_{s,int} - t_w < 0$					0.6203	0.7449	0.7555
0.13	87	$\Delta t_w = t_{s,int} - t_w < 0$				0.669	0.7894	0.879	0.8876
0.2	80	$\Delta t_w = t_{s,int} - t_w < 0$		0.5969	0.7253	0.8237	0.9044	0.968	0.9745
0.4	60	0.6285	0.719876	0.8021	0.8784	0.9502	1.0195	1.080129	1.0863
0.6	40	0.717	0.81396	0.9026	0.9834	1.060514	1.1345	1.1995	1.206
0.8	20	0.81926	0.92435	1.0195	1.1078	1.1911	1.2712	1.3414	1.34836
1	0	0.9399	1.0547	1.1585	1.25455	1.3452	1.43219	1.5085	1.516

Table 14.5 Numerical results $W_{yl,s}(\eta_{l\delta})$ of laminar forced film condensation of water vapour–air mixture on a horizontal flat plate at bulk temperature $t_\infty = 100°C$ (the blank space in the table is corresponds to the region with the wall subcooled temperature $\Delta t_w = t_{s,int} - t_w < 0$)

		$C_{mv,\infty}$							
		0.4	0.5	0.6	0.7	0.8	0.9	0.99	0.999
$(\Delta t_w)_s/t_s$	t_w (°C)	$W_{yl,s}(\eta_{l\delta})$							
0.05	95	Region for $\Delta t_w = t_{s,int} - t_w < 0$						0.00029	0.000305
0.08	92	Region for $\Delta t_w = t_{s,int} - t_w < 0$					0.000324	0.000509	0.000528
0.13	87	Region for $\Delta t_w = t_{s,int} - t_w < 0$				0.000449	0.000722	0.001022	0.001056
0.2	80	Region for $\Delta t_w = t_{s,int} - t_w < 0$		0.000395	0.000743	0.001237	0.001871	0.002509	0.002571
0.4	60	0.002309	0.004443	0.006158	0.007531	0.008658	0.009591	0.0103	0.010365
0.6	40	0.011717	0.013608	0.015081	0.016196	0.017111	0.017849	0.018388	0.018436
0.8	20	0.020864	0.022524	0.023772	0.024733	0.025482	0.026057	0.026455	0.026488
1	0	0.030024	0.031518	0.03261	0.033422	0.034023	0.034454	0.034719	0.03474

Table 14.6 Numerical results $W_{xl,s}(\eta_{l\delta})$ of laminar forced film condensation of water vapour–air mixture on a horizontal flat plate at bulk temperature $t_\infty = 100°C$ (the blank space in the table corresponds to the region with the wall subcooled temperature $\Delta t_w = t_{s,int} - t_w < 0$)

		$C_{mv,\infty}$							
		0.4	0.5	0.6	0.7	0.8	0.9	0.99	0.999
$(\Delta t_w)_s/t_s$	t_w (°C)	$W_{xl,s}(\eta_{l\delta})$							
0.05	95	Region for $\Delta t_w = t_{s,int} - t_w < 0$						0.001896	0.001951
0.08	92	Region for $\Delta t_w = t_{s,int} - t_w < 0$					0.00209	0.002731	0.002791
0.13	87	Region for $\Delta t_w = t_{s,int} - t_w < 0$				0.00268	0.003651	0.004636	0.004745
0.2	80	Region for $\Delta t_w = t_{s,int} - t_w < 0$		0.002641	0.004089	0.005979	0.00822	0.010269	0.010452
0.4	60	0.014547	0.0242	0.02987	0.033148	0.035043	0.036023	0.036385	0.036397
0.6	40	0.06149	0.062318	0.06183	0.060603	0.05909	0.05739	0.05574	0.05557
0.8	20	0.09022	0.08563	0.08145	0.07762	0.07408	0.07075	0.0679	0.06762
1	0	0.10599	0.098552	0.09241	0.08715	0.0825	0.07829	0.07476	0.07442

Table 14.7 Numerical results Φ_s of laminar forced film condensation of water vapour–air mixture on a horizontal flat plate at bulk temperature $t_\infty = 100°C$ (the blank space in the table corresponds to the non-condensable region with the wall subcooled temperature $\Delta t_w = t_{s,int} - t_w < 0$)

			$C_{mv,\infty}$							
			0.4	0.5	0.6	0.7	0.8	0.9	0.99	0.999
$(\Delta t_w)_s$	t_w (°C)		Φ_s							
0.05	95		Region for $\Delta t_w = t_{s,int} - t_w < 0$						0.0005799	0.0006098
0.08	92		Region for $\Delta t_w = t_{s,int} - t_w < 0$					0.0006476	0.0010161	0.0010529
0.13	87		Region for $\Delta t_w = t_{s,int} - t_w < 0$				0.0008955	0.0014379	0.0020311	0.0020989
0.2	80		Region for $\Delta t_w = t_{s,int} - t_w < 0$		0.0007874	0.0014795	0.0024519	0.0036915	0.004923	0.0050428
		(1)	0.0045248	0.0085352	0.0116433	0.0140548	0.0159827	0.0175426	0.0187009	0.0188081
0.4	60	(2)	0.004674	0.008764	0.011898	0.014331	0.016204	0.017603	0.018505	0.018577
		(3)	0.032987	0.026755	0.021865	0.019635	0.013835	0.00346	−0.01049	−0.01227
		(1)	0.0206541	0.0235086	0.0256458	0.027205	0.0284438	0.029411	0.0300841	0.0301454
0.6	40	(2)	0.020391	0.023194	0.025356	0.027018	0.02826	0.029128	0.029617	0.029651
		(3)	−0.01274	−0.01337	−0.01129	−0.00686	−0.00648	−0.00963	−0.01554	−0.01641
		(1)	0.0321856	0.0341041	0.0354943	0.0365214	0.0372727	0.0378234	0.0381711	0.0382001
0.8	20	(2)	0.031542	0.033433	0.034905	0.03602	0.036813	0.037305	0.0375	0.037508
		(3)	−0.01999	−0.01967	−0.01659	−0.01372	−0.01233	−0.01372	−0.01757	−0.01813
		(1)	0.039572	0.0409068	0.041837	0.04249	0.042933	0.0432182	0.0433375	0.0433407
1	0	(2)	0.040192	0.041375	0.042312	0.043003	0.043448	0.043647	0.043616	0.043602
		(3)	0.015668	0.011446	0.011354	0.012073	0.011995	0.009922	0.006421	0.006022

Note: (1) numerical solutions, (2) results evaluated by (14.17) and (3) deviation evaluated by (14.17)

14.8 Prediction Equations of Condensate Mass Transfer

In combination with the reliable formulated equation) for prediction of the condensate mass flow rate parameter, the mass transfer theoretical expression (14.16) will become the following simple and reliable equation for prediction of condensate mass transfer of laminar forced condensation of saturated and superheated water vapour on the horizontal flat plate:

$$G_x = \sqrt{2} b \mu_{l,\mathrm{s}} {\mathrm{Re}_{xl,\mathrm{s}}}^{1/2} \left[A \ln \left(\frac{(\Delta t_\mathrm{w})_\mathrm{s}}{t_\mathrm{s}} \right) + B \right] \qquad \left(0.4 \le \frac{(\Delta t_\mathrm{w})_\mathrm{s}}{t_\mathrm{s}} \le 1 \right) \tag{14.18}$$

where

$$A = 0.0272 \left(C_{\mathrm{mv},\infty} \right)^{-0.3865} \qquad \left(0.4 \le C_{\mathrm{mv},\infty} \le 0.999 \right)$$

$$B = -0.0112 \left(C_{\mathrm{mv},\infty} \right)^2 + 0.0215 \left(C_{\mathrm{mv},\infty} \right) + 0.0334 \qquad \left(0.4 \le C_{\mathrm{mv},\infty} \le 0.999 \right)$$

Obviously, the reliability of (14.18) is same as that of (14.17).

14.9 Equation of Interfacial Vapour Saturation Temperature

14.9.1 For Laminar Forced Film Condensation of Vapour–Gas Mixture

The prediction equation of interfacial vapour saturation temperature $t_{\mathrm{s,int}}$ was reported in Sect. 14.4 for the laminar forced film condensation of water vapour–air mixture. In this section, I will derive the dependent correlation of the interfacial vapour saturation temperature $t_{\mathrm{s,int}}$ in order to investigate another way to predict the interfacial vapour saturation temperature $t_{\mathrm{s,int}}$.

Consulting the condensate mass–energy transformation equation (11.32) with (11.33) for laminar forced film condensation of vapour, the present condensate mass–energy transformation equation for the laminar forced film condensation of vapour–gas mixture becomes

$$\left(-\frac{\mathrm{d}\theta_l}{\mathrm{d}\eta_l} \right)_{\eta_l = 0} = C_{\mathrm{mh}} \Phi_\mathrm{s} \tag{11.32}$$

with

$$C_{\mathrm{mh}} = \frac{\mu_{l,\mathrm{s}} h_{\mathrm{fg}}}{\lambda_{l,\mathrm{w}} \left(t_{\mathrm{s,int}} - t_\mathrm{w} \right)} \tag{14.19}$$

Equation (14.19) describes the condensate mass–energy transformation coefficient C_{mh} for laminar forced film condensation of vapour–gas mixture. Here, $t_{\mathrm{s,int}}$ is the interfacial vapour saturation temperature, $\lambda_{l,\mathrm{w}}$ is the liquid thermal conductivity of the wall, $\mu_{l,\mathrm{s}}$ and h_{fg} are liquid absolute viscosity and latent heat at interfacial vapour saturation temperature $t_{\mathrm{s,int}}$, respectively.

Equation (14.19) is rewritten as

$$C_{\mathrm{mh}}\lambda_{l,\mathrm{w}}\left(t_{\mathrm{s,int}}-t_{\mathrm{w}}\right)-\mu_{l,\mathrm{s}}h_{\mathrm{fg}}=0 \tag{14.20}$$

where from (11.32) the condensate mass–energy transformation coefficient C_{mh} is defined as

$$C_{\mathrm{mh}}=\frac{\left(-\frac{\mathrm{d}\theta_l(\eta_l)}{\mathrm{d}\eta_l}\right)_{\eta_l=0}}{\Phi_{\mathrm{s}}}$$

Equation (14.20) is the equation for interfacial vapour saturation temperature of laminar forced film condensation of vapour–gas mixture and describes the dependent relation of the interfacial vapour saturation temperature $t_{\mathrm{s,int}}$ to the condensate mass–energy transformation coefficient C_{mh}, the liquid thermal conductivity of the wall temperature, $\lambda_{l,\mathrm{w}}$, the liquid thermal conductivity of the wall, $\mu_{l,\mathrm{s}}$ and h_{fg} at interfacial vapour saturation temperature $t_{\mathrm{s,int}}$.

14.9.2 For Laminar Forced Film Condensation of Water Vapour–Air Mixture

Based on the typical experimental data, the condensate latent h_{fg} of water vapour at different interfacial vapour saturation temperature, at an atmospheric pressure as the system pressure of the water vapour–air mixture, can be evaluated by the following equation:

$$h_{\mathrm{fg}}=\left(-2.4355t_{\mathrm{s,int}}+2503.2\right)\times 10^{3} \tag{14.21}$$

According to (10.2) and (10.3), $\lambda_{l,\mathrm{w}}$ and $\mu_{l,\mathrm{s}}$ are expressed as follows for water:

$$\lambda_{l,\mathrm{w}}=-8.01\times 10^{-6}{t_{\mathrm{w}}}^{2}+1.94\times 10^{-3}t_{\mathrm{w}}+0.563 \tag{14.22}$$

$$\mu_{l,\mathrm{s}}=\exp\left[-1.6-\frac{1150}{t_{\mathrm{s,int}}+273}+\left(\frac{690}{t_{\mathrm{s,int}}+273}\right)^{2}\right]\times 10^{-3} \tag{14.23}$$

Then, (14.20) is changed to

$$\begin{aligned}&C_{\mathrm{mh}}\left(-8.01\times 10^{-6}{t_{\mathrm{w}}}^{2}+1.94\times 10^{-3}t_{\mathrm{w}}+0.563\right)\left(t_{\mathrm{s,int}}-t_{\mathrm{w}}\right)\\&-\exp\left[-1.6-\frac{1150}{t_{\mathrm{s,int}}+273}+\left(\frac{690}{t_{\mathrm{s,int}}+273}\right)^{2}\right]\\&\times\left(-2.4355t_{\mathrm{s,int}}+2503.2\right)\times 10^{3}=0\end{aligned} \tag{14.24}$$

with the condensate mass-energy transformation coefficient, defined as

$$C_{\rm mh} = \frac{\left(-\frac{{\rm d}\theta_l(\eta_l)}{{\rm d}\eta_l}\right)_{\eta_l=0}}{\Phi_{\rm s}}$$

Equation (14.24) is the alternative of the condensate mass–energy transformation equation for laminar forced film condensation of water vapour–air mixture, describing the dependent relation of the interfacial vapour saturation temperature $t_{\rm s,int}$ to the condensate mass–energy transformation coefficient. While the latter depends on the solutions, the wall dimensionless temperature gradient

$$\left(-\frac{{\rm d}\theta_l(\eta_l)}{{\rm d}\eta_l}\right)_{\eta_l=0}$$

and condensate mass flow rate parameter Φ_s. With the nonlinear algebraic equation (14.24) for $t_{\rm s,int}$, the interfacial vapour saturation temperature $t_{\rm s,int}$ can be easily found out with iteration approach for any given values of the wall dimensionless temperature gradient, condensate mass flow rate parameter, and interfacial vapour saturation temperature. In fact, (14.24) can also be used for judgement of reliability of numerical solutions of the governing mathematical model.

14.10 Evaluation of Condensate Mass–Energy Transformation Coefficient

It is obvious that the condensate mass–energy transportation coefficient plays an important role in the investigation of the transformation between the data of heat and mass transfer of the film condensation. In this section, the laminar forced film condensation of water vapour–air mixture on a horizontal flat plate at bulk temperature $t_\infty = 100°{\rm C}$ will be taken as an example to introduce the evaluation of the condensate mass–energy transportation coefficient. Here, I will introduce two approaches about such transformation.

The first approach to the evaluation of the condensate mass–energy transportation coefficient is the numerical method. With the systems of key numerical solutions

$$\left(-\frac{{\rm d}\theta_l(\eta_l)}{{\rm d}\eta_l}\right)_{\eta_l=0}$$

and $\Phi_{\rm s}$ listed in Figs. 14.1 and 14.7, respectively, the system of numerical solutions of the condensate mass–energy transformation coefficient

$$C_{\rm mh} = \frac{\left(-\frac{{\rm d}\theta_l(\eta_l)}{{\rm d}\eta_l}\right)_{\eta_l=0}}{\Phi_{\rm s}}$$

Fig. 14.7 Numerical results Φ_s of laminar forced film condensation of water vapour–air mixture on a horizontal flat plate at bulk temperature $t_\infty = 100°C$ (lines 1 → 5 denote the reference wall subcooled grade $(\Delta t_w)_s/t_s = 0.2, 0.4, 0.6, 0.8$ and 1 respectively)

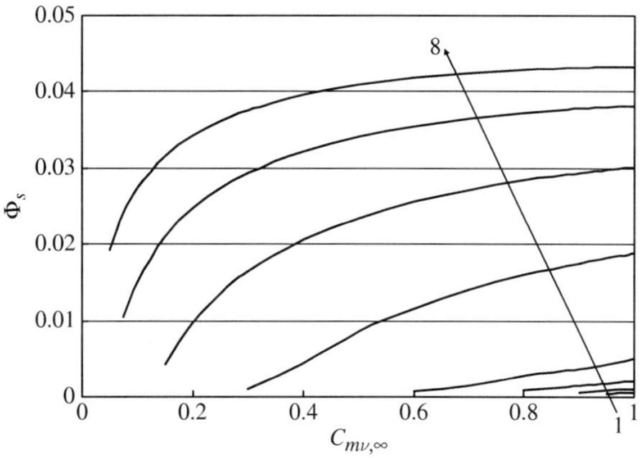

is found out, and their typical values are listed in Table 14.8. On the other hand, for the evaluation of the condensate mass–energy transformation coefficient for laminar forced film condensation of water vapour–air mixture, the formulated equations (14.7) and (14.17) can also be applied for reliable evaluation of the dimensionless wall temperature gradient and the condensate mass flow rate parameter.

The second approach is to use the definition equation (14.19) to evaluate the condensate mass–energy transformation coefficient. In this evaluation, the physical property equations (14.21), (14.22) and (14.23) are used for determination of the related physical properties at the wall surface and the liquid–vapour interface, respectively. The listed numerical solutions of the interfacial vapour saturation temperature in Table 14.8 are taken from Table 14.3 for the laminar forced film condensation of water vapour–air mixture on a horizontal flat plate at bulk temperature $t_\infty = 100\,°C$. On this basis, a set of the evaluated results of condensate mass–energy transportation coefficient are evaluated with (14.19), and their typical values are listed in Table 14.8.

It is seen from Table 14.8 that the results of the condensate mass–energy transformation coefficient evaluated by the two approaches are very well coincident. It follows that the condensate mass–energy transformation equation reported in this book is reliable.

According to the derivation, it is seen that the condensate mass–energy transformation equation (11.32) with (11.33) is universally suitable for any laminar forced film condensation. While, (14.19) is specialized for the laminar forced film condensation of vapour-gas mixture.

14.11 Summary

At present, it is the time to summarize the complete mathematical models of laminar forced film condensation of vapour–gas mixture in Table 14.9, including the two-phase film flows governing partial differential equations, similarity variables, dimensionless similarity governing equations, the physical property factor equations, and heat and mass transfer equations.

Table 14.8 Evaluated results of the condensate mass–energy transformation coefficient

		$C_{mv,\infty}$							
		0.4	0.5	0.6	0.7	0.8	0.9	0.99	0.999
$(\Delta t_w)_s/t_s$	t_w (°C)	$C_{mh} = \frac{\left(-\frac{d\theta_l(\eta_l)}{d\eta_l}\right)_{\eta_l=0}}{\Phi_s}$ $\left(\text{evaluated with the numerical solutions } \left(-\frac{d\theta_l(\eta_l)}{d\eta_l}\right)_{\eta_l=0} \text{ and } \Phi_s\right)$							
0.05	95	The region for $\Delta t_w = t_{s,int} - t_w < 0$						2818.2913	2622.0669
0.08	92	The region for $\Delta t_w = t_{s,int} - t_w < 0$					2489.9777	1321.6446	1257.6082
0.13	87	The region for $\Delta t_w = t_{s,int} - t_w < 0$				1669.7773	881.60154	560.70952	537.38169
0.2	80	The region for $\Delta t_w = t_{s,int} - t_w < 0$		2128.4119	932.66984	495.89163	300.25402	210.56273	203.73655
0.4	60	353.1435	164.01496	108.25085	82.100047	66.884992	56.901328	50.457482	49.8935
0.6	40	69.269439	53.848472	44.67039	38.761271	34.506675	31.216259	28.912254	28.704195
0.8	20	40.079989	33.685109	29.458272	26.431053	24.151733	22.351243	21.026401	20.905703
1	0	29.543616	25.602104	22.886448	20.880191	19.32779	18.080365	17.153751	17.069398

		$C_{mv,\infty}$							
		0.4	0.5	0.6	0.7	0.8	0.9	0.99	0.999
$(\Delta t_w)_s/t_s$	t_w (°C)	$t_{s,int}$,(°C) (numerical solutions listed in Table 14.3)							
0.05	95	The region for $\Delta t_w = t_{s,int} - t_w < 0$						95.3554	95.3818
0.08	92	The region for $\Delta t_w = t_{s,int} - t_w < 0$					92.4173	92.7826	92.8222
0.13	87	The region for $\Delta t_w = t_{s,int} - t_w < 0$				87.6618	88.2448	88.9411	89.0234
0.2	80	The region for $\Delta t_w = t_{s,int} - t_w < 0$		80.5724	81.2944	82.4007	83.8921	85.4456	85.6044
0.4	60	64.4756	69.0113	72.92	76.2787	79.2221	81.8485	83.9902	84.1968

Table 14.8 (continued)

$(\Delta t_w)_s/t_s$	t_w (°C)	$C_{mv,\infty}$ 0.4	0.5	0.6	0.7	0.8	0.9	0.99	0.999
		$C_{mh}=\dfrac{\left(-\dfrac{d\theta_l(\eta_l)}{d\eta_l}\right)_{\eta_l=0}}{\Phi_s}$ $\left(\text{evaluated with the numerical solutions } \left(-\dfrac{d\theta_l(\eta_l)}{d\eta_l}\right)_{\eta_l=0} \text{ and } \Phi_s\right)$							
0.6	40	64.0573	68.833	72.8397	76.1685	79.1254	81.7546	83.8975	84.1027
0.8	20	64.0446	68.8201	72.7723	76.1513	79.106	81.7333	83.8757	84.0789
1	0	64.0409	68.8163	72.7687	76.1453	79.0988	81.7258	83.8666	84.0699

$(\Delta t_w)_s/t_s$	t_w (°C)	$C_{mv,\infty}$ 0.4	0.5	0.6	0.7	0.8	0.9	0.99	0.999
		$C_{mh}=\dfrac{\mu_{l,s}h_{fg}}{\lambda_{l,w}(t_{s,int}-t_w)}$ (evaluated by using (14.19) with (14.21), (14.22) and (14.23) and numerical solutions of $t_{s,int}$ listed in Table 14.3)							
0.05	95	The region for $\Delta t_w = t_{s,int} - t_w < 0$						2813.987	2618.606
0.08	92	The region for $\Delta t_w = t_{s,int} - t_w < 0$					2486.071	1319.855	1255.693
0.13	87	The region for $\Delta t_w = t_{s,int} - t_w < 0$				1667.842	880.2802	559.6544	536.3449
0.2	80	The region for $\Delta t_w = t_{s,int} - t_w < 0$		2127.156	931.6304	495.0115	299.4354	209.7734	203.4164
0.4	60	352.3123981	163.2052	107.4642	81.31813	66.13044	56.16829	49.73873	49.18226
0.6	40	68.37116888	52.99989	43.86424	37.98616	33.71271	30.49497	28.20271	27.99691
0.8	20	39.17334894	32.83423	28.6552	25.66711	23.41056	21.63775	20.33266	20.21485
1	0	28.64684007	24.76763	22.09596	20.12591	18.60122	17.37965	16.46712	16.38404

Table 14.9 Summary on complete mathematical models of laminar forced film condensation of vapour–gas mixture

Governing partial differential equations for liquid film	
Mass equation	$\frac{\partial}{\partial x}(\rho_l w_{xl}) + \frac{\partial}{\partial y}(\rho_l w_{yl}) = 0$
Momentum equation	$\rho_l \left(w_{xl}\frac{\partial w_{xl}}{\partial x} + w_{yl}\frac{\partial w_{xl}}{\partial y} \right) = \frac{\partial}{\partial y}\left(\mu_l \frac{\partial w_{xl}}{\partial y} \right)$
Energy equation	$\rho_l \left[w_x \frac{\partial (c_{p_l} t)}{\partial x} + w_y \frac{\partial (c_{p_l} t)}{\partial y} \right] = \frac{\partial}{\partial y}\left(\lambda_l \frac{\partial t}{\partial y} \right)$
Governing partial differential equations for vapour-gas mixture film	
Mass equation	$\frac{\partial}{\partial x}(\rho_{\mathrm{m}} w_{x\mathrm{m}}) + \frac{\partial}{\partial y}(\rho_{\mathrm{m}} w_{y\mathrm{m}}) = 0$
Momentum equation	$\rho_{\mathrm{m}} \left(w_{x\mathrm{m}}\frac{\partial w_{x\mathrm{m}}}{\partial x} + w_{y\mathrm{m}}\frac{\partial w_{x\mathrm{m}}}{\partial y} \right) = \frac{\partial}{\partial y}\left(\mu_{\mathrm{m}} \frac{\partial w_{x\mathrm{m}}}{\partial y} \right)$
Energy equation	$\rho_{\mathrm{m}} \left[w_{x\mathrm{m}} \frac{\partial (c_{p_{\mathrm{m}}} t_{\mathrm{m}})}{\partial x} + w_{y\mathrm{m}} \frac{\partial (c_{p_{\mathrm{m}}} t_{\mathrm{m}})}{\partial y} \right] = \frac{\partial}{\partial y}\left(\lambda_{\mathrm{m}} \frac{\partial t_{\mathrm{m}}}{\partial y} \right) +$
	$\frac{\partial}{\partial y}\left[\rho_{\mathrm{m}} D_{\mathrm{v}} \left(c_{p_{\mathrm{v}}} - c_{p_{\mathrm{a}}} \right) \frac{\partial C_{\mathrm{mv}}}{\partial y} t_{\mathrm{m}} \right]$
Species equation	$\frac{\partial (w_{x\mathrm{m}} \rho_{\mathrm{m}} C_{\mathrm{mv}})}{\partial x} + \frac{\partial (w_{y\mathrm{m}} \rho_{\mathrm{m}} C_{\mathrm{mv}})}{\partial y} = \frac{\partial}{\partial y}\left(D_{\mathrm{v}} \rho_{\mathrm{m}} \frac{\partial C_{\mathrm{mv}}}{\partial y} \right)$
Boundary conditions of governing partial differential equations	
$y = 0$	$w_{xl} = 0, \quad w_{yl} = 0, \quad t_l = t_{\mathrm{w}}$
$y = \delta_l$	$w_{xl,\mathrm{s}} = w_{x\mathrm{m,s}}$
	$\rho_{l,\mathrm{s}} \left(w_{xl}\frac{\partial \delta_l}{\partial x} - w_{yl} \right)_{l,\mathrm{s}} = \rho_{\mathrm{m,s}} C_{\mathrm{mv,s}} \left(w_{x\mathrm{m}} \frac{\partial \delta_{\mathrm{m}}}{\partial x} - w_{y\mathrm{m}} \right)_{\mathrm{s}} = g_x$
	$\mu_{l,\mathrm{s}} \left(\frac{\partial w_{xl}}{\partial y} \right)_{\mathrm{s}} = \mu_{\mathrm{m,s}} \left(\frac{\partial w_{x\mathrm{m}}}{\partial y} \right)_{\mathrm{s}}$
	$\lambda_{l,\mathrm{s}} \left(\frac{\partial T_l}{\partial y} \right)_{\mathrm{s}_l} = \lambda_{\mathrm{m,s}} \left(\frac{\partial T_{\mathrm{m}}}{\partial y} \right)_{\mathrm{s}} + h_{\mathrm{fg}} \rho_{\mathrm{m,s}} C_{\mathrm{mv,s}} \left(w_{x\mathrm{m}} \frac{\partial \delta_{\mathrm{m}}}{\partial x} - w_{y\mathrm{m}} \right)_{\mathrm{s}}$
	$t = t_{\mathrm{s,int}}$
	$C_{\mathrm{mv}} = C_{\mathrm{mv,s}}$
	$\rho_{\mathrm{m,s}} C_{\mathrm{m,vs}} \left(w_{x\mathrm{m}} \frac{\partial \delta_{\mathrm{m}}}{\partial x} - w_{y\mathrm{m}} \right)_{\mathrm{s}} = D_{\mathrm{v}} \rho_{\mathrm{m,s}} \left(\frac{\partial C_{\mathrm{mv}}}{\partial y} \right)_{\mathrm{s}}$
$y \to \infty$	$w_{x\mathrm{m}} = w_{x\mathrm{m},\infty}, \quad t_{\mathrm{m}} = t_{\infty}, C_{\mathrm{mv}} = C_{\mathrm{mv},\infty}$
Similarity variables for liquid film	
Dimensionless coordinate variable η_l	$\eta_l = \frac{y}{x}\left(\frac{1}{2} \mathrm{Re}_{xl,\mathrm{s}} \right)^{1/2}$
Local Reynolds number	$\mathrm{Re}_{xl,\mathrm{s}} = \frac{w_{x\mathrm{m},\infty} x}{\nu_{l,\mathrm{s}}}$
Dimensionless temperature	$\theta_l(\eta_l) = \frac{t_l - t_{\mathrm{s,int}}}{t_{\mathrm{w}} - t_{\mathrm{s,int}}}$
Dimensionless velocity components	$w_{xl} = w_{x\mathrm{m},\infty} W_{xl}(\eta_l)$
$W_{xl}(\eta_l)$ and $W_{yl}(\eta_l)$	$w_{yl} = w_{x\mathrm{m},\infty} \left(\frac{1}{2} \mathrm{Re}_{xl,\mathrm{s}} \right)^{-1/2} W_{yl}(\eta_l)$

Table 14.9 (continued)

Similarity variables for vapour–gas mixture film	
Dimensionless coordinate variable	$\eta_{\mathrm{m}} = \frac{y}{x}\left(\frac{1}{2}\mathrm{Re}_{x\mathrm{m},\infty}\right)^{1/2}$
Local Reynolds number	$\mathrm{Re}_{x\mathrm{m},\infty} = \frac{w_{x\mathrm{m},\infty} x}{\nu_{\mathrm{m},\infty}}$
Dimensionless temperature	$\theta_{\mathrm{m}}(\eta_{\mathrm{m}}) = \frac{t_{\mathrm{m}} - t_{\infty}}{t_{\mathrm{s,int}} - t_{\infty}}$
Dimensionless velocity components	$w_{x\mathrm{m}} = w_{x\mathrm{m},\infty} W_{x\mathrm{m}}(\eta_{\mathrm{m}})$
$W_{xl}(\eta_l)$ and $W_{yl}(\eta_l)$	$w_{y\mathrm{m}} = w_{x\mathrm{m},\infty}\left(\frac{1}{2}\mathrm{Re}_{x\mathrm{m},\infty}\right)^{-1/2} W_{y\mathrm{m}}(\eta_{\mathrm{m}})$
Vapour-relative mass fraction	$\Gamma_{\mathrm{mv}}(\eta_{\mathrm{m}}) = \frac{C_{\mathrm{mv}} - C_{\mathrm{mv},\infty}}{C_{\mathrm{mv,s}} - C_{\mathrm{mv},\infty}}$
Governing ordinary differential equations for liquid film	
Mass equation	$-\eta_l \frac{\mathrm{d}W_{xl}(\eta_l)}{\mathrm{d}\eta_l} + 2\frac{\mathrm{d}W_{yl}(\eta_l)}{\mathrm{d}\eta_l} + \frac{1}{\rho_l}\frac{\mathrm{d}\rho_l}{\mathrm{d}\eta_l}\left[-\eta_l W_{xl}(\eta_l) + 2W_{yl}(\eta_l)\right] = 0$
Momentum equation	$\frac{\nu_{l,\mathrm{s}}}{\nu_l}\left[-\eta_l W_{xl}(\eta_l) + 2\left(W_{yl}(\eta_l)\right)\right]\frac{\mathrm{d}W_{xl}(\eta_l)}{\mathrm{d}\eta_l} = \frac{\mathrm{d}^2 W_{xl}(\eta_l)}{\mathrm{d}\eta_l^2} + \frac{1}{\mu_l}\frac{\mathrm{d}\mu_l}{\mathrm{d}\eta_l}\frac{\mathrm{d}W_{\mathrm{x}l}(\eta_l)}{\mathrm{d}\eta_l}$
Energy equation	$\mathrm{Pr}_l \frac{\nu_{l,\mathrm{s}}}{\nu_l}\left\{\left[-\eta_l W_{xl}(\eta_l) + 2W_{yl}(\eta_l)\right]\frac{\mathrm{d}\theta_l(\eta_l)}{\mathrm{d}\eta_l} + \left[-\eta_l W_{xl}(\eta_l) + 2W_{yl}(\eta_l)\right]\left(\theta_l(\eta_l) + \frac{t_{\mathrm{s,int}}}{t_{\mathrm{w}} - t_{\mathrm{s,int}}}\right)\frac{1}{c_{pl}}\frac{\mathrm{d}c_{pl}}{\mathrm{d}\eta_l}\right\} = \frac{\mathrm{d}^2\theta_l(\eta_l)}{\mathrm{d}\eta_l^2} + \frac{1}{\lambda_l}\frac{\partial\lambda_l}{\partial\eta_l}\frac{\mathrm{d}\theta_l(\eta_l)}{\mathrm{d}\eta_l}$
Special	
Governing ordinary differential equations for vapour–gas mixture film	
Mass equation	$-\eta_{\mathrm{m}}\frac{\mathrm{d}W_{x\mathrm{m}}(\eta_{\mathrm{m}})}{\mathrm{d}\eta_{\mathrm{m}}} + 2\frac{\mathrm{d}W_{y\mathrm{m}}(\eta_{\mathrm{m}})}{\mathrm{d}\eta_{\mathrm{m}}} + \frac{1}{\rho_{\mathrm{m}}}\frac{\mathrm{d}\rho_{\mathrm{m}}}{\mathrm{d}\eta_{\mathrm{m}}}\left[-\eta_{\mathrm{m}}\cdot W_{x\mathrm{m}}(\eta_{\mathrm{m}}) + 2W_{y\mathrm{m}}(\eta_{\mathrm{m}})\right] = 0$
Momentum equation	$\frac{\nu_{\mathrm{m},\infty}}{\nu_{\mathrm{m}}}\left[-\eta_{\mathrm{m}}\cdot W_{x\mathrm{m}}(\eta_{\mathrm{m}}) + 2W_{y\mathrm{m}}(\eta_{\mathrm{m}})\right]\frac{\mathrm{d}W_{x\mathrm{m}}(\eta_{\mathrm{m}})}{\mathrm{d}\eta_{\mathrm{m}}} = \frac{\mathrm{d}^2 W_{x\mathrm{m}}(\eta_{\mathrm{m}})}{\mathrm{d}\eta_{\mathrm{m}}^2} + \frac{1}{\mu_{\mathrm{m}}}\frac{\mathrm{d}\mu_{\mathrm{m}}}{\mathrm{d}\eta_l}\frac{\mathrm{d}W_{x\mathrm{m}}(\eta_{\mathrm{m}})}{\mathrm{d}\eta_{\mathrm{m}}}$
Energy equation	$\frac{\nu_{\mathrm{m},\infty}}{\nu_{\mathrm{m}}}\left\{\left[-\eta_{\mathrm{m}} W_{x\mathrm{m}}(\eta_{\mathrm{m}}) + 2W_{y\mathrm{m}}(\eta_{\mathrm{m}})\right]\frac{\mathrm{d}\theta_{\mathrm{m}}(\eta_{\mathrm{m}})}{\mathrm{d}\eta_{\mathrm{m}}} + \left[-\eta_{\mathrm{m}} W_{x\mathrm{m}}(\eta_{\mathrm{m}}) + 2W_{y\mathrm{m}}(\eta_{\mathrm{m}})\right]\frac{1}{c_{p\mathrm{m}}}\frac{\mathrm{d}c_{p\mathrm{m}}}{\mathrm{d}\eta_{\mathrm{m}}}\left(\theta_{\mathrm{m}} + \frac{t_{\infty}}{t_{\mathrm{s,int}} - t_{\infty}}\right)\right\} = \frac{1}{\mathrm{Pr}_{\mathrm{m}}}\left[\frac{\mathrm{d}^2\theta_{\mathrm{m}}(\eta_{\mathrm{m}})}{\mathrm{d}^2\eta_{\mathrm{m}}} + \frac{1}{\lambda_{\mathrm{m}}}\frac{\mathrm{d}\lambda_{\mathrm{m}}}{\mathrm{d}\eta_{\mathrm{m}}}\cdot\frac{\mathrm{d}\theta_{\mathrm{m}}(\eta_{\mathrm{m}})}{\mathrm{d}\eta_{\mathrm{m}}}\right] + \frac{1}{Sc_{\mathrm{m},\infty}}\frac{\nu_{\mathrm{m},\infty}}{\nu_{\mathrm{m}}}\frac{c_{p\mathrm{v}} - c_{p\mathrm{g}}}{c_{p\mathrm{m}}}\left(C_{\mathrm{mv,s}} - C_{\mathrm{mv},\infty}\right)\left[\frac{\mathrm{d}\Gamma_{\mathrm{m,v}}(\eta_{\mathrm{m}})}{\mathrm{d}\eta_{\mathrm{v}}}\frac{\mathrm{d}\theta_{\mathrm{m}}(\eta_{\mathrm{m}})}{\mathrm{d}\eta_{\mathrm{m}}} + \left(\theta_{\mathrm{m}} + \frac{t_{\infty}}{t_{\mathrm{s,int}} - t_{\infty}}\right)\frac{\mathrm{d}^2\Gamma_{\mathrm{mv}}(\eta_{\mathrm{m}})}{\mathrm{d}\eta_{\mathrm{v}}^2} + \left(\theta_{\mathrm{m}} + \frac{t_{\infty}}{t_{\mathrm{s,int}} - t_{\infty}}\right)\frac{1}{\rho_{\mathrm{m}}}\frac{\mathrm{d}\rho_{\mathrm{m}}}{\mathrm{d}\eta_l}\frac{\mathrm{d}\Gamma_{\mathrm{mv}}(\eta_{\mathrm{m}})}{\mathrm{d}\eta_{\mathrm{v}}}\right]$
Species equation	$\left[-\eta_{\mathrm{m}} W_{x\mathrm{m}}(\eta_{\mathrm{m}}) + 2W_{y\mathrm{m}}(\eta_{\mathrm{m}})\right]\frac{\mathrm{d}\Gamma_{\mathrm{mv}}(\eta_{\mathrm{m}})}{\mathrm{d}\eta_{\mathrm{m}}} = \frac{1}{Sc_{\mathrm{m},\infty}}\left[\frac{\mathrm{d}^2\Gamma_{\mathrm{mv}}(\eta_{\mathrm{m}})}{d\eta_{\mathrm{v}}^2} + \frac{1}{\rho_{\mathrm{m}}}\frac{\mathrm{d}\rho_{\mathrm{m}}}{\mathrm{d}\eta_{\mathrm{m}}}\frac{\mathrm{d}\Gamma_{\mathrm{mv}}(\eta_{\mathrm{m}})}{\mathrm{d}\eta_{\mathrm{m}}}\right]$

Table 14.9 (continued)

Boundary conditions of governing ordinary differential equations

$\eta_l = 0:$ $\quad W_{xl} = 0, \quad W_{yl} = 0, \quad \theta_l = 1$

$\eta_l = \eta_{l\delta}$ $\quad W_{x\mathrm{m},\infty}(\eta_\mathrm{m}) = W_{xl,\mathrm{s}}(\eta_l)$

$$W_{y\mathrm{m,s}}(\eta_\mathrm{m}) = \frac{1}{2C_\mathrm{mv,s}}\frac{\rho_{l,\mathrm{s}}}{\rho_\mathrm{m,s}}\left(\frac{\nu_{l,\mathrm{s}}}{\nu_{\mathrm{m},\infty}}\right)^{1/2}\left[-\eta_{l\delta}\cdot W_{xl,\mathrm{s}}(\eta_l) + 2W_{yl,\mathrm{s}}(\eta_l)\right]$$

$$\left(\frac{\mathrm{d}W_{x\mathrm{m}}(\eta_\mathrm{m})}{\mathrm{d}\eta_\mathrm{m}}\right)_\mathrm{s} = \frac{\mu_{l,\mathrm{s}}}{\mu_\mathrm{m,s}}\left(\frac{\nu_{\mathrm{m},\infty}}{\nu_{l,\mathrm{s}}}\right)^{1/2}\left(\frac{\mathrm{d}W_x(\eta_l)}{\mathrm{d}\eta l}l\right)_\mathrm{s}$$

$$\left(\frac{\mathrm{d}\theta_\mathrm{m}(\eta_\mathrm{m})}{\mathrm{d}\eta_\mathrm{m}}\right)_\mathrm{s} = \frac{\lambda_{l,\mathrm{s}}\left(\frac{\nu_{\mathrm{m},\infty}}{\nu_{l,\mathrm{s}}}\right)^{1/2}(t_\mathrm{w} - t_\mathrm{s,int})\left(\frac{\mathrm{d}\theta_l(\eta_l)}{\mathrm{d}\eta_l}\right)_\mathrm{s} + 2h_\mathrm{fg}\rho_\mathrm{m,s}\nu_{\mathrm{m},\infty}C_\mathrm{mv,s}W_{y\mathrm{m,s}}(\eta_\mathrm{m})}{\lambda_\mathrm{m,s}\left(t_\mathrm{s,int} - t_\infty\right)}$$

$\theta_{l,\mathrm{s}}(\eta_{l\delta}) = 0, \quad \theta_\mathrm{m,s}(\eta_\mathrm{m0}) = 1$

$\Gamma_\mathrm{mv,s}(\eta_\mathrm{m0}) = 1, \quad C_\mathrm{mv} = C_\mathrm{mv,s}$

$$\left(\frac{\mathrm{d}\Gamma_\mathrm{mv}(\eta_\mathrm{m})}{\mathrm{d}\eta_\mathrm{m}}\right)_\mathrm{s} = -2Sc_{\mathrm{m},\infty}\frac{C_\mathrm{mv,s}}{C_\mathrm{mv,s} - C_{\mathrm{mv},\infty}}W_{y\mathrm{m,s}}(\eta_\mathrm{m0})$$

$\eta_\mathrm{m} \to \infty$ $\quad W_{x\mathrm{m},\infty}(\eta_{\mathrm{m}\infty}) = 1, \quad \theta_{\mathrm{m},\infty} = 0, \quad \Gamma_{\mathrm{mv},\infty} = 0$

Physical property factor equations of condensate liquid film

For condensate water film medium

$$\mathrm{Pr}_l = \frac{\mu_l \cdot c_{pl}}{\lambda_l} = \frac{\left[\exp\left(-1.6 - \frac{1150}{T} + \left(\frac{690}{T}\right)^2\right) \times 10^{-3}\right] \times 4200}{-8.01 \times 10^{-6}t^2 + 1.94 \times 10^{-3}t + 0.563}$$

$$\frac{1}{c_{pl}}\frac{\mathrm{d}c_{pl}}{\mathrm{d}\eta_l} = 0$$

$$\frac{1}{\rho_l}\frac{\mathrm{d}\rho_l}{\mathrm{d}\eta_l} = \frac{\left[-2 \times 4.48 \times 10^{-3}t\,(t_\mathrm{w} - t_\mathrm{s})\frac{\mathrm{d}\theta_l(\eta_l)}{\mathrm{d}\eta_l}\right]}{\left(-4.48 \times 10^{-3}t^2 + 999.9\right)}$$

$$\frac{1}{\lambda_l}\frac{\mathrm{d}\lambda_l}{\mathrm{d}\eta_l} = \frac{\left[\left(-2 \times 8.01 \times 10^{-6} + 1.94 \times 10^{-3}\right)(t_\mathrm{w} - t_\mathrm{s})\frac{\mathrm{d}\theta_l(\eta_l)}{\mathrm{d}\eta_l}\right]}{-8.01 \times 10^{-6}t^2 + 1.94 \times 10^{-3}t + 0.563}$$

$$\frac{1}{\mu_l}\frac{\mathrm{d}\mu_l}{\mathrm{d}\eta_l} = \left(\frac{1150}{T^2} - 2 \times \frac{690^2}{T^3}\right)(t_\mathrm{w} - t_\mathrm{s})\frac{\mathrm{d}\theta_l(\eta_l)}{\mathrm{d}\eta_l}$$

Physical property factor equations of vapour–gas mixture film

Concentration-dependent physical property factors

$$\frac{1}{\rho_\mathrm{m}}\frac{\mathrm{d}\rho_\mathrm{m}}{\mathrm{d}\eta_\mathrm{m}} = \frac{1}{\rho_\mathrm{g}}\frac{\mathrm{d}\rho_\mathrm{g}}{\mathrm{d}\eta_\mathrm{m}} + \frac{1}{\rho_\mathrm{v}}\frac{\mathrm{d}\rho_\mathrm{v}}{\mathrm{d}\eta_\mathrm{m}} - \frac{C_\mathrm{mv}\left(\rho_\mathrm{v} - \rho_\mathrm{g}\right)}{(1 - C_\mathrm{mv})\rho_\mathrm{v} + C_\mathrm{mv}\rho_\mathrm{g}}$$

$$\left[\frac{(1 - C_\mathrm{mv})}{C_\mathrm{mv}}\frac{\rho_\mathrm{v}}{\left(\rho_\mathrm{v} - \rho_\mathrm{g}\right)}\frac{1}{\rho_\mathrm{v}}\frac{\mathrm{d}\rho_\mathrm{v}}{\mathrm{d}\eta_\mathrm{m}} + \frac{\rho_\mathrm{g}}{\left(\rho_\mathrm{v} - \rho_\mathrm{g}\right)}\frac{1}{\rho_\mathrm{g}}\frac{\mathrm{d}\rho_\mathrm{g}}{\mathrm{d}\eta_\mathrm{m}} - \frac{C_\mathrm{mv,s} - C_{\mathrm{mv},\infty}}{C_\mathrm{mv}}\frac{\mathrm{d}\Gamma_\mathrm{mv}(\eta_\mathrm{m})}{\mathrm{d}\eta_\mathrm{m}}\right]$$

$$\frac{1}{\mu_\mathrm{m}}\frac{\mathrm{d}\mu_\mathrm{m}}{\mathrm{d}\eta_\mathrm{m}} = C_\mathrm{mv}\frac{\mu_\mathrm{v}}{\mu_\mathrm{m}}\frac{1}{\mu_\mathrm{v}}\frac{\mathrm{d}\mu_\mathrm{v}}{\mathrm{d}\eta_\mathrm{m}} + (1 - C_\mathrm{mv})\frac{\mu_\mathrm{g}}{\mu_\mathrm{m}}\frac{1}{\mu_\mathrm{g}}\frac{\mathrm{d}\mu_\mathrm{g}}{\mathrm{d}\eta_\mathrm{m}} + \frac{\mu_\mathrm{v} - \mu_\mathrm{g}}{\mu_\mathrm{m}}\left(C_\mathrm{mv,s} - C_{\mathrm{mv},\infty}\right)\frac{\mathrm{d}\Gamma_\mathrm{mv}}{\mathrm{d}\eta_\mathrm{m}}$$

$$\frac{1}{\lambda_\mathrm{m}}\frac{\mathrm{d}\lambda_\mathrm{m}}{\mathrm{d}\eta_\mathrm{m}} = C_\mathrm{mv}\frac{\lambda_\mathrm{v}}{\lambda_\mathrm{m}}\frac{1}{\lambda_\mathrm{v}}\frac{\mathrm{d}\lambda_\mathrm{v}}{\mathrm{d}\eta_\mathrm{m}} + \left(1 - C_\mathrm{m,v}\right)\frac{\lambda_\mathrm{g}}{\lambda_\mathrm{m}}\frac{1}{\lambda_\mathrm{g}}\frac{\mathrm{d}\lambda_\mathrm{g}}{\mathrm{d}\eta_\mathrm{m}} + \frac{\lambda_\mathrm{v} - \lambda_\mathrm{g}}{\lambda_\mathrm{m}}\left(C_\mathrm{mv,s} - C_{\mathrm{mv},\infty}\right)\frac{\mathrm{d}\Gamma_\mathrm{mv}}{\mathrm{d}\eta_\mathrm{m}}$$

Table 14.9 (continued)

Temperature dependent physical property factors	For density factors $\frac{1}{\rho_\mathrm{v}}\frac{\mathrm{d}\rho_\mathrm{v}}{\mathrm{d}\eta_\mathrm{m}} = -\frac{\left(T_\mathrm{s,int}/T_\infty - 1\right)\mathrm{d}\theta_\mathrm{m}/\mathrm{d}\eta_\mathrm{m}}{\left(T_\mathrm{s,int}/T_\infty - 1\right)\theta_\mathrm{m} + 1}$, $\frac{1}{\rho_\mathrm{g}}\frac{\mathrm{d}\rho_\mathrm{g}}{\mathrm{d}\eta_\mathrm{m}} = -\frac{\left(T_\mathrm{s,int}/T_\infty - 1\right)\mathrm{d}\theta_\mathrm{m}/\mathrm{d}\eta_\mathrm{m}}{\left(T_\mathrm{s,int}/T_\infty - 1\right)\theta_\mathrm{m+1}}$ For absolute viscosity factors $\frac{1}{\mu_\mathrm{v}}\frac{\mathrm{d}\mu_\mathrm{v}}{\mathrm{d}\eta_\mathrm{m}} = \frac{n_{\mu,\mathrm{v}}\left(T_\mathrm{s,int}/T_\infty - 1\right)\mathrm{d}\theta_\mathrm{m}/\mathrm{d}\eta_\mathrm{m}}{\left(T_\mathrm{s,int}/T_\infty - 1\right)\theta_\mathrm{m} + 1}$, $\frac{1}{\mu_\mathrm{g}}\frac{\mathrm{d}\mu_\mathrm{g}}{\mathrm{d}\eta_\mathrm{m}} = \frac{n_{\mu,\mathrm{g}}\left(T_\mathrm{s,int}/T_\infty - 1\right)\mathrm{d}\theta_\mathrm{m}/\mathrm{d}\eta_\mathrm{m}}{\left(T_\mathrm{s,int}/T_\infty - 1\right)\theta_\mathrm{m} + 1}$ For thermal conductivity factors $\frac{1}{\lambda_\mathrm{v}}\frac{\mathrm{d}\lambda_\mathrm{v}}{\mathrm{d}\eta_\mathrm{m}} = \frac{n_{\lambda,\mathrm{v}}\left(T_\mathrm{s,int}/T_\infty - 1\right)\mathrm{d}\theta_\mathrm{m}/\mathrm{d}\eta_\mathrm{m}}{\left(T_\mathrm{s,int}/T_\infty - 1\right)\theta_\mathrm{m} + 1}$, $\frac{1}{\lambda_\mathrm{g}}\frac{\mathrm{d}\lambda_\mathrm{g}}{\mathrm{d}\eta_\mathrm{m}} = \frac{n_{\lambda,\mathrm{g}}\left(T_\mathrm{s,int}/T_\infty - 1\right)\mathrm{d}\theta_\mathrm{m}/\mathrm{d}\eta_\mathrm{m}}{\left(T_\mathrm{s,int}/T_\infty - 1\right)\theta_\mathrm{m} + 1}$
Condensate heat transfer	
Wall dimensionless temperature grade	For laminar forced film condensation of water vapour–air mixture on a horizontal plate at $t_\infty = 100°\mathrm{C}$ $-\left(\frac{\mathrm{d}\theta_l}{\mathrm{d}\eta_l}\right)_{\eta_l=0} = \mathrm{A}C_{\mathrm{mv},\infty}{}^{B}$ $(0.4 \le C_{\mathrm{mv},\infty} \le 0.999$ where $A = 0.5868\left[\frac{(\Delta t_\mathrm{w})_\mathrm{s}}{t_\mathrm{s}}\right]^{-0.3272}$ $\left(0.05 \le \frac{(\Delta t_\mathrm{w})_\mathrm{s}}{t_\mathrm{s}} \le 0.2\right)$ $A = -0.334\frac{(\Delta t_\mathrm{w})_\mathrm{s}}{t_\mathrm{s}} + 1.0766$ $\left(0.2 \le \frac{(\Delta t_\mathrm{w})_\mathrm{s}}{t_\mathrm{s}} \le 1\right)$ $B = -0.2486\left[\frac{(\Delta t_\mathrm{w})_\mathrm{s}}{t_\mathrm{s}}\right]^{-0.8082}$ $\left(0.05 \le \frac{(\Delta t_\mathrm{w})_\mathrm{s}}{t_\mathrm{s}} \le 0.2\right)$ $B = 2.8813\left[\frac{(\Delta t_\mathrm{w})_\mathrm{s}}{t_\mathrm{s}}\right]^{3} - 6.3245\left[\frac{(\Delta t_\mathrm{w})_\mathrm{s}}{t_\mathrm{s}}\right]^{2} + 4.5528\frac{(\Delta t_\mathrm{w})_\mathrm{s}}{t_\mathrm{s}} - 1.6063$ $\left(0.2 \le \frac{(\Delta t_\mathrm{w})_\mathrm{s}}{t_\mathrm{s}} \le 1\right)$
Local values of heat transfer, $\mathrm{Nu}_{xl,\mathrm{w}} = \frac{\alpha_{x,\mathrm{w}}x}{\lambda_{l,\mathrm{w}}}$	$q_{x,\mathrm{w}} = \lambda_{l,\mathrm{w}}\left(t_\mathrm{w} - t_\mathrm{s,int}\right)\left(\frac{1}{2}\mathrm{Re}_{xl,\mathrm{s}}\right)^{1/2} x^{-1}\left(-\frac{\mathrm{d}\theta_l(\eta_l)}{\mathrm{d}\eta_l}\right)_{\eta_l=0}$ $\alpha_{x,\mathrm{w}} = \lambda_{l,\mathrm{w}}\left(\frac{1}{2}\mathrm{Re}_{xl,\mathrm{s}}\right)^{1/2} x^{-1}\left(-\frac{\mathrm{d}\theta_l(\eta_l)}{\mathrm{d}\eta_l}\right)_{\eta_l=0}$ $\mathrm{Nu}_{xl,\mathrm{w}} = \left(\frac{1}{2}\mathrm{Re}_{xl,\mathrm{s}}\right)^{1/2}\left(-\frac{\mathrm{d}\theta_l(\eta_l)}{\mathrm{d}\eta_l}\right)_{\eta_l=0}$

Table 14.9 (continued)

Average values of heat transfer, $\overline{\mathrm{Nu}}_{xl,\mathrm{w}} = \dfrac{\overline{\alpha}_{x,\mathrm{w}} x}{\lambda_{l,\mathrm{w}}}$	$\overline{Q}_{x,\mathrm{w}} = \sqrt{2}x^{-1}\lambda_{l,\mathrm{w}}\left(t_\mathrm{w} - t_{\mathrm{s,int}}\right)\left(\mathrm{Re}_{x\mathrm{l,s}}\right)^{1/2}\left(-\dfrac{\mathrm{d}\theta_l(\eta_l)}{\mathrm{d}\eta_l}\right)_{\eta_l=0}$ $\overline{\alpha}_{x,\mathrm{w}} = \sqrt{2}\lambda_{l,\mathrm{w}}\mathrm{Re}_{xl,\mathrm{s}}^{1/2}x^{-1}\left(-\dfrac{\mathrm{d}\theta_l(\eta_l)}{\mathrm{d}\eta_l}\right)_{\eta_l=0}$ $\overline{\mathrm{Nu}}_{xl,\mathrm{w}} = \sqrt{2}\left(\mathrm{Re}_{x\mathrm{l,s}}\right)^{1/2}\left(-\dfrac{\mathrm{d}\theta_l(\eta_l)}{\mathrm{d}\eta_l}\right)_{\eta_l=0}$
Total value of heat transfer	$Q_{x,\mathrm{w}} = \sqrt{2}b\lambda_{l,\mathrm{w}}\left(t_\mathrm{w} - t_{\mathrm{s,int}}\right)\left(\mathrm{Re}_{xl,\mathrm{s}}\right)^{1/2}\left(-\dfrac{\mathrm{d}\theta_l(\eta_l)}{\mathrm{d}\eta_l}\right)_{\eta_l=0}$
Prediction equation of heat transfer	For laminar forced film condensation of water vapour–air mixture on a horizontal plate at $t_\infty = 100°\mathrm{C}$ $Q_x = \sqrt{2}b\lambda_{l,\mathrm{w}}\left(t_\mathrm{w} - t_{\mathrm{s,int}}\right)\left(\mathrm{Re}_{xl,\mathrm{s}}\right)^{1/2} A\,\exp\left[B\dfrac{(\Delta t_\mathrm{w})_\mathrm{s}}{t_\mathrm{s}}\right]$ $\left(0.4 \le \dfrac{(\Delta t_\mathrm{w})_\mathrm{s}}{t_\mathrm{s}} \le 1\right)$ $\overline{\mathrm{Nu}}_{xl,\mathrm{w}} = \sqrt{2}\left(\mathrm{Re}_{xl,\mathrm{s}}\right)^{1/2} A\,\exp\left[B\dfrac{(\Delta t_\mathrm{w})_\mathrm{s}}{t_\mathrm{s}}\right]$ $\left(0.4 \le \dfrac{(\Delta t_\mathrm{w})_\mathrm{s}}{t_\mathrm{s}} \le 1\right)$ where $A = 1.0977{C_{\mathrm{mv},\infty}}^{-0.6304}$ $(0.4 \le C_{\mathrm{mv},\infty} \le 0.999)$ $B = 0.1333\,ln\left(C_{\mathrm{mv},\infty}\right) - 0.3936$ $(0.4 \le C_{\mathrm{mv},\infty} \le 0.999)$
Condensate mass flow rate	
Mass flow rate, G_x	$G_x = \sqrt{2}b\mu_{l,\mathrm{s}}{\mathrm{Re}_{xl,\mathrm{s}}}^{1/2}\Phi_\mathrm{s}$ where mass flow rate parameter $\Phi_\mathrm{s} = \left(\eta_{l\delta}W_{xl,\mathrm{s}} - 2W_{yl,\mathrm{s}}\right)$
Mass flow rate parameter	For laminar forced convection of water vapour–air mixture on a horizontal plate at $t_\infty = 100°\mathrm{C}$ $\Phi_\mathrm{s} = A\,ln\left(\dfrac{(\Delta t_\mathrm{w})_\mathrm{s}}{t_s}\right) + B$ $\left(0.4 \le \dfrac{(\Delta t_\mathrm{w})_\mathrm{s}}{t_\mathrm{s}} \le 1\right)$ where $A = 0.0272\left(C_{\mathrm{mv},\infty}\right)^{-0.3865}$ $(0.4 \le C_{\mathrm{mv},\infty} \le 0.999)$ $B = -0.0112\left(C_{\mathrm{mv},\infty}\right)^2 + 0.0215\left(C_{\mathrm{mv},\infty}\right) + 0.0334$ $(0.4 \le C_{\mathrm{mv},\infty} \le 0.999)$

Table 14.9 (continued)

Evaluation of interfacial vapour saturation equations $t_{s,int}$	
With prediction equation	For laminar forced convection of water vapour–air mixture on a horizontal plate at $t_\infty = 100°C$ $t_{s,int} = AC_{mv,\infty}{}^{B}$ $(0.4 \le C_{mv,\infty} \le 0.999)$ $A = 177.5\left[\frac{(\Delta t_w)_s}{t_s}\right]^2 - 112.12\frac{(\Delta t_w)_s}{t_s} + 100.59$ $\left(0.05 \le \frac{(\Delta t_w)_s}{t_s} \le 0.4\right)$ $B = -4.9374\left[\frac{(\Delta t_w)_s}{t_s}\right]^3 + 3.6734\left[\frac{(\Delta t_w)_s}{t_s}\right]^2 - 0.0445\frac{(\Delta t_w)_s}{t_s} + 0.0243$ $\left(0.05 \le \frac{(\Delta t_w)_s}{t_s} \le 0.4\right)$ $t_{s,int} = 21.79\,ln\left(C_{mv,\infty}\right) + 84.076$ $(0.4 \le C_{mv,\infty} \le 0.999)$ and $(0.4 \le C_{mv,\infty} \le 1)$
With mass–energy transformation equation	For laminar forced convection of vapour–gas mixture $t_{s,int}$ is a solution of the condensate mass–energy transformation equation: $\frac{\left(-\frac{d\theta_l(\eta_l)}{d\eta_l}\right)_{\eta_l=0}}{\Phi_s}\lambda_{l,w}\left(t_{s,int} - t_w\right) - \mu_{l,s}h_{fg} = 0$ For laminar forced convection of water vapour–air mixture $t_{s,int}$ is a solution of the condensate mass–energy transformation equation: $\frac{\left(-\frac{d\theta_l(\eta_l)}{d\eta_l}\right)_{\eta_l=0}}{\Phi_s}\left(-8.01\times10^{-6}t_w^2 + 1.94\times10^{-3}t_w + 0.563\right)\left(t_{s,int} - t_w\right)$ $-\exp\left[-1.6 - \frac{1150}{t_{s,int}+273} + \left(\frac{690}{t_{s,int}+273}\right)^2\right]$ $(-2.4355t_{s,int} + 2503.2) = 0$

14.12 Remarks

Through the heat and mass transfer analyses, the equations of condensate heat and mass transfer of laminar forced film condensation of vapour–gas mixture are provided, where only dimensionless wall temperature gradient, interfacial vapour saturation temperature, and defined condensate mass flow rate parameter are non-given conditions respectively.

The laminar forced film condensation of water vapour in presence of air on a horizontal flat plate is taken as example for the numerical calculation of condensate heat and mass transfer. A system of numerical results of the dimensionless temperature gradient and mass flow rate parameter are obtained, and the following phenomena are found.

With decreasing bulk vapour mass fraction, the wall dimensionless temperature gradient will increase at accelerative pace. For special bulk vapour mass fraction, increasing the reference wall reference subcooled grade will cause decrease of the wall dimensionless temperature gradient. Decreasing the bulk vapour mass fraction will cause decrease of the condensate mass flow rate parameter, and such decrease of the mass flow rate parameter will be even obvious with increasing wall subcooled grade. For a special bulk vapour mass fraction, increasing the reference wall subcooled grade will cause increase of the condensate mass flow rate parameter. These phenomena demonstrate decisive effect of the non-condensable gas on condensate heat and mass transfer of the laminar forced film condensation of vapour–gas mixture.

The system of the rigorous key solutions of the wall dimensionless temperature gradient and the condensate mass flow rate parameter is formulated the simple and reliable equations for the laminar film condensation of water vapour–air mixture. In combination with the provided theoretical heat and transfer equations, these formulated equations of the wall dimensionless temperature gradient and condensate mass flow rate parameter can be used for simple and reliable evaluation of the heat and mass transfer of laminar film condensation of water vapour–air mixture.

The interfacial vapour saturation temperature necessary for correct prediction of laminar film condensation of vapour–gas mixture is deeply investigated here. By taking the laminar forced film condensation of water vapour–air mixture as an example, three methods are reported for evaluation of the interfacial vapour saturation temperature: (1) the numerical calculation method, (2) prediction with the related formulation equation and (3) prediction by solving the modified condensate mass–energy transformation equation for describing the dependent relation for the interfacial vapour saturation temperature. The calculated results of the interfacial vapour saturation temperature related to the different methods are well coincident. It proves that the system of investigations in this chapter on similarity analysis method, the similarity mathematical model, the numerical calculation and treatment method of temperature- and concentration- dependent physical properties is valid for extensive study of heat and mass transfer of laminar forced film condensation of vapour–gas mixture.

The author believes that the analysis and calculation methods reported in this work can be conveniently extended to other different types of laminar forced film condensation from vapour–gas mixture.

14.13 Calculation Examples

Example 1 A flat plate, 0.3 m in width and 0.3 m in length, is located horizontally in the flowing water vapour–air mixture with 2.9 m/s as the bulk flow velocity. The wall temperature is $t_w = 85\,°C$, reference vapour saturation temperature is $t_s = 100°C$, the vapour bulk temperature is $t_\infty = 100°C$ and the bulk vapour mass fraction is $C_{m,v\infty} = 0.95$. Please solve following questions:

(1) Calculate heat transfer rate on the plate.
(2) From the heat transfer rate obtained in solution (1) for laminar forced film condensation of water vapour–air mixture on the horizontal flat plate and the corresponding result of pure vapour calculated in Example 1 of Chap. 11, which point can you get?

(3) Calculate the condensate mass flow rate on the plate.
(4) From the condensate mass flow rate in solution (3) for laminar forced film condensation of water vapour–air mixture on the horizontal flat plate and the corresponding result of pure vapour calculated in Example 1 of Chap. 11, which point can you get?

Solution 1: First of all, it is very important to predict the interfacial vapour saturation temperature for prediction of heat transfer of laminar forced film condensation of water vapour–air mixture. With $t_w = 85$, the reference wall subcooled grade

$$\frac{(\Delta t_w)_s}{t_s} = \frac{t_s - t_w}{t_s} = \frac{100 - 85}{100} = 0.15$$

Then, for evaluation of interfacial vapour saturation temperature, the following equation is applied:

$$t_{s,\text{int}} = aC_{\text{mv},\infty}{}^{b} \qquad (0.4 \le C_{\text{mv},\infty} \le 0.999) \tag{14.8}$$

where

$$a = 177.5\left[\frac{(\Delta t_w)_s}{t_s}\right]^2 - 112.12\frac{(\Delta t_w)_s}{t_s} + 100.59 \qquad \left(0.05 \le \frac{(\Delta t_w)_s}{t_s} \le 0.4\right)$$

$$b = -4.9374\left[\frac{(\Delta t_w)_s}{t_s}\right]^3 + 3.6734\left[\frac{(\Delta t_w)_s}{t_s}\right]^2 - 0.0445\frac{(\Delta t_w)_s}{t_s} + 0.0243 \qquad \left(0.05 \le \frac{(\Delta t_w)_s}{t_s} \le 0.4\right)$$

Furthermore,

$$a = 177.5 \times 0.15^2 - 112.12 \times 0.15 + 100.59 = 87.76575$$

$$b = -4.9374 \times 0.15^3 + 3.6734 \times 0.15^2 - 0.0445 \times 0.15 + 0.0243 = 0.082713$$

$$t_{s,\text{int}} = 87.76575 \times 0.95^{0.082713} = 87.394^\circ\text{C}$$

Then, the related water physical properties are determined.
With (14.22), we have

$$\begin{aligned}\lambda_{l,w} &= -8.01 \times 10^{-6} t_w{}^2 + 1.94 \times 10^{-3} t_w + 0.563 \\ &= -8.01 \times 10^{-6} \times 85^2 + 1.94 \times 10^{-3} \times 85 + 0.563 \\ &= 0.67\,\text{W}/\left(\text{m}\,^\circ\text{C}\right)\end{aligned}$$

With (14.23), we have

$$\mu_{l,\mathrm{s}} = \exp\left(-1.6 - \frac{1150}{t_{\mathrm{s,int}} + 273} + \left(\frac{690}{t_{\mathrm{s,int}} + 273}\right)^2\right) \times 10^{-3}$$
$$= \exp\left(-1.6 - \frac{1150}{87.394 + 273} + \left(\frac{690}{87.394 + 273}\right)^2\right) \times 10^{-3}$$
$$= 324.5 \times 10^{-6} \quad \mathrm{kg/(m\,s)}$$

With (14.21), we have

$$h_{\mathrm{fg}} = (-2.4355 t_{\mathrm{s,int}} + 2503.2) \times 10^3$$
$$= (-2.4355 \times 87.394 + 2503.2) \times 10^3$$
$$= 2290.35 \times 10^3\ \mathrm{kJ/kg}$$
$$t_{\mathrm{f}} = \frac{t_{\mathrm{w}} + t_{\mathrm{s,int}}}{2} = \frac{85 + 87.394}{2} = 86.197^\circ\mathrm{C}$$

The related physical properties of water and water vapour are listed in Table 14.10.

The local Reynolds number with variable physical properties is calculated as

$$\mathrm{Re}_{x\mathrm{m},\infty} = \frac{w_{x\mathrm{v},\infty}x}{\nu_{\mathrm{m},\infty}} = \frac{2.9 \times 0.3}{20.12 \times 10^{-6}} = 43240.6 < 5 \times 10^5$$

then, the condensate flow can be regarded as laminar.

From (14.4) and (14.7), the following equation can be used for evaluation of condensate heat transfer rate on the plate:

$$Q_x = \sqrt{2} b \lambda_{l,\mathrm{w}} \left(t_{\mathrm{w}} - t_{\mathrm{s,int}}\right) \left(\mathrm{Re}_{xl,\mathrm{s}}\right)^{1/2} \left(-\frac{\mathrm{d}\theta_l(\eta_l)}{\mathrm{d}\eta_l}\right)_{\eta_l=0} \qquad (14.4)$$

and

$$\left(-\frac{\mathrm{d}\theta_l\,(\eta_l)}{\mathrm{d}\eta_l}\right)_{\eta_l=0} = A C_{\mathrm{mv},\infty}{}^B \qquad (0.4 \le C_{\mathrm{mv},\infty} \le 0.999) \qquad (14.7)$$

Table 14.10 The related physical properties of water and water vapour

Materials	Water					
Temperature	t_{w} (°C)	$t_{\mathrm{s,int}}$ (°C)			t_{f} (°C)	t_∞ (°C)
	85	87.394			86.197	100
Physical properties	$\lambda_{l,\mathrm{w}}$, (W/(m °C))	$\nu_{l,\mathrm{s}}$ (m²/s)	$\mu_{l,\mathrm{s}}$ (kg/(m s))	h_{fg} (kJ/kg)	$\nu_{l,\mathrm{f}}$ (m²/s)	$\nu_{\mathrm{m},\infty}$ (m²/s)
	0.6700	0.3326×10^{-6}	324.5×10^{-6}	2290.352	0.3372×10^{-6}	20.12×10^{-6}

With $\frac{(\Delta t_{\mathrm{w}})_{\mathrm{s}}}{t_{\mathrm{s}}} = 0.15$ we have

$$A = 0.5868\left(\frac{(\Delta t_{\mathrm{w}})_{\mathrm{s}}}{t_{\mathrm{s}}}\right)^{-0.3272} \qquad \left(0.05 \le \frac{(\Delta t_{\mathrm{w}})_{\mathrm{s}}}{t_{\mathrm{s}}} \le 0.2\right)$$

$$B = -0.2486\left(\frac{(\Delta t_{\mathrm{w}})_{\mathrm{s}}}{t_{\mathrm{s}}}\right)^{-0.8082} \qquad \left(0.05 \le \frac{(\Delta t_{\mathrm{w}})_{\mathrm{s}}}{t_{\mathrm{s}}} \le 0.2\right)$$

Here,

$$A = 0.5868 \times (0.15)^{-0.3272} = 1.09162$$

$$B = -0.2486 \times (0.15)^{-0.8082} = -1.15182$$

$$\left(-\frac{\mathrm{d}\theta_l(\eta_l)}{\mathrm{d}\eta_l}\right)_{\eta_l=0} = 1.09162 \times 0.95^{-1.15182} = 1.158057$$

$$\mathrm{Re}_{xl,\mathrm{s}} = \frac{w_{x\mathrm{m},\infty}x}{\nu_{l,\mathrm{s}}} = \frac{2.9 \times 0.3}{0.3326 \times 10^{6}} = 2615754.7$$

Thus, the heat transfer rate on the plate is calculated as

$$Q_x = \sqrt{2} \times 0.3 \times 0.67 \times (85 - 87.394) \times 2615754.7^{1/2} \times 1.158057$$
$$= -1274.6\,\mathrm{W}$$

where the negative value of the result of heat transfer rate shows that the heat flux is from the vapour–gas mixture bulk to the wall plate surface.

Here, it should be indicated that the wall subcooled temperature of this example is only equal to

$$t_{\mathrm{s,int}} - t_{\mathrm{w}} = 87.394 - 85 = 2.394°\mathrm{C}$$

so small wall subcooled temperature requires a very accurate determination of the interfacial vapour saturation temperature. Otherwise, a big deviation for the wall subcooled temperature $t_{\mathrm{s,int}} - t_{\mathrm{w}}$ will be easily produced, which will cause a big deviation for the calculated result of the heat transfer of the laminar forced film condensation of vapour–gas mixture.

Solution 2: If we take $(Q_x)_{\mathrm{m}}$ and $(Q_x)_{\mathrm{p}}$ as heat transfer rates for laminar film condensation from water vapour–air mixture and pure water vapour, respectively, their ratio will be

$$\frac{(Q_x)_{\mathrm{m}}}{(Q_x)_{\mathrm{p}}} = \frac{1274.6}{5225.4} = 0.2439$$

where $(Q_x)_{\mathrm{p}} = 5225.4$ comes from the calculated result of example 1 of Chap. 11. The very small ratio $(Q_x)_{\mathrm{m}}/(Q_x)_{\mathrm{p}}$ demonstrates that the non-condensable air in the vapour–gas mixture strongly reduces the condensate heat transfer.

Solution 3: With (14.16), the condensate mass flow rate on the plate can be calculated as

$$G_x = \sqrt{2} b \mu_{l,s} \mathrm{Re}_{xl,s}{}^{1/2} \Phi_s \tag{14.16}$$

where only the mass flow rate parameter Φ_s is the unknown variable. Evaluation of the mass flow rate parameter Φ_s can be done by using the condensate mass–energy equation (11.32) with (14.19) as follows:

$$\left(-\frac{d\theta_l}{d\eta_l}\right)_{\eta_l=0} = C_{mh}\Phi_s \tag{11.32}$$

with

$$C_{mh} = \frac{\mu_{l,s} h_{fg}}{\lambda_{l,w}\left(t_{s,int} - t_w\right)} \tag{14.19}$$

With the physical property data listed in Table 14.9, we have

$$C_{mh} = \frac{\mu_{l,s} h_{fg}}{\lambda_{l,w}\left(T_{s,int} - T_w\right)} = \frac{324.5 \times 10^{-6} \times 2290352}{0.67 \times (87.394 - 85)} = 463.36$$

From solution (1) we have $\left(-\frac{d\theta_l(\eta_l)}{d\eta_l}\right)_{\eta_l=0} = 1.158057$

Then,

$$\Phi_s = \frac{\left(-\frac{d\theta_l}{d\eta_l}\right)_{\eta_l=0}}{C_{mh}} = \frac{1.158057}{463.36} = 0.002499$$

In this case, the condensate mass flow rate on the plate can be calculated as

$$\begin{aligned} G_x &= \sqrt{2} b \mu_{l,s} \mathrm{Re}_{xl,s}{}^{1/2} \Phi_s \\ &= \sqrt{2} \times 0.3 \times 324.5 \times 10^{-6} \times 2615754.7^{1/2} \times 0.002499 \\ &= 0.0005564\,\mathrm{kg/s} \\ &= 2.003\,\mathrm{kg/h} \end{aligned}$$

Solution 4: If we take $(G_x)_m$ and $(G_x)_p$ as condensate mass flow rate on the plate for laminar film condensation from water vapour–air mixture and pure water vapour, respectively, their ratio will be

$$\frac{(G_x)_m}{(G_x)_p} = \frac{0.0005564}{0.002251} = 0.2472$$

where $(G_x)_\mathrm{p} = 0.002251$ comes from the calculated result of example 1 of Chap. 11. The very small ratio $(G_x)_\mathrm{m}/(G_x)_\mathrm{p}$ demonstrates that the non-condensable air in the vapour–gas mixture strongly reduces the condensate mass transfer. On the other hand, it is seen that the evaluated ratio $(G_x)_\mathrm{m}/(G_x)_\mathrm{p}$ is very close to the evaluated ratio $(Q_x)_\mathrm{m}/(Q_x)_\mathrm{p}$, which demonstrates that the related theoretical analysis and calculation methods are reliable.

Exercises

1. Please point out at least two approaches to evaluate the interfacial vapour saturation temperature $t_{\mathrm{s,int}}$.
2. Follow example 1, in which only the wall temperature is changed to 65°C and all other conditions are kept. Please solve following questions:

   (1) Calculate heat transfer rate on the plate.
   (2) From the heat transfer rate obtained in solution (1) for laminar forced film condensation of water vapour–air mixture on the horizontal flat plate and the corresponding result of pure vapour calculated in exercise 2 of Chap. 11, which point can you get?
   (3) Calculate the condensate mass flow rate on the plate.
   (4) From the condensate mass flow rate in solution (3) for laminar forced film condensation of water vapour–air mixture on the horizontal flat plate and the corresponding result of pure vapour calculated in example 1 of Chap. 11, which point can you get?

3. Please explain why the condensate mass–energy transformation equation

$$\left(-\frac{\mathrm{d}\theta_l}{\mathrm{d}\eta_l}\right)_{\eta_l=0} = C_\mathrm{mh}\Phi_\mathrm{s} \tag{11.32}$$

   is universally eligible for laminar forced film condensation of both pure vapour and vapour–gas mixture.
4. The following two equations describe the condensate mass–energy transformation coefficient, respectively, for laminar forced film condensation of vapour and that of vapour–gas mixture:

$$C_\mathrm{mh} = \frac{\mu_{l,\mathrm{s}} h_\mathrm{fg}}{\lambda_{\mathrm{l,w}}\,(t_\mathrm{s} - t_\mathrm{w})} \tag{11.33}$$

$$C_\mathrm{mh} = \frac{\mu_{l,\mathrm{s}} h_\mathrm{fg}}{\lambda_{l,\mathrm{w}}\left(t_{\mathrm{s,int}} - t_\mathrm{w}\right)} \tag{14.19}$$

   (1) Please identify the related physical significances of the physical property variables $\mu_{l,\mathrm{s}}$, h_fg and $\lambda_{l,\mathrm{w}}$ in the two equations.
   (2) Explain the coincidence of the two equations and analyze the physical significance of the condensate mass–energy transformation equation.

Part IV
Appendix

Appendix A
Tables with Physical Properties

Physical Properties of Gases at Atmospheric Pressure

Air [1]							
T K	ρ kg/m^3	c_p kJ/(kg °C)	$\mu \times 10^6$ kg/(m s)	$\nu \times 10^6$ m^2/s	λ W/(m °C)	$a \times 10^6$ m^2/s	Pr
100	3.5562	1.032	7.11	1.999	0.00934	2.54	0.786
150	2.3364	1.012	10.34	4.426	0.0138	5.84	0.758
200	1.7458	1.007	13.25	7.59	0.0181	10.3	0.737
250	1.3947	1.006	15.96	11.443	0.0223	15.9	0.720
300	1.1614	1.007	18.46	15.895	0.0263	22.5	0.707
350	0.9950	1.009	20.82	20.925	0.0300	29.9	0.7
400	0.8711	1.014	23.01	26.415	0.0338	38.3	0.69
450	0.7740	1.021	25.07	32.39	0.0373	47.2	0.686
500	0.6964	1.030	27.01	38.785	0.0407	56.7	0.684
550	0.6329	1.040	28.84	45.568	0.0439	66.7	0.683
600	0.5804	1.051	30.58	52.688	0.0469	76.9	0.685
650	0.5356	1.063	32.25	60.213	0.0497	87.3	0.69
700	0.4975	1.075	33.88	68.101	0.05240	98	0.695
750	0.4643	1.087	35.46	76.373	0.0549	109	0.702
800	0.4354	1.099	36.98	84.933	0.0573	120	0.709
850	0.4097	1.11	38.43	93.8	0.0596	131	0.716
900	0.3868	1.121	39.81	102.921	0.0620	143	0.720
950	0.3666	1.131	41.13	112.193	0.0643	155	0.723
1000	0.3482	1.141	42.44	121.884	0.0667	168	0.726
1100	0.3166	1.159	44.9	141.819	0.0715	195	0.728
1200	0.2920	1.175	47.3	161.986	0.0763	224	0.728

D.Y. Shang, *Theory of Heat Transfer with Forced Convection Film Flows*,
Heat and Mass Transfer, DOI 10.1007/978-3-642-12581-2_15,

Monoxide, CO [1]

T K	ρ kg/m^3	c_p kJ/(kg °C)	$\mu \times 10^6$ kg/(m s)	$\nu \times 10^6$ m^2/s	λ W/(m °C)	$a \times 10^6$ m^2/s	Pr
200	1.6888	1.045	12.7	7.5201	0.017	9.63	0.781
220	1.5341	1.044	13.7	8.9303	0.0190	11.9	0.753
240	1.4055	1.043	14.7	10.4589	0.0206	14.1	0.744
260	1.2967	1.043	15.7	12.1077	0.0221	16.3	0.741
280	1.2038	1.042	16.6	13.7897	0.0236	18.8	0.733
300	1.1233	1.043	17.5	15.5791	0.025	21.3	0.730
320	1.0529	1.043	18.4	17.4755	0.0263	23.9	0.730
340	0.9909	1.044	19.3	19.4772	0.0278	26.9	0.725
360	0.9357	1.045	20.2	21.5881	0.0291	29.8	0.729
380	0.8864	1.047	21	23.6913	0.0305	32.9	0.719
400	0.8421	1.049	21.8	25.8877	0.0318	36.0	0.719
450	0.7483	1.055	23.7	31.6718	0.0350	44.3	0.714
500	0.67352	1.065	25.4	37.7123	0.0381	53.1	0.710
550	0.61226	1.076	27.1	44.2622	0.0411	62.4	0.710
600	0.56126	1.088	28.6	50.9568	0.0440	72.1	0.707
650	0.51806	1.101	30.1	58.1014	0.0470	82.4	0.705
700	0.48102	1.114	31.5	65.4858	0.0500	93.3	0.702
750	0.44899	1.127	32.9	73.2756	0.0528	104	0.702
800	0.42095	1.140	34.3	81.4824	0.0555	116	0.705

Helium, He [1]

T K	ρ kg/m^3	c_p kJ/(kg °C)	$\mu \times 10^6$ kg/(m s)	$\nu \times 10^6$ m^2/s	λ W/(m °C)	$A \times 10^6$ m^2/s	Pr
100	0.4871	5.193	9.63	19.77	0.073	28.9	0.686
120	0.406	5.193	10.7	26.36	0.0819	38.8	0.679
140	0.3481	5.193	11.8	33.90	0.0907	50.2	0.676
160	0.30945	5.193	12.9	41.69	0.0992	63.2	0.6745
180	0.2708	5.193	13.9	51.33	0.1072	76.2	0.673
200	0.2462	5.193	15	60.93	0.1151	91.6	0.674
220	0.2216	5.193	16	72.20	0.1231	107	0.675
240	0.20455	5.193	17	83.11	0.13	124	0.6785
260	0.1875	5.193	18	96	0.137	141	0.682
280	0.175	5.193	19	108.57	0.145	160.5	0.681
300	0.1625	5.193	19.9	122.46	0.152	180	0.68
350	0.1422	5.193	22.1	155.42	0.17	237.5	0.6775
400	0.1219	5.193	24.3	199.34	0.187	295	0.675
450	0.10972	5.193	26.3	239.70	0.204	364.5	0.6715
500	0.09754	5.193	28.3	290.14	0.22	434	0.668
600	0.083615	5.193	32	382.71	0.252	601	0.661
700	0.06969	5.193	35	502.22	0.278	768	0.654
800		5.193	38.2		0.304		
900		5.193	41.4		0.33		
1000	0.04879	5.193	44.6	914.12	0.354	1400	0.654

Hydrogen, H_2 [1]

T K	ρ kg/m^3	c_p kJ/(kg °C)	$\mu \times 10^6$ kg/(m s)	$\nu \times 10^6$ m^2/s	λ W/(m °C)	$a \times 10^6$ m^2/s	Pr
100	0.24255	11.23	4.21	19.77	0.067	24.6	0.707
200	0.12115	13.54	6.81	26.35	0.131	79.9	0.704
300	0.08078	14.31	8.96	33.90	0.183	158	0.701
400	0.06059	14.48	10.82	41.69	0.226	258	0.695
500	0.04848	14.52	12.64	51.33	0.266	378	0.691
600	0.0404	14.55	14.24	60.93	0.305	519	0.678
700	0.03463	14.61	15.78	72.20	0.342	676	0.675
800	0.0303	14.7	17.24	83.11	0.378	849	0.67
900	0.02694	14.83	18.65	96	0.412	1030	0.671
1000	0.02424	14.99	20.13	108.57	0.448	1230	0.673
1100	0.02204	15.17	21.3	122.46	0.488	1460	0.662
200	0.0202	15.37	22.62	155.41	0.528	1700	0.659
1300	0.01865	15.59	23.85	199.34	0.569	1955	0.655
1400	0.01732	15.81	25.07	239.70	0.61	2230	0.65
1500	0.01616	16.02	26.27	290.14	0.655	2530	0.643
1600	0.0152	16.28	27.37	382.70	0.697	2815	0.639

Nitrogen, N_2 [1]

T K	ρ kg/m^3	c_p kJ/(kg °C)	$\mu \times 10^6$ kg/(m s)	$\nu \times 10^6$ m^2/s	λ W/(m °C)	$a \times 10^6$ m^2/s	Pr
100	3.4388	1.07	6.88	2.000	0.0958	2.6	0.768
150	2.2594	1.05	10.06	4.45	0.0139	5.86	0.759
200	1.6883	1.043	12.02	7.126	0.0183	10.4	0.736
250	1.3488	1.042	15.49	11.48	0.0222	15.8	0.727
300	1.1233	1.041	17.82	15.86	0.0259	22.1	0.716
350	0.9625	1.042	20	20.78	0.0293	29.2	0.711
400	0.8425	1.045	22.04	26.16	0.0327	37.1	0.704
450	0.7485	1.05	23.96	32.01	0.0358	45.6	0.703
500	0.6739	1.056	25.77	38.24	0.0389	54.7	0.7
550	0.6124	1.065	27.47	44.86	0.0417	63.9	0.702
600	0.5615	1.075	29.08	51.79	0.0416	73.9	0.701
700	0.4812	1.098	32.1	66.71	0.0499	94.4	0.706
800	0.4211	1.122	34.91	82.91	0.0548	116	0.715
900	0.3743	1.146	37.53	100.27	0.0597	139	0.721
1000	0.3368	1.167	39.99	118.74	0.0647	165	0.721
1100	0.3062	1.187	42.32	138.21	0.07	193	0.718
1200	0.2807	1.204	44.53	158.64	0.0758	224	0.707
1300	0.2591	1.219	46.62	179.93	0.081	256	0.701

Oxygen, O_2 [1]

T K	ρ kg/m^3	c_p kJ/(kg °C)	$\mu \times 10^6$ kg/(m s)	$\nu \times 10^6$ m^2/s	λ W/(m °C)	$a \times 10^6$ m^2/s	Pr
100	3.945	0.962	7.64	1.936629	0.00925	2.44	0.796
150	2.585	0.921	11.48	4.441006	0.0138	5.8	0.766
200	1.93	0.915	14.75	7.642487	0.0183	10.4	0.737
250	1.542	0.915	17.86	11.58236	0.0226	16	0.723
300	1.284	0.92	20.72	16.13707	0.0268	22.7	0.711
350	1.1	0.929	23.35	21.22727	0.0296	29	0.733
400	0.962	0.942	25.82	26.83992	0.033	36.4	0.737
450	0.8554	0.956	28.14	32.89689	0.0363	44.4	0.741
500	0.7698	0.972	30.33	39.39984	0.0412	55.1	0.716
550	0.6998	0.988	32.4	46.29894	0.0441	63.8	0.726
600	0.6414	1.003	34.37	53.58591	0.0473	73.5	0.729
700	0.5498	1.031	38.08	69.26155	0.0523	93.1	0.744
800	0.481	1.054	41.52	86.32017	0.0589	116	0.743
900	0.4275	1.074	44.72	104.6082	0.0649	141	0.74
1000	0.3848	1.09	47.7	123.9605	0.071	169	0.733
1100	0.3498	1.103	50.55	144.5111	0.0758	196	0.736
1200	0.3206	1.115	53.25	166.0948	0.0819	229	0.725
1300	0.206	1.125	58.84	285.6311	0.0871	262	0.721

Carbon dioxide, CO_2 [1]

T K	ρ kg/m^3	c_p kJ/(kg °C)	$\mu \times 10^6$ kg/(m s)	$\nu \times 10^6$ m^2/s	λ W/(m °C)	$A \times 10^6$ m^2/s	Pr
220	2.4733	0.783	11.105	4.490	0.010805	5.92	0.818
250	2.1675	0.804	12.59	5.809	0.012884	7.401	0.793
300	1.7973	0.871	14.958	8.322	0.016572	10.588	0.77
350	1.5362	0.9	17.205	11.200	0.02047	14.808	0.755
400	1.3424	0.942	19.32	14.392	0.02461	19.463	0.738
450	1.1918	0.98	21.34	17.906	0.02897	24.813	0.721
500	1.0732	1.013	23.26	21.67	0.03352	30.84	0.702
550	0.9739	1.047	25.08	25.752	0.03821	37.5	0.695
600	0.8938	1.076	26.83	30.018	0.04311	44.83	0.668

Ammonia, NH_3 [2]

T K	ρ kg/m^3	c_p kJ/(kg °C)	$\mu \times 10^6$ kg/(m s)	$\nu \times 10^6$ m^2/s	λ W/(m °C)	$A \times 10^6$ m^2/s	Pr
220	0.3828	2.198	7.255	18.952	0.0171	20.54	0.93
273	0.7929	2.177	9.353	11.796	0.022	13.08	0.9
323	0.6487	2.177	11.035	17.011	0.027	19.2	0.88
373	0.559	2.236	12.886	23.052	0.0327	26.19	0.87
423	0.4934	2.315	14.672	29.736	0.0391	34.32	0.87
473	0.4405	2.395	16.49	37.435	0.0476	44.21	0.84

Water vapour [2]

T K	ρ kg/m^3	c_p kJ/(kg °C)	$\mu \times 10^6$ kg/(m s)	$\nu \times 10^6$ m^2/s	λ W/(m °C)	$a \times 10^6$ m^2/s	Pr
380	0.5863	2.06	1.271	2.168	0.0246	20.36	1.06
400	0.5542	2.014	1.344	2.425	0.0261	23.38	1.04
450	0.4902	1.98	1.525	3.111	0.0299	30.7	1.01
500	0.4405	1.985	1.704	3.868	0.0339	38.7	0.996
550	0.4005	1.997	1.884	4.704	0.0379	47.5	0.991
600	0.3652	2.026	2.067	5.660	0.0422	57.3	0.986
650	0.338	2.056	2.247	6.648	0.0464	66.6	0.995
700	0.314	2.085	2.426	7.726	0.0505	77.2	1
750	0.2931	2.119	2.604	8.884	0.0549	88.3	1.05
800	0.2739	2.152	2.786	10.172	0.0592	100.1	1.01
850	0.2579	2.186	2.969	11.51221	0.0637	113	1.019

Gas mixture [3]

t °C	ρ kg/m^3	c_p kJ/(kg °C)	$\mu \times 10^6$ kg/(m s)	$\nu \times 10^6$ m^2/s	λ W/(m °C)	$a \times 10^6$ m^2/s	Pr
0	1.295	1.042	15.8	12.2	0.0228	12.2	0.72
100	0.95	1.068	20.4	21.47	0.0313	21.54	0.69
200	0.748	1.097	24.5	32.75	0.0401	32.8	0.67
300	0.617	1.122	28.2	45.71	0.0484	45.81	0.65
400	0.525	1.151	31.7	60.38	0.057	60.38	0.64
500	0.457	1.185	34.8	76.15	0.0656	76.3	0.63
600	0.405	1.214	37.9	93.58	0.0742	93.61	0.62
700	0.363	1.239	40.7	112.12	0.0827	112.1	0.61
800	0.33	1.264	43.4	131.52	0.0915	131.8	0.6
900	0.301	1.29	45.9	152.49	0.1	152.5	0.59
1000	0.275	1.306	48.4	176	0.109	174.3	0.58
1100	0.257	1.323	50.7	197.28	0.1175	197.1	0.57
1200	0.24	1.34	53	220.83	0.1262	221	0.56

Water vapour [2]

T K	ρ kg/m^3	c_p kJ/(kg °C)	$\mu \times 10^6$ m^2/s	$\nu \times 10^6$ m^2/s	λ W/(m °C)	$a \times 10^6$ m^2/s	Pr
380	0.5863	2.06	12.71	21.68	0.0246	20.36	1.06
400	0.5542	2.014	13.44	24.25	0.0261	23.38	1.04
450	0.4902	1.98	15.25	31.11	0.0299	30.7	1.01
500	0.4405	1.985	17.04	38.68	0.0339	38.7	0.996
550	0.4005	1.997	18.84	47.04	0.0379	47.5	0.991
600	0.3652	2.026	20.67	56.6	0.0422	57.3	0.986
650	0.338	2.056	22.47	66.48	0.0464	66.6	0.995
700	0.314	2.085	24.26	77.26	0.0505	77.2	1
750	0.2931	2.119	26.04	88.84	0.0549	88.3	1.05
800	0.2739	2.152	27.86	101.72	0.0592	100.1	1.01
850	0.2579	2.186	29.69	115.12	0.0637	113	1.019
380	0.5863	2.06	12.71	21.68	0.0246	20.36	1.06
400	0.5542	2.014	13.44	24.25	0.0261	23.38	1.04
450	0.4902	1.98	15.25	31.11	0.0299	30.7	1.01
500	0.4405	1.985	17.04	38.68	0.0339	38.7	0.996
550	0.4005	1.997	18.84	47.04	0.0379	47.5	0.991
600	0.3652	2.026	20.67	56.6	0.0422	57.3	0.986
650	0.338	2.056	22.47	66.48	0.0464	66.6	0.995
700	0.314	2.085	24.26	77.26	0.0505	77.2	1
750	0.2931	2.119	26.04	88.84	0.0549	88.3	1.05
800	0.2739	2.152	27.86	101.72	0.0592	100.1	1.01
850	0.2579	2.186	29.69	115.12	0.0637	113	1.019

Physical Properties of Some Saturated Liquid

Ammonia, NH_3 [4]

t °C	ρ kg/m^3	c_p kJ/(kg °C)	$\mu \times 10^6$ kg/(m s)	$\nu \times 10^6$ m^2/s	λ W/(m °C)	$a \times 10^7$ m^2/s	Pr
–50	703.69	4.463	306.11	0.435	0.547	1.742	2.6
–40	691.68	4.467	280.82	0.406	0.547	1.775	2.28
–30	679.34	4.467	262.9	0.387	0.549	1.801	2.15
–20	666.69	4.509	254.01	0.381	0.547	1.819	2.09
–10	653.55	4.564	247.04	0.378	0.543	1.825	2.07
0	640.1	4.635	238.76	0.373	0.54	1.819	2.05
10	626.16	4.714	230.43	0.368	0.531	1.801	2.04
20	611.75	4.798	219.62	0.359	0.521	1.775	2.02
30	596.37	4.89	208.13	0.349	0.507	1.742	2.01
40	580.99	4.999	197.54	0.34	0.493	1.701	2
50	564.33	5.116	186.23	0.33	0.476	1.654	1.99

Carbon dioxide, CO_2 [4]

t °C	ρ kg/m^3	c_p kJ/(kg °C)	$\mu \times 10^6$ kg/(m s)	$\nu \times 10^6$ m^2/s	λ W/(m °C)	$a \times 10^7$ m^2/s	Pr
–50	1156.34	1.84	137.61	0.119	0.085	0.4021	2.96
–40	1117.77	1.88	131.9	0.118	0.1011	0.481	2.45
–30	1076.76	1.97	125.98	0.117	0.1116	0.5272	2.22
–20	1032.39	2.05	118.72	0.115	0.1151	0.5445	2.12
–10	983.38	2.18	111.12	0.113	0.1099	0.5133	2.2
0	926.99	2.47	100.11	0.108	0.1045	0.4578	2.38
10	860.03	3.14	86.86	0.101	0.0971	0.3608	2.8
20	772.57	5	70.3	0.091	0.0872	0.2219	4.1
30	597.81	36.4	47.82	0.08	0.0703	0.0279	28.7

Sulphur dioxide, SO_2 [4]

t °C	ρ kg/m^3	c_p kJ/(kg °C)	$\mu \times 10^6$ kg/(m s)	$\nu \times 10^6$ m^2/s	λ W/(m °C)	$a \times 10^7$ m^2/s	Pr
–50	1560.84	1.3595	755.45	0.484	0.242	1.141	4.24
–40	1536.81	1.3607	651.61	0.424	0.235	1.130	3.74
–30	1520.64	1.3616	564.16	0.371	0.230	1.117	3.31
–20	1488.60	1.3624	482.31	0.324	0.225	1.107	2.93
–10	1463.61	1.3628	421.52	0.288	0.218	1.097	2.62
0	1438.46	1.3636	369.68	0.257	0.211	1.081	2.38
10	1412.51	1.3645	327.7	0.232	0.204	1.066	2.18
20	1386.40	1.3653	291.14	0.210	0.199	1.050	2.00
30	1359.33	1.3662	258.27	0.190	0.192	1.035	1.83
40	1329.22	1.3674	229.96	0.173	0.185	1.019	1.70
50	1299.10	1.3683	210.45	0.162	0.177	0.999	1.61

Freon 12, CCl_2F_2 [4]

t °C	ρ kg/m^3	c_p kJ/(kg °C)	$\mu \times 10^6$ kg/(m s)	$\nu \times 10^6$ m^2/s	λ W/(m °C)	$a \times 10^7$ m^2/s	Pr
–50	1546.75	0.8750	479.49	0.310	0.067	0.501	6.2
–40	1518.71	0.8847	423.72	0.279	0.069	0.514	5.4
–30	1489.56	0.8956	376.86	0.253	0.069	0.526	4.8
–20	1460.57	0.9073	343.23	0.235	0.071	0.539	4.4
–10	1429.49	0.9203	315.92	0.221	0.073	0.550	4.0
0	1397.45	0.9345	299.05	0.214	0.073	0.557	3.8
10	1364.30	0.9496	276.95	0.203	0.073	0.560	3.6
20	1330.18	0.9659	263.38	0.198	0.073	0.560	3.5
30	1295.10	0.9835	251.25	0.194	0.071	0.560	3.5
40	1257.13	1.019	240.11	0.191	0.069	0.555	3.5
50	1215.96	1.0216	231.03	0.190	0.067	0.545	3.5

$C_2H_4(OH)_2$ [4]

t °C	ρ kg/m^3	c_p kJ/(kg °C)	$\mu \times 10^6$ kg/(m s)	$\nu \times 10^6$ m^2/s	λ W/(m °C)	$a \times 10^7$ m^2/s	Pr
0	1130.75	2.294	65052.05	57.53	0.242	0.934	615
20	1116.65	2.382	21417.35	19.18	0.249	0.939	204
40	1101.43	2.474	9571.427	8.69	0.256	0.939	93
60	1087.66	2.562	5166.385	4.75	0.260	0.932	51
80	1077.56	2.650	3211.129	2.98	0.261	0.921	32.4
100	1058.50	2.742	2148.755	2.03	0.263	0.908	22.4

Mercury, Hg [4]

t °C	ρ kg/m^3	c_p kJ/(kg °C)	$\mu \times 10^6$ kg/(m s)	$\nu \times 10^6$ m^2/s	λ W/(m °C)	$a \times 10^7$ m^2/s	Pr
0	13628.22	0.1403	1689.9	0.124	8.2	42.99	0.0288
20	13579.04	0.1394	1548.01	0.114	8.69	46.04	0.0249
50	13505.84	0.1386	1404.61	0.104	9.40	50.22	0.0207
100	13384.58	0.1373	1242.09	0.0928	10.51	57.16	0.0162
150	13264.28	0.1365	1131.44	0.0853	11.49	63.54	0.0134
200	13144.94	0.1570	1054.22	0.0802	12.34	69.08	0.0116
250	13025.60	0.1357	996.46	0.0765	13.07	74.06	0.0103
315.5	12847.00	0.134	864.6	0.0673	14.02	81.50	0.0083

Water, H_2O [2]

t °C	ρ kg/m^3	c_p kJ/(kg °C)	$\mu \times 10^6$ kg/(m s)	$\nu \times 10^6$ m^2/s	λ W/(m °C)	$a \times 10^6$ m^2/s	Pr
0	999.9	4.217	1752.5	1.7527	0.569	0.13494	12.99
10	999.7	4.193	1299.2	1.2996	0.586	0.13980	9.30
20	998.2	4.182	1001.5	1.0033	0.602	0.1442	6.96
30	995.7	4.179	797	0.8004	0.617	0.14828	5.4
40	992.2	4.179	651.3	0.6564	0.630	0.14953	4.32
50	988.1	4.181	544	0.5450	0.643	0.15408	3.54
60	983.2	4.185	460	0.4679	0.653	0.15870	2.97
70	977.8	4.190	400.5	0.4014	0.662	0.15834	2.54
80	971.8	4.197	351	0.3612	0.669	0.16403	2.20
90	965.3	4.205	311.3	0.3225	0.675	0.16629	1.94
100	958.4	4.216	279	0.2911	0.680	0.16829	1.73
110	951.0	4.229	252.2	0.2652	0.683	0.16983	1.56
120	943.1	4.245	230	0.2439	0.685	0.17110	1.43
130	934.8	4.263	211	0.2257	0.687	0.17239	1.31
140	926.1	4.285	195	0.2106	0.687	0.17312	1.22
150	917.0	4.310	181	0.1974	0.686	0.17357	1.14
160	907.4	4.339	169	0.1862	0.684	0.17373	1.07
170	897.3	4.371	158.5	0.1766	0.681	0.17363	1.02
180	886.9	4.408	149.3	0.1683	0.676	0.17291	0.97
190	876.0	4.449	141.2	0.1612	0.671	0.17217	0.94
200	863.0	4.497	133.8	0.1550	0.664	0.17109	0.91
210	852.3	4.551	127.3	0.1494	0.657	0.16938	0.88
220	840.3	4.614	121.5	0.1446	0.648	0.16713	0.86
230	827.3	4.686	119.7	0.145	0.639	0.16483	1.185
240	813.6	4.770	111.4	0.1369	0.629	0.16208	0.85
250	799.0	4.869	107	0.1339	0.617	0.15860	0.84
260	784.0	4.985	103	0.1314	0.604	0.15455	0.85
270	767.9	5.13	99.4	0.1294	0.589	0.14952	0.86
280	750.7	5.3	96.1	0.1280	0.573	0.14402	0.89
290	732.3	5.51	93	0.1270	0.558	0.14300	0.92
300	712.5	5.77	90.1	0.1265	0.540	0.13136	0.96

Temperature Parameters of Gases [5–7]

Gas	n_μ	n_λ	$n_{\mu\lambda}$	n_{c_p}	Temperature range, k	Recommended Pr
Ar	0.72	0.73	0.7255	0.01	220–1500	0.622
He	0.66	0.725	0.69575	0.01	273–873	0.675
H_2	0.68	0.8	0.746	0.042	220–700	0.68
Air	0.68	0.81	0.7515	0.078	230–1000	0.7
CO	0.71	0.83	0.776	0.068	220–600	0.72
N_2	0.67	0.76	0.7195	0.07	220–1200	0.71
O_2	0.694	0.86	0.7853	0.108	230–600	0.733
Water vapour	1.04	1.185	1.11975	0.003	380–800	1
Gas mixture	0.75	1.02	0.8985	0.134	273–1173	0.63
CO_2	0.88	1.3	1.111	0.34	220–700	0.73
CH_4	0.78	1.29	1.0605	0.534	273–1000	0.74
CCl_4	0.912	1.29	1.1199	0.28	260–400	0.8
SO_2	0.91	1.323	1.13715	0.257	250–900	0.81
H_2S	1	1.29	1.1595	0.18	270–400	0.85
NH_3	1.04	1.375	1.22425	0.34	250–900	0.87

Notes:

1. Component of the gas mixture: $CO_2 = 0.13$, $N_2 = 0.76$, and water vapour $= 0.11$
2. The temperature parameters n_μ, n_λ, and n_{c_p} are defined by simple power-law of gases as follows:

$$\frac{\mu}{\mu_\infty} = \left(\frac{T}{T_\infty}\right)^{n_\mu}, \quad \frac{\lambda}{\lambda_\infty} = \left(\frac{T}{T_\infty}\right)^{n_\lambda} \text{ and } \frac{c_p}{c_{p\infty}} = \left(\frac{T}{T_\infty}\right)^{n_{c_p}},$$

   where T_∞ is a related reference temperature
3. The overall temperature parameter $n_{\mu\lambda}$ is defined as $n_{\mu\lambda} = 0.45 n_\mu + 0.55 n_\lambda$

References

1. F.P. Incropera, D.P. Dewitt, *Fundamen300tals of Heat Transfer* (Wiley, New York, NY, 1981)
2. T. Cebeci, P. Brabshaw, *Physical and Computational Aspects of Convective Heat Transfer* (Springer, New York, NY, 1984)
3. Shiming Yang, Heat Transfer, 2nd Ed., (Higher Education, Beijing, 1987)
4. E.R.G. Eckert, R.M. Drake, *Heat and Mass Transfer*, 2nd edn. (McGraw-Hill Book, New York, NY, 1959)
5. D.Y. Shang, B.X. Wang, Effect of variable thermophysical properties on laminar free convection of gas. Int. J. Heat Mass Transf. **33**(7), 1387–1395 (1990)
6. D.Y. Shang, B.X. Wang, Effect of variable thermophysical properties on laminar free convection of polyatomic gas. Int. J. Heat Mass Transf. **34**(3), 749–755 (1991)
7. D.Y. Shang, *Free Convection Film Flows and Heat Transfer* (Springer, Berlin, Heidelberg and New York, NY, 2006)

Index

D.Y. Shang, *Theory of Heat Transfer with Forced Convection Film Flows*, Heat and Mass Transfer, DOI 10.1007/978-3-642-12581-2,